Novel Proteins for Food, Pharmaceuticals, and Agriculture

Novel Proteins for Food, Pharmaceuticals, and Agriculture

Sources, Applications, and Advances

Edited by

Dr Maria Hayes
Teagasc Food Research Centre
Ashtown
Dublin
Ireland

WILEY Blackwell

This edition first published 2019

Registered Office
John Wiley & Sons, Inc., 111 River Street, Hoboken, NJ 07030, USA

Editorial Office
111 River Street, Hoboken, NJ 07030, USA

For details of our global editorial offices, customer services, and more information about Wiley products visit us at www.wiley.com.

Wiley also publishes its books in a variety of electronic formats and by print-on-demand. Some content that appears in standard print versions of this book may not be available in other formats.

Library of Congress Cataloging-in-Publication Data

Names: Hayes, Maria, editor.
Title: Novel proteins for food, pharmaceuticals, and agriculture : sources, applications, and advances / edited by Maria Hayes.
Description: First edition. | Hoboken, NJ, USA : Wiley-Blackwell, 2018. | Includes bibliographical references. |
Identifiers: LCCN 2018034376 (print) | LCCN 2018034929 (ebook) | ISBN 9781119385356 (Adobe PDF) | ISBN 9781119385325 (ePub) | ISBN 9781119385301 (hardcover)
Subjects: | MESH: Dietary Proteins
Classification: LCC QP551 (ebook) | LCC QP551 (print) | NLM QU 55.4 | DDC 615.1/9—dc23
LC record available at https://lccn.loc.gov/2018034376

Cover Design: Wiley
Cover Images: Background © Pam Susemiehl/Getty Images; (Inset L-R) © wrangel/Getty Images, © Santiago Urquijo/Getty Images, © magnez2/Getty Images

Set in 10/12pt WarnockPro by SPi Global, Chennai, India

10 9 8 7 6 5 4 3 2 1

Contents

About the Editor

Dr Maria Hayes has an Honours degree in Science (Industrial Microbiology and Chemistry) from University College Dublin and a PhD in Microbiology and Chemistry from University College Cork, the topic of which concerned the generation of antimicrobial and heart health beneficial peptides from dairy processing waste streams. Following her PhD, Maria worked with the Centre of Marine Biotechnology (CAMBIO), an Enterprise Ireland-funded centre based in Donegal, Ireland. This work involved understanding, isolating, and purifying chitinolytic bacteria from marine processing by-products, specifically crab and prawn shell material, and was carried out in conjunction with an industry partner.

In 2008, Maria joined Teagasc as the programme manager and full-time researcher on the Marine Functional Foods Research Initiative (NutraMara project). The aim of this project was to utilise marine resources for the recovery of valuable and healthy food ingredients with enhanced health benefits (functional foods). Maria works extensively on method development for the recovery of food ingredients and biomolecules from marine, dairy, plant, and animal by-products or co-products generated during food processing. She has published over 100 academic papers, many of which concern the utilisation of by-products/co-products and rest raw materials of marine and meat processing, and has collaborated with the main Irish universities and institutes of technology as well as with international partners including NIH in the US, the University of Nottingham UK and others. She is also a member of the European Chitin Society (EUCHIS) and the WG leader on EUALGAE and is a member of the EU COST action imPARAS. Her research interests include utilisation of rest raw materials from marine processing and algal research for food. She enjoys and maintains several industry collaborations within Ireland and internationally. She is currently the chief co-ordinator of several by-product utilisation projects in Ireland, including BRAVO, FISHBOWL, MUSSELS, and BRAVO 2, and one EU project (The IDEA project) which deals with development of economically viable algae-based value chains. She has published over 100 academic research papers concerning proteins from different sources including cereal, dairy, meat, marine, and algae.

List of Contributors

Joana Abelho
Allmicroalgae
Lisbon
Portugal

Carlos Álvarez Garcia
Department of Food Quality and Sensory Science
Dublin
Ireland

Leen Bastiaens
VITO
Mol
Belgium

Stephen Bleakley
Teagasc Food Research Centre
Food BioSciences Department
Dublin
Ireland

Graziella Chini Zittelli
National Research Council
Institute of Ecosystem Study
Florence
Italy

Alan T. Critchley
Verschuren Centre for Sustainability in Energy and the Environment
Cape Breton University
Sydney
Nova Scotia
Canada

Lakshmi A. Dave
Riddet Institute
Massey University
Palmerston North
New Zealand

Victória del Pino
Necton S.A.
Olhão
Portugal

John Dodd
AlgaeCytes Ltd.
Discovery Parkhouse
Sandwich
UK

Christine Edwards
School of Pharmacy and Life Sciences
Robert Gordon University
Aberdeen
UK

Shane Feeney
Teagasc Food Research Centre
Moorepark
Fermoy, Co. Cork
Ireland

and

Advanced Glycoscience Research Cluster
National Centre for Biomedical Engineering Science
National University of Ireland Galway
Galway
Ireland

Marco García-Vaquero
University College Dublin (UCD)
School of Veterinary Medicine
Dublin
Ireland

Spyros Gkelis
School of Biology
Aristotle University of Thessaloniki
Department of Botany
Thessaloniki
Greece

Luisa Gouveia
LNEG, National Laboratory of Energy and Geology, Bioenergy Unit
Lisbon
Portugal

Maria Hayes
Teagasc Food Research Centre
Food BioSciences Department
Dublin
Ireland

Rita M. Hickey
Teagasc Food Research Centre
Moorepark
Fermoy, Co. Cork
Ireland

Muge Isleten Hosoglu
Canakkale Onsekiz Mart University
Canakkale
Turkey

Lokesh Joshi
Advanced Glycoscience Research Cluster
National Centre for Biomedical Engineering Science
National University of Ireland Galway
Galway
Ireland

Despoina Konstantinou
School of Biology
Aristotle University of Thessaloniki
Department of Botany
Thessaloniki
Greece

Morten Laake
Sigtun Innovation AS
HUB Lillehammer Regional Innovation Center
Lillehammer
Norway

Tomas Lafarga
Institut de Recerca i Tecnologia Agroalimentàries (IRTA)
Postharvest Programme
Processed Fruits and Vegetables
Lleida
Spain

Joana Gabriela Laranjeira da Silva
Allmicroalgae
Lisbon
Portugal

Catherine Lefranc-Millot
Roquette
Lestrem
France

Ismael Marcet Manrique
University of Oviedo
Department of Chemical and Environmental Engineering
Oviedo
Spain

Beth Mason
Verschuren Centre for Sustainability in Energy and the Environment
Cape Breton University
Sydney
Nova Scotia
Canada

Leticia Mora
Instituto de Agroquímica y Tecnología de Alimentos (CSIC)
Valencia
Spain

Sinead T. Morrin
Teagasc Food Research Centre
Moorepark
Fermoy, Co. Cork
Ireland

Patrick Murray
Limerick Institute of Technology
Limerick City
Ireland

Koenraad Muylaert
KU, Leuven campus Kortrijk
Laboratory Aquatic Biology
Kortrijk
Belgium

Chigozie Louis Okolie
Verschuren Centre for Sustainability in Energy and the Environment
Cape Breton University
Sydney
Nova Scotia
Canada

and

Department of Plant, Food, and Environmental Sciences
Faculty of Agriculture
Dalhousie University
Truro
Canada

Hugo Pereira
Allmicroalgae
Lisbon
Portugal

Milagro Reig
Universidad Politécnica de Valencia
Instituto de Ingeniería de Alimentos para el Desarrollo
Valencia
Spain

Vytas Rimkus
Spila, UAB
Vilnius
Lithuania

Ivo Safarik
Department of Nanobiotechnology
Biology Centre, ISB
Academy of Sciences
Ceske Budejovice
Czech Republic

Kari Skjanes
Norwegian Institute of Bioeconomy Research - NIBIO
Ås
Norway

Hanne Skomedal
Norwegian Institute of Bioeconomy Research - NIBIO
Ås
Norway

Virginie Teichman-Dubois
Roquette
Lestrem
France

Fidel Toldrá
Instituto de Agroquímica y Tecnología de Alimentos (CSIC)
Valencia
Spain

Fidel Toldrá-Reig
Instituto de Tecnología Química (CSIC-UPV)
Valencia
Spain

Preface

The aim of this book is to highlight the various sources, applications, and advancements that exist, and that are currently being researched, concerning proteins from novel and traditional sources. Sources discussed include marine, by-product proteins, plant, dairy and meat and novel sources including insect, rapeseed/canola as well as cereal sources. Applications discussed in the book include food, functional foods, feed, chemical, and pharmaceutical as well as niche applications. Novel marine proteins from macroalgae and microalgae as well as insect protein are examined. Protein-derived bioactive and technofunctional ingredients and their applications feature. Bioactive compounds and functional foods represent a major market application in food and other industries. This book discusses sustainable protein resources and gives examples of bioactive compounds isolated from these and other resources that have or could achieve high market value in specific markets. Functional food formulations and pharmaceutical applications, and how they alter biological activity to provide therapeutic benefits, nutritional values, and health protection are covered. Hydrolysate generation and applications feature. Technofunctional applications of proteins and how technologies, such as those used in the dairy processing industry, are examined and how these technologies may be applied to non-dairy protein sources are discussed. The book also looks at the screening process for identification of bioactive molecules derived from protein sources and the use of a biorefinery concept during protein extraction processes for protein. Furthermore, computing methods for efficient *in silico* analysis of proteins and hydrolysate products, including peptides, are discussed. Different protein production methods used for bioactive protein and peptide isolation and identification are highlighted. Furthermore, the market opportunities that exist for novel proteins such as insect, macroalgal-derived and others are discussed, as are the identification and commercialisation of new proteins for different markets. This book also highlights regulations in the US, Europe, Japan, and China regarding health or novel food claims for functional food products made with novel proteins.

This work has evolved from my current research interests in protein from all resources. By 2050, the world population will require 70% more food than currently consumed. Total global consumption of protein, per person, has increased to 36 g and this is largely driven by the consumption of alternative proteins to dairy and meat which has increased by 15% since 1960. I believe that alternative protein streams of marine origin and utilisation of total raw material will increasingly contribute to food supply requirements, through production of marine-derived protein ingredients and advancement of technologies and methods to deal with by-products from meat, milk, and plant processing. I have a keen interest in the potential health effects of proteins and protein hydrolysates and peptides and also the functional application of these. I think there is currently no book that covers protein, hydrolysates of protein and their applications for the health and technofunctional attributes as well as processing technologies that could be applied to various protein sources to improve product production and expand markets for the clear consumer demand that exists. This book also takes the sustainability of protein resources into account and discusses the restrictions that exist concerning development.

It is clear that novel protein sources, including insect protein, will contribute to the global protein requirement in the coming years. This book also focuses on how processing technologies that are applied today to resources such as dairy could be applied to novel protein sources such as insect, macroalgal, and others. It is suitable for those who work in a multidisciplinary environment and brings together protein chemistry, process engineering, and food biotechnology disciplines. It also looks at the potential to exploit key market opportunities in sports, nutrition, and technofunctional protein applications.

1

Biological Roles and Production Technologies Associated with Bovine Glycomacropeptide

Shane Feeney[1,2], Lokesh Joshi[2], and Rita M. Hickey[1]

[1] *Teagasc Food Research Centre, Moorepark, Fermoy, Co. Cork, Ireland*
[2] *Advanced Glycoscience Research Cluster, National Centre for Biomedical Engineering Science, National University of Ireland Galway, Galway, Ireland*

1.1 Introduction

Glycomacropeptide (GMP) is a casein-derived whey peptide found in 'sweet' whey. The addition of chymosin to milk during cheese making enzymatically hydrolyses or cleaves the milk protein (kappa-casein) into two peptides, an insoluble peptide (para-kappa-casein) and a soluble hydrophilic glycopeptide (GMP), as shown in Figure 1.1. The larger peptide, para-kappa-casein, contains the amino acid residues 1–105 and becomes coagulated and incorporated into the cheese curd. The smaller peptide, which contains the amino acid residues 106–169 (GMP), becomes soluble and is incorporated into the whey (Walstra et al. 2006). GMP is the third most abundant whey protein, after beta-lactoglobulin and alpha-lactalbumin, accounting for approximately 15–25% (1.2–1.5 $g L^{-1}$) of the total whey protein (Thomä-Worringer et al. 2006). GMP is highly polar and has unique characteristics due to the absence of phenylalanine, tryptophan, tyrosine, histidine, arginine or cysteine residues (Neelima et al. 2013). The peptide is rich, however, in branched chain amino acids, such as isoleucine and valine (Marshall 2004; Krissansen 2007).

At least 13 genetic variants of bovine kappa-casein have been identified which have different post-translational modifications (PTMs) and vary in their level of phosphorylation and glycosylation (Thomä-Worringer et al. 2006). The average molecular weight for GMP is 7500 Da, whereas the highest recorded molecular weight is 9631 Da (Mollé and Léonil 2005). It has been suggested that the peptide has the ability to associate and dissociate under certain pH conditions, possibly explaining why molecular weights of between 14 and 30 kDa are observed via SDS-PAGE (Galindo-Amaya 2006; Farías et al. 2010).

Given the heterogeneity of GMP, there is no single isoelectric point (pI) assigned to GMP but the pI of the peptide portion is approximately 4, varying with PTM. Approximately 60% of GMP consists of O-linked carbohydrate chains which are composed of mainly galactose (gal), *N*-acetyl galactosamine (GalNAc) and *N*-neuraminic acid (Neu5Ac) attached at threonine residues. Saito et al. (1991) determined via high-performance liquid chromatography (HPLC) the distribution of monosaccharide,

Novel Proteins for Food, Pharmaceuticals, and Agriculture: Sources, Applications, and Advances, First Edition. Edited by Maria Hayes.

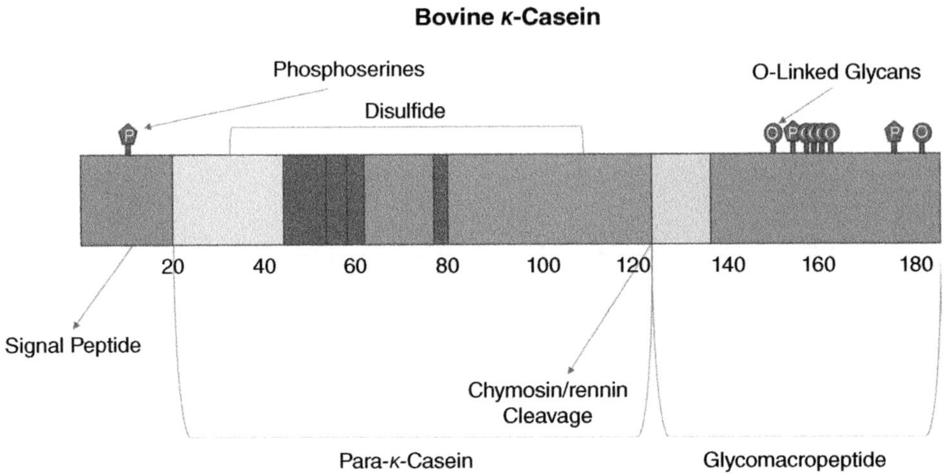

Figure 1.1 Bovine kappa-casein structure which varies depending on its post-translational modifications (phosphorylation and glycosylation). During cheese making, hydrolysis by chymosin releases the water-soluble fragment para-kappa-casein and the hydrophilic glycomacropeptide.

disaccharide, trisaccharide (straight and branched) and tetrasaccharide chains as 0.8%, 6.3%, 18.4%, 18.5% and 56.0%, respectively, while Mollé and Léonil (1995) identified five potential glycosylation sites using electrospray-ionisation mass spectrometry (ESI–MS) (Saito et al. 1991; Molle and Leonil 1995). Glycosylation influences the physical properties of GMP such as solubility (Taylor and Woonton 2009) and its emulsifying and foaming properties (Kreuß and Kulozik 2009). Moreover, variations in glycosylation can occur over the course of lactation (Recio et al. 2009; Neelima et al. 2013). For instance, colostrum GMP has an elevated glycan content (Guerin et al. 1974). Only GalNAc, Gal and Neu5Ac have been identified in GMP glycans from mature milk, but glycans from colostrum samples in addition contain *N*-acetylglucosamine (GlcNAc) and fucose (Fuc). Furthermore, a greater number of glycans and more complex structures have been identified in colostrum GMP (Fiat et al. 1988). A disialylated tetrasaccharide is the most abundant glycan present in mature GMP (Saito and Itoh 1992), and this high level of sialylation is vital for some of GMP's biological activities, as will be discussed later. Commercially available forms of GMP contain approximately 8% sialic acid (Arla Food Ingredients and Agropur Ingredients).

The aim of this chapter is to provide an overview of the state of the art in research regarding the functional role of GMP in maintaining and improving human health which is summarised in Table 1.1 and providing better knowledge on the isolation and detection of GMP as an ingredient in functional or medical foods.

1.2 Biological Properties Associated with Glycomacropeptide

1.2.1 Management of Phenylketonuria

Phenylketonuria (PKU) (OMIM 261600) is an autosomal recessive disorder caused by mutations in the phenylalanine hydroxylase (PAH) gene that encodes the enzyme which

Table 1.1 Biofunctional roles of GMP in improving human health.

Bioactivity	Reference
Management of PKU	Etzel (2004); Ney et al. (2008); Ney et al. (2016)
Ability to bind cholera toxin and *E. coli* enterotoxins	Kawasaki et al. (1992) – cholera toxin Isoda et al. (1999) – *E. coli* heat labile enterotoxins
Inhibition of bacterial and viral adhesion	Neeser et al. (1988) – cariogenic bacteria Neeser et al. (1994) – cariogenic bacteria Neeser et al. (1995) – cariogenic bacteria Schüpbach et al. (1996) – cariogenic bacteria Bruck et al. (2006a) – *E. coli, Salmonella typhimurium, Shigella flexneri* Nakajima et al. (2005) – *E. coli* Rhoades et al. (2005) – *E. coli* Feeney et al. (2017) – *E. coli* Kawasaki et al. (1993a) – human influenza virus Dosako et al. (1992) – Epstein–Barr virus
Suppression of gastric secretions	Beucher et al. (1994) – gastric secretions Yvon et al. (1994) – gastric secretions Stan EYa et al. (1983) – gastric secretions
Promotion of bifidobacterial growth	Brody (2000) – review has several examples Thomä-Worringer et al. (2006) – review has several examples Recio et al. (2009) – review has several examples O'Riordan et al. (2014) – review has several examples
Reduction in intestinal epithelial cell barrier dysfunction	Rong et al. (2015) – barrier function Feeney et al. (2017) – barrier function
Modulation of immune system responses	Brody (2000) – review has several examples Daddaoua et al. (2005) – anti-inflammatory activity Requena et al. (2008) – anti-inflammatory activity Requena et al. (2010) – anti-inflammatory activity Lopez-Posadas et al. (2010) – anti-inflammatory activity Cui et al. (2017) – ulcerative colitis Jimenez et al. (2012) – control of allergic diseases

catalyses the conversion of phenylalanine (Phe) to tyrosine (Tyr) in a reaction dependent on the essential PAH co-factor tetrahydrobiopterin (Blau et al. 2010). Tyr is an essential amino acid in PKU. Normal intake of dietary protein in untreated PKU causes Phe to accumulate in blood, leading to toxic concentrations of Phe in the brain and intellectual disability (Vockley et al. 2014). The main therapy for PKU is long-term adherence to a low-Phe diet that limits Phe intake from natural foods that contain protein, and supplements with special medical formulas that supply vitamins, minerals and all essential amino acids except Phe (MacLeod et al. 2009; Singh et al. 2014). The absence of Phe in GMP makes this peptide a valuable dietary ingredient for patients who are suffering from PKU. GMP can be made into a variety of palatable GMP medical foods that are low in Phe and high in protein content (Etzel 2004; Lim et al. 2007).

A number of preclinical studies in a PKU mouse model demonstrated that GMP supplemented with limiting amino acids supports growth and reduces concentrations of Phe in plasma and brain, improves bone status and reduces metabolic stress compared with an amino acid diet (Ney et al. 2008; Solverson et al. 2012a,b). Clinical evaluation of GMP found that in 11 PKU subjects, safety, acceptability, improved satiety and greater protein retention were observed with GMP medical formulas compared with amino acid medical formulas (Ney et al. 2008; MacLeod et al. 2009; van Calcar et al. 2009). Based on the results of these studies, GMP medical formulas first became available in the United States in 2010, with Cambrooke Therapeutics, Agropur Ingredients and Nestlé supplying formulas containing GMP for the treatment of PKU.

1.2.2 Anti-Infective Properties

It is now accepted that mucosal surface adherence of bacteria is required for colonisation and subsequent development of disease. When in the adherent state, these bacteria are more likely to survive as their resistance to cleansing mechanisms, immune factors, bacteriolytic enzymes and antibiotics is higher (Ofek et al. 2003). Bacterial surface components that mediate adherence are collectively known as adhesins (Moran et al. 2009). Several bacterial species utilise specific adhesins, or proteinaceous lectins, that bind glycan structures on the surface of host tissues to facilitate attachment. Milk glycans, such as those associated with GMP, have also been shown to obstruct specific host–pathogen interactions, including bacterial adhesion to the host ligands (Cravioto et al. 1991; Simon et al. 1997; Coppa et al. 2006). The glycans can structurally mimic epithelial cell surface glycans and thus function as decoys that pathogens can bind to instead of the host and thereby prevent infection (Sharon and Ofek 2000). GMP-derived peptides have been found in the intestinal lumen and blood of human (Chabance et al. 1998; Ledoux et al. 1999) and animal subjects (Fosset et al. 2002) after ingestion of the peptide and milk products, suggesting that GMP survives digestion and can be produced in the gastrointestinal tract where it can be absorbed by intestinal cells. It is believed that because GMP is O-glycosylated, some protein fragments are non-digestible, and so they reach the distal segment of the gastrointestinal tract intact where they can exert their anti-infective properties (Boutrou et al. 2008).

In terms of inhibiting bacterial adhesion, GMP has been shown to reduce the adherence of pathogens such as *Salmonella typhimurium*, *Shigella flexneri* and *E. coli* to certain intestinal cell lines (Nakajima et al. 2005; Rhoades et al. 2005; Bruck et al. 2006a,b). Strömqvist et al. (1995) demonstrated that GMP inhibited adhesion of *Helicobacter* to sections of stomach tissue. GMP has also been shown to inhibit binding of cholera toxin to Chinese hamster ovary cells at concentrations as low as 20 ppm (Kawasaki et al. 1992). Nakajima et al. (2005) found that GMP inhibits the association of EHEC O157 with Caco-2 cells and the association of EPEC with Caco-2 cells based on pathogen binding to its sialic acid component. The glycopeptide was also found to inhibit the adhesion of certain strains of EPEC to HT-29 cells (Rhoades et al. 2005) and the ETEC strain K88 to porcine intestinal cells and porcine mucus (Gonzalez-Ortiz et al. 2013, 2014). Recently, Feeney et al. (2017) found that GMP reduced intestinal epithelial cell barrier dysfunction and adhesion of enterohemorrhagic and enteropathogenic *E.coli in vitro*.

Another important property associated with GMP is the ability to inhibit the adhesion of cariogenic bacteria such as *Streptococcus mutans*, *S. sanguis* and *S. sobrinus* to oral surfaces, therefore modifying the composition of plaque bacteria to control acid production and, in turn, reducing the demineralization of enamel and promoting remineralization (Moynihan et al. 2000; Kashket and DePaola 2002; Janer et al. 2004). In this respect, GMP as an ingredient in dental hygiene products such as toothpaste and mouthwash to protect against tooth decay and plaque formation has received much attention in recent years.

Glycomacropeptide has also been shown to possess antibacterial properties and can inhibit the growth of both gram-positive, such as *Streptococcus mutans,* and gram-negative bacterial species, such as *Porphyromonas gingivalis* and *E. coli.* (Malkoski et al. 2001). In addition, GMP is effective in preventing haemagglutination by *Actinomyces viscosus, Streptococcus sanguis* and *Streptococcus mutans* (Neeser et al. 1988, 1994, 1995). Furthermore, bioactive peptides released by the pepsin treatment of GMP have been shown to have an antibacterial effect on *E. coli* in acidic media and also improve the resistance of *Lactobacillus rhamnosus* to acid stress (Robitaille et al. 2012). GMP is also known to have antiviral activity against human rotavirus (HRV) infection *in vitro* (Inagaki et al. 2014). Desialylated kappa-casein obtained by neuraminidase treatment exhibited anti-HRV activity, whereas deglycosylated kappa-casein obtained by *O*-glycosidase treatment lacked antiviral activity, indicating that glycans other than sialic acid were responsible for the activity. Kawaski et al. (1993a,b) demonstrated that GMP also inhibits haemagglutination by four strains of human influenza virus while Dosako et al. (1992) demonstrated that GMP prevents Epstein–Barr virus from inducing morphological transformations in peripheral lymphocytes.

1.2.3 Prebiotic

There are contradictory data on the effects of GMP on the growth promotion of *Bifidobacterium* and *Lactobacillus* strains (Azuma et al. 1984; Poch and Bezkorovainy 1991; Idota et al. 1994; Bruck et al. 2006a,b; Cicvárek et al. 2010; Hernandez-Hernandez et al. 2011). The differences observed between studies may be as a result of the quality and/or purity of the GMP used. Robitaille et al. (2012) demonstrated that highly purified GMP exhibits dose-dependent growth-promoting activity for lactic acid bacteria in a minimal culture medium. This study also concluded that the presence of glycans linked to caseinomacropeptide is not required for the growth-promoting activity. When added to diets given to infants (Bruck et al. 2006b) and piglets (Gustavo Hermes et al. 2013), caseinomacropeptide also increased lactobacilli populations in faeces and in ileal and proximal colonic digesta, respectively, suggesting that caseinomacropeptide could also be a growth promoter for lactic acid bacteria *in vivo*.

To address the influence of hydrolysis on the growth-promoting activity of caseinomacropeptide, effects of peptic and tryptic digests of the peptide on probiotic lactic acid bacteria growth were investigated (Robitaille and Champagne 2014). Pepsin treatment was effective in promoting the growth in milk of all probiotic bacteria tested, with biomass levels being improved significantly, by 1.7 to 2.6 times ($P < 0.05$), depending on the strain. Another study by Tian et al. (2014) demonstrated that GMP significantly improved the growth of probiotic bacteria supplemented in yogurt. GMP increased growth of *Streptococcus thermophilus* ($P < 0.05$), while it had little effect on the growth of *Lactobacillus bulgaricus* ($P > 0.05$). An addition of 1.5% GMP increased

Bifidobacterium animalis subsp. *lactis* (Bb12) growth in the yogurt fourfold relative to the control (no GMP). The authors concluded that the growth-promoting effect of GMP was not linked to its sialic acid content but might be related to its high Glutamic acid, Leucine, and Alanine content.

Recently, Sawin et al. (2015) reported that GMP feeding resulted in a significant decrease in the abundance of Proteobacteria and the genus *Desulfovibrio* in the caecal and faecal microbiota of wild-type and phenylketonuric mice. Increased short chain fatty acids (SCFA) and lower indices of inflammation were also observed in comparison to the casein and amino acid-based control diets. Ntemiri et al. (2017) used an *in vitro* batch fermentation (artificial colon model) to simulate colonic fermentation processes of two GMP products, a commercially available GMP concentrate and a semi-purified GMP concentrate, and lactose. Faecal samples were collected from healthy and frail older people. Sequencing analysis revealed that the commercial GMP preparation had a positive effect on the growth of health-promoting taxa such as *Coprococcus* and *Clostridium* cluster XIVb. GMP also increased SCFA production and sustained the diversity of the microbiota from healthy elderly inocula and to a lesser extent in inocula from frail elderly subjects under *in vitro* conditions. Taken together, a unifying feature and potential mechanism for the reported health-promoting properties associated with GMP may be its role in the development of a more host-friendly flora that may increase our ability to resist acute infection.

1.2.4 Immunomodulatory Activities Associated with GMP

1.2.4.1 Inflammation and Allergy

Most studies which examine the effects of pure GMP on cells of the immune system are performed *in vitro* and focus on lymphocytes. Allergic disorders can be identified by screening for regulation of allergen-specific immunoglobulin antibodies and T helper type 2 cells. There have been several *in vivo* studies which demonstrate GMP's ability to inhibit splenocyte propagation and how it can be used to suppress immune responses such as allergic reactions. GMP has been shown to inhibit lipopolysaccharide-induced splenocyte proliferation (Otani and Monnai 1993; Mikkelsen et al. 2005), to suppress interleukin (IL)-2 receptor expression in mouse CD4+ T-cells (Otani et al. 1996) and to block serum IgG antibody production by mouse lymphocytes (Monnai et al. 1998).

Investigations carried out on the effect of GMP on macrophages have shown that GMP increases the release of IL-1 receptor while leaving IL-1-beta unaffected, in a mouse monocytic cell line (Monnai and Otani 1997). GMP also increases the phagocytic activity and propagation of human macrophage cells (U937) at $10\,\mu g\,mL^{-1}$ (Li and Mine 2004a). Li and Mine (2004a) found that pepsin hydrolysis of GMP increased cell proliferation and phagocytic action while sialidase treatment decreased it; however, 70% of the activity was still retained suggesting that both the terminal sialic acids and polypeptide portions of GMP are required for optimal stimulatory effects.

Requena et al. (2009) demonstrated that GMP induced cytokine production in human monocytes via stimulation of the MAPK and the NF-kappa-B signal transduction pathways by upregulating the secretion of TNF, IL-1-beta, and IL-8 in a concentration-dependent fashion. The effect of GMP on cytokine secretion was confirmed using human primary blood monocytes (Requena et al. 2009). Jimenez et al. (2012) demonstrated that orally consumed GMP prevented allergic sensitization and reduced severity of the early phase reaction induced by antigen in cutaneous hypersensitivity and in anaphylaxis in ovalbumin-sensitized rats. GMP also displays immunoregulatory activity in allergic asthma models, as it effectively suppresses blood and lung eosinophilia, goblet cell hyperplasia, and collagen deposition in airways. The beneficial effect of GMP in asthma is associated with downregulation of IL-5 and IL-13 and upregulation of IL-10 expression in asthmatic lung tissue (Roldan et al. 2016). A study in rats demonstrated that the prebiotic action of GMP leads to an allergy-protective microbiota, resulting in an increase in TGF-beta production and a reduction in mast cell response to allergens (Jimenez et al. 2016). Results from a separate rat trial have recently indicated that GMP has an inhibitory effect on atopic dermatitis through downregulating the Th2-dominant immune response (Munoz et al. 2017).

1.2.4.2 Colitis

In ulcerative colitis (UC), both innate and adaptive immunity dysfunctions may contribute to disease pathogenesis (Baumgart and Carding 2007; Abraham and Cho 2009). With the exception of 5-aminosalicylate (5ASA)-containing drugs, most medications used for UC have side effects and significant complication risks. Therefore, a novel strategy for UC is highly desired. GMP is a promising candidate for UC treatment because of its ability to modulate gut microbiota and regulate immune responses (Daddaoua et al. 2005). Hvas et al. (2016) used GMP as a nutritional therapy to treat patients with active distal UC and found that it was well tolerated and accepted by patients, and the disease-modifying effect of GMP was similar to that of 5ASA. Ortega-Gonzalez et al. (2014) found that GMP exhibited intestinal anti-inflammatory effects in a lymphocyte transfer mouse model of colitis by reducing the activity of colonic myeloperoxidase and the percentage of $CD4^+$ interferon (IFN)-gamma$^+$ cells in mesenteric lymph nodes. Cui et al. (2017), in a recent study of famoxadone (OXZ)-induced mouse experimental UC, found that it could significantly improve morphological injury to intestinal mucosa in OXZ-induced UC mice to the same extent as regular treatment of UC. The study found that GMP could significantly reduce the expression of human mucosal addressin cell adhesion molecule-1 (MAdCAM-1), Cluster of differentiation 4 (CD4) and Cluster of differentiation 8 (CD8) in the lamina propria of the intestinal mucosa and significantly stimulate the secretion of sIgA to increase intestinal immunity. Furthermore, GMP was found to be directly involved in inhibiting the MAPK pathway and activating the TGF-beta-1/Smad signal transduction cascade, which could maintain immunological regulation of the intestinal mucosa and protect the function of the intestinal mucosal barrier.

1.2.5 Satiety

Glycomacropeptide is believed to stimulate the release of the hormone CCK in the gastrointestinal tract. CCK slows gastric emptying which may in turn promote satiety (Keogh et al. 2010). However, studies investigating the effect of GMP on food intake and satiety have resulted in mixed findings. Degen et al. (2001) demonstrated that oral GMP stimulates CCK hormone and increased satiety in human test subjects. Royle et al. (2008) found that GMP was associated with reduced fat mass in Wistar rats fed *ad libitum* for seven weeks with diets differing in protein type and amount. In addition, Veldhorst et al. (2009) showed a decrease in food energy intake at a subsequent meal 180 minutes after consumption of a test breakfast containing whey protein with GMP compared to whey protein without GMP. However, Burton-Freeman (2008) found that GMP had no effect on satiety or on food intake 75 minutes after consumption but did reduce daily food intake.

More recently, Keogh et al. (2010) compared the ability of GMP and a GMP-depleted whey protein concentrate to stimulate CCK, by making subjective measures of satiety and food intake for 20 overweight/obese male test subjects. Blood samples, CCK levels record, and subjective measures of satiety were collected before and 15, 30, 60, 90, 120, and 180 minutes after GMP consumption. A lunchtime meal of hot food was provided from which subjects ate *ad libitum* until satisfied. Energy and nutrient intakes from the food consumed were calculated. There was no significant difference in CCK levels, subjective measures of satiety, food intake, and energy intake between treatments across all GMP concentrations.

Chungchunlam et al. (2009), however, showed that the ingestion of preload drinks enriched with whey protein containing naturally present GMP resulted in subjects consuming a lower energy intake at a subsequent meal and reporting a greater feeling of fullness compared with a maltodextrin carbohydrate-enriched control beverage. However, the natural presence of GMP in whey hinders investigation of the satiating effect of whey protein, and the influence of GMP alone on satiety and food intake. The group (Chungchunlam et al. 2014) later found that whey GMP alone did not reduce subsequent food intake compared with a drink enriched with maltodextrin, but whey protein had a greater satiating effect than maltodextrin. Therefore, this study suggests that the presence of GMP in whey does not appear to be the cause of the observed effect of whey protein on satiety. Despite this controversy, several companies still claim that GMP promotes weight loss.

1.2.6 Anticarcinogenic

The development of colorectal cancer is a complex pathological process which involves multiple steps and stages, with changes from normal crypt foci to aberrant crypt foci (ACF), adenoma formation, expansion and eventual development to colorectal cancer (Belinsky et al. 1998; Takayama et al. 1998). Chen et al. (2014) utilised dimethylhydrazine (DMH)-induced colorectal cancer (CC) model rats to explore the effects of GMP on colorectal cancer. Rats with CC were orally given various concentrations of GMP or the same volume of phosphate-buffered saline for 15 weeks. The total numbers of ACF and crypts per focus in colon were scored, the methylation level of DNA extracted from colon was detected and the expression of p16 and mucin 2 (MUC2) proteins was measured. The results showed that although ACF were found in rats treated with GMP, their number was significantly decreased compared to that of the control rats. In addition, methylation and expression levels of p16 and MUC2 were also inhibited by GMP, which were more obvious in rats treated with higher concentrations of GMP. The study highlights the potential of GMP as nutritional therapy for preventing colorectal cancer.

1.3 Glycomacropeptide Production

Several methodologies have been employed to isolate and purify GMP from cheese whey, with the main challenge arising from separating GMP from the other whey proteins (Abd El-Salam (2006); Tullio et al. (2007); Neelima et al. (2013); Nakano and

Ozimek (2014). GMP is known to be soluble in 8% (w/v) trichloroacetic acid (TCA) solution, while all other whey proteins are precipitated. Therefore, GMP can be separated from sweet whey proteins by deproteinization with TCA (Nakano et al. 2002). GMP can also be separated from other whey proteins by cellulose acetate electrophoresis in pyridine/acetic acid because of pI differences between the proteins (Nakano et al. 2009). These methods are useful for preparation of GMP at laboratory scale, but are not suitable for large-scale production of GMP for commercial use. For this reason, isolation methods to generate highly pure GMP for use in humans have become an area of interest for many whey producers. Techniques such as thermal treatment, ultrafiltration (UF), ethanol precipitation, complexation, supercritical carbon dioxide processing, various chromatographic methods, and combinations thereof have been used to isolate GMP.

1.3.1 Thermal Treatment and Ethanol Precipitation

Thermal treatment and ethanol precipitation may be useful to eliminate significant amounts of proteins/peptides from GMP-containing streams (Saito et al. 1991; Berrocal and Neeser 1993; Martín-Diana et al. 2002; Li and Mine 2004b). Rojas and Torres (2013) evaluated the isolation and recovery on whey GMP by means of thermal treatment (90 °C). Eighteen samples (2 L each) of sweet whey, resuspended commercial whey (positive control), and acid whey (negative control) were processed. The indirect presence of GMP was verified using chemical tests and 15% SDS-PAGE. At 90 °C, bands of 14, 20, and 41 kDa bands were observed in sweet whey (Rojas and Torres 2013). These bands may correspond to oligomers of GMP. Peptide recovery showed an average of 1.5 $g L^{-1}$ (34.08%). The results indicate that industrial-scale GMP production is feasible. However, it should be noted that the yield and glycosylation characteristics of GMP are significantly influenced by the severity of milk heat treatment. The greater the severity of heating, the lower the quantity of GMP recovered, and the lower the quantity of sialic acid associated with the soluble GMP fraction (Taylor and Woonton 2009). Previously, Saito et al. (1991) reported complete loss of sialic acid from GMP by heating sweet whey at pH 3.0 and 98 °C for one hour. Since a considerable number of biological activities associated with GMP can be attributed to sialic acid (see above), determination of this glycan residue during isolation of GMP is important.

1.3.2 Complexation

Another approach for the isolation of GMP involves the use of an enzymatic cross-linking technique (Tolkach and Kulozik 2005). This method involves the pretreatment of WPC with the enzyme transglutaminase (Tgase) followed by microfiltration. GMP can be cross-linked to Tgase due to the presence of two glutamine and three lysine residues in its amino acid sequence. The native whey proteins show much less sensitivity to cross-linking by this enzyme due to their globular structure despite the presence of glutamine and lysine. The covalent linked GMP aggregates can be removed be means of microfiltration or diafiltration. However, the procedure only separates native whey proteins from GMP and may not be applicable to heated whey. As cross-linking of other milk proteins by Tgase changes their functional properties (Bönisch et al. 2007; Czernicka et al. 2009), it is possible that the same may occur for GMP isolated by Tgase.

Chitosan, a polysaccharide comprising co-polymers of glucosamine and *N*-acetyl-glucosamine, is a derivative of the naturally abundant biopolymer chitin (Singh 2015). The polycationic character of chitosan at acidic pH values allows the formation of complexes with negatively charged GMP molecules, inducing their flocculation (Nakano et al. 2004; Casal et al. 2005). Glycosylated GMP was found to have a higher affinity for chitosan when compared to non-glycosylated forms. The carboxylic groups in the carbohydrate moiety of the GMP increase the negative charge of the molecule and may play a role in the selective precipitation. The authors found that at pH 5.0, 0.08 mg mL^{-1} of chitosan completely removed the GMP whereas 70% of non-glycosylated GMP remained in solution. As the pH increased, the amount of chitosan required to ensure complete removal of GMP also increased by up to 0.19 and 0.34 mg mL^{-1} for pH 6.0 and 6.6, respectively (Casal et al. 2005). In another study by Nakano et al. (2006a), chitosan was mixed with sweet whey at pH 3.0 to form soluble chitosan-GMP complex. This complex was then separated from the other whey proteins and peptides by UF (molecular weight cut-off (MWCO) 100 000). The chitosan–GMP complex recovered in the retentate was then dialysed to give an insoluble complex, from which GMP was separated by elution with sodium chloride. The GMP fraction obtained accounted for an average 6.4% of dry non-dialyzable fraction of sweet whey, and contained 7.9% (w/w) sialic acid. The adsorption capacity and selectivity of these chitosan materials, however, are generally low.

A further study by Li et al. (2010) used beta-cyclodextrin which was immobilised to native chitosan beads by cross-linking with 1,6-hexamethylene di-isocyanate (HMDI). The resultant modified beads had a superior adsorption affinity for GMP. At pH 3.0, 90.23% of GMP was adsorbed, with a maximum adsorption capacity corresponding to 12.87 mg of sialic acid/g-adsorbent. Subsequent desorption experiments demonstrated that the modified beads could be regenerated and used for further cycles without significant decreases in capacity and selectivity.

1.3.3 Aqueous Two-Phase Systems

Aqueous two-phase extraction has been used widely as a mild separation method in many research fields (Oliveira et al. 2002). An aqueous two-phase system (ATPS) is formed when two water-soluble polymers, such as polyethylene glycol (PEG) and dextran, or a polymer and a salt are dissolved in water beyond a critical concentration at which two immiscible phases are formed (Gu and Glatz 2007). Da Silva et al. (2009) explored the possibility of partitioning of GMP using ATPS. In their study, the use of PEG and sodium citrate as ATPS for the partitioning of GMP was proposed. The results demonstrated that the partitioning of GMP depends on PEG molar mass, tie line length, pH, sodium chloride concentration, and temperature. The data indicated that GMP is preferentially partitioned into the PEG phase without addition of sodium chloride at pH 8.0. Larger tie line lengths and higher temperatures favour GMP partition to the PEG phase. Furthermore, it was verified that PEG molar mass and concentration have a slight effect on GMP partition. The increase in the molar mass of PEG induces a reduction of the protein solubility in the top PEG-rich phase, showing that the use of PEG1500 is beneficial for the extraction of GMP. A protein recovery higher than 85% was obtained in the top phase (PEG-rich phase) of these systems, demonstrating its suitability as a starting point for the separation of GMP.

A more recent study by Wu et al. (2012), using PEG/ammonium sulfate to separate GMP, found also that molecular weight of PEG and concentration of both the phases influence the partitioning of GMP. The study showed that 18% (w/w) PEG6000 and 15% (w/w) ammonium sulfate were the optimum conditions for protein recovery (69.2%) (Wu et al. 2012). However, the use of this system still requires another purification step, as do most of the methods described above.

1.3.4 Ultrafiltration

Ultrafiltration is a commonly used process for concentrating a dilute product in dairy streams, particularly in whey streams. UF is used to separate different sized whey proteins in a solution based on the membrane pore size or MWCO. The molecular weight of GMP is pH dependent and at neutral pH, self-association occurs, which forms oligomers through non-covalent bonds, whereas at low pH this is prevented through partial disassociation (Xu et al. 2000). Kawasaki et al. (1993b) used this molecular weight-pH dependency to influence the separation of GMP from other whey proteins including bovine serum albumin, immunoglobulin, beta-lactoglobulin, and alpha-lactalbumin in a UF system with a cut-off range of 20–50 kDa. This resulted in a GMP purity of between 18% and 63%. Once the permeate containing the GMP was adjusted to pH 7, it was passed through the same membrane where the purity of the GMP recovered was reported to be between 81% and 86%.

Li and Mine (2004c) compared the efficiency of three techniques of GMP isolation from WPI: TCA, ethanol precipitation, and UF. The TCA pretreatment recovered only sialo-GMP (glycosylated) and eliminated all contaminated proteins; however, the recovery rate was the lowest (6.7% of the initial WPI). Ethanol precipitation recovered 20.4% of GMP from WPI and 75.7% was glycosylated, but the heating process may lead to degradation of the glycans. UF was found to be the most effective in recovering GMP. The recovery rate was 33.9% with 81.6% sialo-GMP. The authors concluded that the carbohydrate profile of GMP varied widely and depended on the isolation method. Based on the high recovery of sialo-GMP, a combination of UF and anionic chromatography was deemed to be a suitable approach on an industrial scale.

Javanmard et al. (2012) isolated GMP for use as a source of protein for PKU patients from a whey protein solution (10% protein w/v) using dual UF disc membranes with 50 and 10 kDa cut-offs. The whey was passed through the membranes at pH 3.5, 4, and 4.5 with an ambient temperature of 25 °C. Diafiltration was used for purification of GMP in both UF phases. After UF, the Phe content of the whey at pH 4 was lowest, indicating a high purity and recovery rate of GMP. The disadvantage of employing some of these processes is that salts, solvents, buffers, and acids may be required, which introduce contaminants into the GMP product that must be removed through additional processing steps.

Supercritical carbon dioxide (SCO2) is an established method for the fractionation of alpha-lactalbumin and beta-lactoglobulin from WPC and WPI (Bonnaillie and Tomasula 2012). Bonnaillie et al. (2014) used SCO_2 to further fractionate WPI via acid precipitation of alpha-lactalbumin, beta-lactoglobulin, and the minor whey proteins to obtain GMP-enriched solutions. The process was optimised and alpha-lactalbumin precipitation maximised at low pH and a temperature (T) ≥65 °C, where beta-lactoglobulin with 84% purity and GMP with 58% purity were obtained, after UF and diafiltration to separate beta-lactoglobulin from the GMP solution. At 70 °C, beta-lactoglobulin also precipitated with alpha-lactalbumin, leaving a GMP-rich solution with up to 94% purity after UF.

1.3.5 Chromatography

1.3.5.1 Gel Filtration

Gel filtration chromatography separates molecules based on their size by filtration through a gel. In 1998, Nakano and Ozimek performed a study to determine the molecular weight of GMP and whether it is in fact influenced by pH (Nakano and Ozimek 1998). The authors found that GMP could be purified from sweet whey using a Sephacryl S-200 gel. To purify GMP from a solution of hydrolysed caseinate, the same group (Nakano and Ozimek 2000) used a number of processes involving (i) exclusion chromatography with Sephacryl S-200 at pH 7, (ii) addition of acid solution, pH 3.5 to precipitate contaminating proteins and peptides, and (iii) chromatography in Sephacryl S-200 at pH 3.5. The resulting GMP preparation was considered highly pure based on subsequent sialic acid determination, amino acid analysis, and SDS-PAGE. A follow-up study by the group (Nakano and Ozimek 2002) demonstrated that GMP could also be isolated from sweet whey by taking advantage of the pH-dependent difference in elution patterns between GMP and other whey proteins using Sephacryl S-200 gel chromatography. The isolated fraction contained a high molecular weight GMP accounting for 0.02% (w/v) of sweet whey. The GMP fraction contained 106 μg sialic acid/mg dry weight. Again, a high level of purity was observed through the use of gel electrophoresis and amino acid analysis.

1.3.5.2 Affinity

Affinity chromatography (AC) is a selective purification, separation, and enrichment technique in which the compound of interest must have a specific property that can be exploited during the affinity procedure (Hernández-Ledesma et al. 2014). Morr and Seo (1988) were among the first to use affinity chromatography for the isolation of GMP. The method involved a combination of chromatography steps including Con A-Sepharose affinity chromatography (Morr and Seo 1988). Saito et al. (1991) fractionated GMP into sialo- and asialo-GMP using peanut (*Arachis hypogoea*) lectin-affinity chromatography. Recently, Baieli et al. (2017) developed a GMP purification process based on the affinity of sialic acid for wheat germ agglutinin (WGA). After formation of chitosan beads and adsorption of WGA, the agglutinin was covalently attached with glutaraldehyde. Two matrices with different WGA density were assayed for GMP adsorption. GMP was recovered as an aggregate with an overall yield of 64%.

1.3.5.3 Hydrophobic Interaction

Hydrophobic interaction chromatography (HIC) relies on hydrophobic interactions between immobilised hydrophobic ligands and non-polar regions on the surface of proteins (Hernández-Ledesmaet al. 2013). Nakano and Ozimek (2000) investigated purification of GMP from the non-dialysable fraction of sweet whey by chromatography on phenyl-agarose and found that most of the whey proteins were adsorbed on the column, while the GMP was not. The estimated yield of the GMP fraction was approximately 1.6 g L^{-1} of sweet whey. Another study by the group, (Silva-Hernandez et al. 2002) used a combination of anion exchange and HIC to purify GMP from goat sweet whey. Approximately 0.06% (w/v) of sweet whey was recovered as GMP. Amino acid analysis of the GMP preparation demonstrated that the GMP was of high purity.

1.3.5.4 Ion Exchange

Ion exchange chromatography (IEX) separates proteins on the basis of their ionic charge. In anion exchange chromatography, negatively charged proteins are attracted to a positively charged group. Conversely, in cation exchange chromatography, positively charged proteins bind to a negatively charged medium. For preparation of GMP for use in food, ion exchange is one of the most common techniques used, considering, as mentioned, that GMP has a pI ~4, which is lower than the pI (>4.1) of the other major whey proteins (beta-lactoglobulin, alpha-lactalbumin, serum albumin, immunoglobulins, etc.) (Eigel et al. 1984) and therefore can be separated from these whey proteins based on the difference in pI.

Both cation and anion exchange chromatography have been used to purify GMP. Léonil and Mollé (1991) isolated GMP from sweet whey using cation exchange HPLC on a Mono-S column. More recently, LaClair et al. (2009), in their experiment on PKU diet preparation, reported that the Phe concentration ($5\,mg\,g^{-1}$ of product) in a commercially available GMP was too high, and thus, refined the product by using cation exchange chromatography to reduce the Phe level to $2.7\,mg\,g^{-1}$ protein equivalent (43% reduction) (Léonil and Mollé 1991; LaClair et al. 2009).

Glycomacropeptide has more commonly been purified from sweet whey and its fractions by anion exchange chromatography (Kawasaki et al. 1992; Nakano and Ozimek 1999b; Ayers et al. 2003). In 1992, Tanimoto et al. used whey prepared from lactic casein for large-scale production of GMP (Tanimoto et al. 1992). The whey was filtered, desalted, and lyophilised, resulting in a GMP preparation with a Phe content of 2.4% (w/w). The GMP was then purified by ion exchange chromatography, using QSepharose®, presenting Phe levels of 0.9% (w/w). In 1995, Outinen et al. developed a simple method for isolating a peptide fraction, consisting largely of GMP from emmental cheese whey (Outinen et al. 1995). The whey was first clarified by microfiltration, the pH was then adjusted to 5.0 and it was then passed through a polystyrene basic anion exchange resin column. The GMP which was selectively adsorbed on the column was released with dilute sodium chloride solution, subsequently desalted, concentrated by UF and then dried. Other whey components remained intact. About 70% of GMP originally present in whey was recovered in the prepared GMP fraction with a yield of 253 mg from 200 mL of clarified whey. The authors claimed a GMP purity of 70–80%.

The GMP purification process for scaled-up production proposed by Ayers et al. (2003) also involves using anion exchange (QA GiboCel TM, Life Techonologies Ltd) under conditions in which GMP is adsorbed, followed by elution, removal of impurities by numerous processes and recovery of purified GMP (Ayers et al. 2003). At pH 5.1, there was elution of 91% of the GMP. Anion exchange chromatography was also used by Nakano and Ozimek (1999a,b) to purify GMP from the non-dialysable fraction of whey using a DEAE-Sephacel column at two pH values (6.4 and 3.0). The column at pH 3.0 resulted in GMP of high purity and yielded $1\,g\,L^{-1}$ of whey (Nakano and Ozimek 1999b). The authors concluded that chromatography with DEAE Sephacel at pH 3.0 is an easy method that can be applied for large-scale production of GMP. The same authors (Nakano and Ozimek 1999b) also purified GMP from sweet whey dialysed in water by anion exchange chromatography on DEAE Sephacel at pH 2.0–4.5. Dialysis of sweet whey was shown to be important to maximise the yield of GMP adsorbed to the anion exchanger. Only highly sialylated GMP, accounting for approximately 55% of total sialic acid content, was adsorbed on the anion exchanger from non-dialysed sweet whey.

Xu et al. (2000) reported that GMP was selectively adsorbed from Cheddar cheese whey at pH 4.7 on a polystyrene anion exchange resin IRA93. The adsorbed material was then released with dilute sodium chloride solution, desalted and concentrated by UF using an Amicon YM 100 membrane. Tek et al. (2005) investigated the effect of whey conductivity, pH, and salt concentration of the elution buffer on GMP recovery and its extent of contamination using anion exchange chromatography. The results demonstrated that GMP recovery increased substantially with decreasing conductivity and increasing pH of the whey feed stream. Increasing the pH, but not increasing the conductivity, increased contamination of the GMP by primarily beta-lactoglobulin. A salt concentration of at least 0.1 M was required for complete elution of bound GMP. This study helped define conditions required for GMP recovery by a process chromatography system that uses food-grade buffers, operates at industrially relevant flow rates, and achieves up to 98% recovery.

A direct-capture anion exchange membrane adsorption process for the separation of a pure glycosylated GMP fraction of caseinomacropeptide was successfully developed at pilot plant scale (Kreuß and Kulozik 2009). The process included a desalting and concentration step, which was performed by a 10 kDa UF/diafiltration. The efficiency of the UF was strongly influenced by the pH of the solutions and displayed optimum performance at pH 4.1 for the eluate. A recent study by Nakano and Ozimek (2016) was undertaken to determine whether commercially available GMP can be refined by anion exchange chromatography with high reproducibility to provide GMP with no contaminating amino acids. Anion exchange chromatography on DEAE Sephacel was employed to remove impurities. The results demonstrated that Phe-containing proteins or peptides do not bind to the column, while most GMP, accounting for 93% of total recovered sialic acid, can bind to the column.

Overall, it may be that a combination of approaches for the isolation of GMP from whey may provide the best results. Etzel (2001) developed a process for producing a substantially pure GMP from whey using two steps. First, the GMP was recovered from whey using a strong anion exchanger (quaternary amino ethyl cellulose ion exchanger). The column was washed with water to remove contaminants and the GMP was then eluted with 0.5 M sodium chloride. The elute was adjusted to pH 7.15 using 1 M sodium hydroxide. In the second step, the crude GMP solution was then chromatographed using metal affinity chromatography (iminodiacetic acid agarose beads containing immobilised Cu sup2+ metal ion). The beads were washed with 0.02 M sodium phosphate and 0.5 M sodium chloride pH 7.15. The effluent up to this point was the substantially purified GMP product. The invention also describes a method for large-scale production of GMP in a substantially pure form using fewer steps than methods of similar capability in purity.

Doultani et al. (2003) also used a two-step approach to purify GMP through the use of cation and anion exchange. A cation exchange resin was used to recover WPI from sweet whey and the effluent was fed to an anion exchange resin to recover GMP. The authors reported that nearly all of the major whey proteins (alpha-lactalbumin, beta-lactoglobulin, immunoglobulin G, and serum albumin) and about half of the total Kjeldahl nitrogen were recovered by the cation exchange resin. The anion exchange resin recovered nearly all the GMP from the effluent of the cation exchanger. This was considered the first process to simultaneously manufacture WPI and GMP from a single stream of whey, increasing the value obtained from whey.

Combining ion exchange and UF, however, remains the most popular choice for large-scale production of GMP (Kawasaki et al. 1994; Etzel 2004). In view of the different processes available for the production of GMP and the importance of its structure (in particular glycosylation) in its various biological activities, it is necessary to know whether the methodology used to obtain GMP influences its structure and activity (Recio et al. 2009). Similarly, storage and technological processes, such as heating, may affect the glycosylation degree or the chemical stability of GMP (Lieske et al. 2004) and should be considered when selecting conditions for large-scale production of GMP.

1.4 Detection of Glycomacropeptide

Isolation of GMP requires assays or methods to track its presence at each step of purification. Monitoring the presence of GMP is also important in the detection of cheese whey in milk as a marker of adulteration (Neelima et al. 2013). The section below summarises the various methods employed for measuring GMP. However, the reader is referred to Neelima et al. (2013) and Nakano and Ozimek (2014) for further detail.

The quantification of GMP in dairy protein fractions and products, including enzymatically treated casein, cheese whey, WPC or WPI, is generally performed through the identification and measurement of characteristic peaks with HPLC or capillary zone electrophoresis, coupled with mass spectrometry or gel electrophoresis to determine the molecular weight and identity of isolated peptides. Colorimetric, immunological, and biosensor-based methods are also available. The content of GMP can also be monitored by measuring ultraviolet (UV) absorbance at wavelength 205–230 nm to determine its peptide amide bond (Peterson 1983). UV absorbance at 280 nm can be used to estimate the amount of contaminating protein/peptide other than GMP (El-Salam et al. 1996; Nakano and Ozimek 1999b, 2000) as GMP has no aromatic amino acids and therefore displays no UV absorbance at 280 nm (Léonil and Mollé 1991; Abd El-Salam 2006).

1.4.1 Chromatography

Kawasaki et al. (1992) described a method for determination of glycosylated and non-glycosylated GMP by size exclusion chromatography using two coupled TSK gel G3000 PW columns and anion exchange chromatography on a Mono Q HR 5/5 column. The method was able to distinguish heterogeneous GMP by the variations in carbohydrate moieties and to determine GMP content without pretreatment of milk samples with thiobarbituric acid. In 2003, Ferreira and Oliveira used RP-HPLC with a polystyrene divinylbenzene column on adulterated samples (Ferreira and Oliveira 2003). The authors used pH 4.6 filtrate from the samples, and eluted using the gradient of two solvents. Solvent A was 0.1% trifluoroacetic acid (TFA) in water and solvent B was 95% acetonitrile–5% water–0.1% TFA. The eluate was monitored by a UV detector at 214 nm and enabled the separation of three peaks, two of which corresponded to the non-glycosylated forms of GMP and the third to the less well-resolved glycosylated GMP components.

1.4.2 Capillary Electrophoresis

In capillary electrophoresis (CE), small amounts of fluids are injected into a separation channel using plug injection. Substances are separated based on their electrophoretic mobility, which is proportional to their charge to size ratio in the interior of a small capillary filled with an electrolyte (Neelima et al. 2013).

In 1995, Otte described a CE method for the separation of the major whey proteins using acidic and basic conditions (Otte et al. 1995). Under acidic conditions (70 mM phosphate, pH 2.5), a GMP peak was separated from the other whey proteins. Also in 1995, van Riel and Olieman used the CE method for the detection of rennet whey solids in skim milk and buttermilk powder, based on the non-glycosylated variants of GMP (van Riel and Olieman 1995). Cherkaoui et al. (1997) then validated a method for the analysis and quantitative determination of a GMP mixture. Recio et al. (1996, 2000) demonstrated the suitability of the CE method to identify GMP in liquid milk samples and observed the presence of detectable GMP in stored UHT milk samples incubated with psychrotrophs.

1.4.3 SDS-PAGE

Galindo-Amaya (2006) developed a SDS-PAGE-based method for the detection of GMP in milk. The authors isolated GMP from milk and adulterated milk samples with sequential precipitation in TCA, treatment with ethanol-ether, and re-suspension in Tris–HCl buffer. Precipitates were analysed by SDS-PAGE and GMP was evidenced as a trimer of 20.8 kDa in samples of sweet whey, and mixtures of whey and milk (1%, 5%, 10%, and 50%), but absent in samples of acid whey and raw milk. The authors claim that the detection of GMP in milk using SDS-PAGE is a sensitive and specific method of detecting milk adulteration in whey, to levels as low as 1%.

1.4.4 Colorimetric

Other methods to measure GMP are based on determination of its glycan moieties, namely sialic acid, which are reviewed extensively by (Nakano and Ozimek 2014). The use of a sialic acid assay to track GMP isolation and purification is important considering the numerous biological activities of GMP which can be attributed to sialic acid (Fukuda et al. 2004). GMP sialic acid has been analysed using the acidic ninhydrin reaction (Yao et al. 1989), a fluorimetric method involving 14% (w/v) TCA and sodium metaperiodate (Rao et al. 2012), and also by HPLC (Fernando and Woonton 2010). Nakano and Ozimek (2014) describe the thiobarbituric acid reaction with 1-propanol as a chromophore extracting solvent as an inexpensive, practical and specific technique. Sephacryl S-200 gel filtration chromatography, cellulose acetate electrophoresis, and SDS-PAGE are the major techniques used to identify sialic acid specific to GMP (Nakano and Ozimek 1999a; Nakano et al. 2006b, 2007, 2009).

1.4.5 Immunological

A number of methods based on immunochemical assays have been developed to detect and quantify GMP, such as biosensor-based immunoassays (Haasnoot et al. 2004), immunoblot (Chávez et al. 2008), immunochromatographic testing (Oancea 2009), and inhibition enzyme-linked immunosorbent assay (ELISA) (Bitri et al. 1993; Picard et al. 1994; Bremer et al. 2008). Immunoblot and immunochromatographic tests have low sensitivity and biosensor methods require expensive equipment. ELISA is the most frequently used immunoassay in routine test quantification because it is a simple, sensitive, and reliable assay system that allows the use of small sample volumes and high sample numbers. In 2012, Chávez et al. developed and validated a sandwich ELISA system using polyclonal antibodies toward pure bovine GMP that affords a more sensitive analysis of raw milk for the routine detection of GMP as an indicator of adulteration with cheese whey (Chávez et al. 2012).

1.5 Conclusion

In summary, GMP is a bioactive peptide from cheese whey with unique chemical properties in terms of glycosylation, absence of aromatic amino acids and self-association. It is well known that GMP exhibits several useful biological activities, including protection from toxins, bacteria, and viruses and regulation of the immune system. In addition to these biological properties, the peptide has a number of functional physical properties (emulsification and foaming properties, gel formation, wide pH range, solubility, wide heat stability, etc.) that make it a valuable ingredient. For these reasons, dairy companies have invested heavily in methods of selectively isolating intact GMP from whey in order to produce a premium functional ingredient in a commercially viable manner. The main aim is to market GMP for application in the diet of hepatic and PKU patients and also for inclusion in infant formula and sports supplements.

Currently, anion exchange is the technique of choice for large-scale production of GMP from whey. However, commercial production of Phe-free GMP still remains a challenge. Agropur Ingredients recently received the Breakthrough Award for Dairy Ingredient Innovation for its development of highly pure GMP. The product contains just 1–2 mg of Phe/g protein, making it highly functional for PKU products. PKU sphere™ is a powdered, low-phenylalanine protein substitute developed by Nestlé which also contains approximately 1 mg of Phe/g protein. Following on from this, further research should explore ways of developing efficient techniques to purify GMP with undetectable levels of Phe on a commercial scale.

The addition of GMP to infant formula has also raised concern due to its high threonine content (12–13 threonine residues) which increases the occurrence of hyperthreoninaemia in infants fed formula containing GMP versus breastfed infants (Rigo et al. 2001). However, more recent studies suggest this increased occurrence of hyperthreoninaemia is due to differences in threonine metabolism among the infants tested (Sandström et al. 2008). Taking this into consideration, future studies are required to increase our knowledge of the biological and structural functions of GMP while also focusing on the safety of its inclusion as a food ingredient.

References

Abd El-Salam, M. (2006). Separation of casein glycomacropeptide from whey: methods of potential industrial application. *International Journal of Dairy Science* 1 (1): 93–99.

Abraham, C. and Cho, J.H. (2009). Inflammatory bowel disease. *New England Journal of Medicine* 361 (21): 2066–2078.

Ayers, J. S., K. P. Coolbear, D. F. Elgar, and M. Pritchard. 2003. Process for isolating glycomacropeptide from dairy products with a phenylalanine impurity of 0.5% w/w. Google Patents.

Azuma, N., Yamauchi, K., and Mitsuoka, T. (1984). Bifidus growth-promoting activity of a glycomacropeptide derived from human K-casein. *Agricultural and Biological Chemistry* 48 (8): 2159–2162.

Baieli, M.F., Urtasun, N., Martinez, M.J. et al. (2017). Affinity chromatography matrices for depletion and purification of casein glycomacropeptide from bovine whey. *Biotechnology Progress* 33 (1): 171–180.

Baumgart, D.C. and Carding, S.R. (2007). Inflammatory bowel disease: cause and immunobiology. *Lancet* 369 (9573): 1627–1640.

Belinsky, S.A., Nikula, K.J., Palmisano, W.A. et al. (1998). Aberrant methylation of p16(INK4a) is an early event in lung cancer and a potential biomarker for early diagnosis. *Proceedings of the National Academy of Sciences of the United States of America* 95 (20): 11891–11896.

Berrocal, R. and J.-R. Neeser. 1993. Production of kappa-caseino-glycomacropeptide. Google Patents.

Beucher, S., Levenez, F., Yvon, M., and Corring, T. (1994). Effects of gastric digestive products from casein on CCK release by intestinal cells in rat. *Journal of Nutritional Biochemistry* 5 (12): 578–584.

Bitri, L., Rolland, M., and Besançon, P. (1993). Immunological detection of bovine caseinomacropeptide in ovine and caprine dairy products. *Milchwissenschaft* 48 (7): 367–371.

Blau, N., van Spronsen, F.J., and Levy, H.L. (2010). Phenylketonuria. *Lancet* 376 (9750): 1417–1427.

Bönisch, M.P., Huss, M., Weitl, K., and Kulozik, U. (2007). Transglutaminase cross-linking of milk proteins and impact on yoghurt gel properties. *International Dairy Journal* 17 (11): 1360–1371.

Bonnaillie, L.M., Qi, P., Wickham, E., and Tomasula, P.M. (2014). Enrichment and purification of casein glycomacropeptide from whey protein isolate using supercritical carbon dioxide processing and membrane ultrafiltration. *Foods* 3 (1): 94–109.

Bonnaillie, L.M. and Tomasula, P.M. (2012). Fractionation of whey protein isolate with supercritical carbon dioxide to produce enriched α-lactalbumin and β-lactoglobulin food ingredients. *Journal of Agricultural and Food Chemistry* 60 (20): 5257–5266.

Boutrou, R., Jardin, J., Blais, A. et al. (2008). Glycosylations of kappa-casein-derived caseinomacropeptide reduce its accessibility to endo- but not exointestinal brush border membrane peptidases. *Journal of Agricultural and Food Chemistry* 56 (17): 8166–8173.

Bremer, M.G., Kemmers-Voncken, A.E., Boers, E.A. et al. (2008). Enzyme-linked immunosorbent assay for the detection of bovine rennet whey powder in milk powder and buttermilk powder. *International Dairy Journal* 18 (3): 294–302.

Brody, E.P. (2000). Biological activities of bovine glycomacropeptide. *British Journal of Nutrition* 84 ((Suppl 1): S39–S46.

Bruck, W.M., Kelleher, S.L., Gibson, G.R. et al. (2006a). The effects of alpha-lactalbumin and glycomacropeptide on the association of CaCo-2 cells by enteropathogenic

Escherichia coli, Salmonella typhimurium and *Shigella flexneri. FEMS Microbiology Letters* 259 (1): 158–162.
Bruck, W.M., Redgrave, M., Tuohy, K.M. et al. (2006b). Effects of bovine alpha-lactalbumin and casein glycomacropeptide-enriched infant formulae on faecal microbiota in healthy term infants. *Journal of Pediatric Gastroenterology and Nutrition* 43 (5): 673–679.
Burton-Freeman, B.M. (2008). Glycomacropeptide (GMP) is not critical to whey-induced satiety, but may have a unique role in energy intake regulation through cholecystokinin (CCK). *Physiology and Behavior* 93 (1–2): 379–387.
Casal, E., Corzo, N., Moreno, F., and Olano, A. (2005). Selective recovery of glycosylated caseinmacropeptide with chitosan. *Journal of Agricultural and Food Chemistry* 53 (4): 1201–1204.
Chabance, B., Marteau, P., Rambaud, J.C. et al. (1998). Casein peptide release and passage to the blood in humans during digestion of milk or yogurt. *Biochimie* 80 (2): 155–165.
Chávez, N.A., Jauregui, J., Palomares, L.A. et al. (2012). A highly sensitive sandwich ELISA for the determination of glycomacropeptide to detect liquid whey in raw milk. *Dairy Science and Technology* 92 (2): 121–132.
Chávez, N.A., Salinas, E., Jauregui, J. et al. (2008). Detection of bovine milk adulterated with cheese whey by western blot immunoassay. *Food and Agricultural Immunology* 19 (4): 265–272.
Chen, Q., Wang, H., Zhu, C. et al. (2014). Anti-apoptotic effects of milk-derived casein glycomacropeptide on mice with ulcerative colitis. *Food and Agricultural Immunology* 25 (4): 453–466.
Cherkaoui, S., Doumenc, N., Tachon, P. et al. (1997). Development of a capillary zone electrophoresis method for caseinoglycomacropeptide determination. *Journal of Chromatography A* 790 (1): 195–205.
Chungchunlam, S.M., Moughan, P.J., Awati, A., and Morton, H.R. (2009). The influence of whey protein and glycomacropeptide on satiety in adult humans. *Physiology and Behavior* 96 (1): 162–168.
Chungchunlam, S.M., Henare, S.J., Ganesh, S., and Moughan, P.J. (2014). Effect of whey protein and glycomacropeptide on measures of satiety in normal-weight adult women. *Appetite* 78: 172–178.
Cicvárek, J., Čurda, L., Elich, O. et al. (2010). Effect of caseinomacropeptide concentrate addition on the growth of bifidobacteria. *Czech Journal of Food Sciences* 28 (6): 485–494.
Coppa, G.V., Zampini, L., Galeazzi, T. et al. (2006). Human milk oligosaccharides inhibit the adhesion to Caco-2 cells of diarrheal pathogens: *Escherichia coli, Vibrio cholerae*, and *Salmonella fyris. Pediatric Research* 59 (3): 377–382.
Cravioto, A., Tello, A., Villafan, H. et al. (1991). Inhibition of localized adhesion of enteropathogenic *Escherichia coli* to HEp-2 cells by immunoglobulin and oligosaccharide fractions of human colostrum and breast milk. *Journal of Infectious Diseases* 163 (6): 1247–1255.
Cui, Y., Zhu, C., Ming, Z. et al. (2017). Molecular mechanisms by which casein glycomacropeptide maintains internal homeostasis in mice with experimental ulcerative colitis. *PLoS One* 12 (7): e0181075.
Czernicka, M., Domagała, J., Sady, M., and Wieteska, I. (2009). Functional properties of milk proteins modified by transglutaminase depending on incubation conditions with the enzyme. *Biotechnology in Animal Husbandry* 25 (5–6-2): 737–743.

Da Silva, C., Coimbra, J.S.d.R., Rojas, E.E.G., and Teixeira, J. (2009). Partitioning of glycomacropeptide in aqueous two-phase systems. *Process Biochemistry* 44 (11): 1213–1216.

Daddaoua, A., Puerta, V., Zarzuelo, A. et al. (2005). Bovine glycomacropeptide is anti-inflammatory in rats with hapten-induced colitis. *Journal of Nutrition* 135 (5): 1164–1170.

Degen, L., Matzinger, D., Drewe, J., and Beglinger, C. (2001). The effect of cholecystokinin in controlling appetite and food intake in humans. *Peptides* 22 (8): 1265–1269.

Dosako, S., H. Kusano, E. Deya, and T. Idota. 1992. Infection protectant. Google Patents.

Doultani, S., Turhan, K., and Etzel, M. (2003). Whey protein isolate and glycomacropeptide recovery from whey using ion exchange chromatography. *Journal of Food Science* 68 (4): 1389–1395.

Eigel, W., Butler, J., Ernstrom, C. et al. (1984). Nomenclature of proteins of cow's milk: fifth revision1. *Journal of Dairy Science* 67 (8): 1599–1631.

El-Salam, M.A., El-Shibiny, S., and Buchheim, W. (1996). Characteristics and potential uses of the casein macropeptide. *International Dairy Journal* 6 (4): 327–341.

Etzel, M. R. 2001. Production of substantially pure kappa casein macropeptide. Google Patents.

Etzel, M.R. (2004). Manufacture and use of dairy protein fractions. *Journal of Nutrition* 134 (4): 996S–1002S.

Farías, M., Martinez, M., and Pilosof, A. (2010). Casein glycomacropeptide pH-dependent self-assembly and cold gelation. *International Dairy Journal* 20 (2): 79–88.

Feeney, S., Ryan, J.T., Kilcoyne, M. et al. (2017). Glycomacropeptide reduces intestinal epithelial cell barrier dysfunction and adhesion of entero-hemorrhagic and entero-pathogenic *Escherichia coli* in vitro. *Foods* 6 (11): ii.

Fernando, S.F. and Woonton, B.W. (2010). Quantitation of N-acetylneuraminic (sialic) acid in bovine glycomacropeptide (GMP). *Journal of Food Composition and Analysis* 23 (4): 359–366.

Ferreira, I.M. and Oliveira, M. (2003). Determination of caseinomacropeptide by an RP-HPLC method and monitoring of the addition of rennet whey to powdered milk. *Journal of Liquid Chromatography and Related Technologies* 26 (1): 99–107.

Fiat, A.M., Chevan, J., Jolles, P. et al. (1988). Structural variability of the neutral carbohydrate moiety of cow colostrum kappa-casein as a function of time after parturition. Identification of a tetrasaccharide with blood group I specificity. *European Journal of Biochemistry* 173 (2): 253–259.

Fosset, S., Fromentin, G., Gietzen, D.W. et al. (2002). Peptide fragments released from Phe-caseinomacropeptide in vivo in the rat. *Peptides* 23 (10): 1773–1781.

Fukuda, S.P., Roig, S.M., and Prata, L.F. (2004). Correlation between acidic ninhydrin and HPLC methods to evaluate fraudulent addition of whey in milk. *Le Lait* 84 (5): 501–512.

Galindo-Amaya, L.M. (2006). Standardization of glycomacropeptide detection with SDS-PAGE as a milk adulteration index. *Revista Cientifica* 16 (3): 308–314.

Gonzalez-Ortiz, G., Hermes, R.G., Jimenez-Diaz, R. et al. (2013). Screening of extracts from natural feed ingredients for their ability to reduce enterotoxigenic *Escherichia coli* (ETEC) K88 adhesion to porcine intestinal epithelial cell-line IPEC-J2. *Veterinary Microbiology* 167 (3–4): 494–499.

Gonzalez-Ortiz, G., Perez, J.F., Hermes, R.G. et al. (2014). Screening the ability of natural feed ingredients to interfere with the adherence of enterotoxigenic *Escherichia coli*

(ETEC) K88 to the porcine intestinal mucus. *British Journal of Nutrition* 111 (4): 633–642.

Gu, Z. and Glatz, C.E. (2007). Aqueous two-phase extraction for protein recovery from corn extracts. *Journal of Chromatography B* 845 (1): 38–50.

Guerin, J., Alais, C., Jolles, J., and Jolles, P. (1974). Kappa-casein from bovine colostrum. *Biochimica et Biophysica Acta* 351 (2): 325–332.

Gustavo Hermes, R., Molist, F., Francisco Perez, J. et al. (2013). Casein glycomacropeptide in the diet may reduce *Escherichia coli* attachment to the intestinal mucosa and increase the intestinal lactobacilli of early weaned piglets after an enterotoxigenic *E. coli* K88 challenge. *British Journal of Nutrition* 109 (6): 1001–1012.

Haasnoot, W., Smits, N.G., Kemmers-Voncken, A.E., and Bremer, M.G. (2004). Fast biosensor immunoassays for the detection of cows' milk in the milk of ewes and goats. *Journal of Dairy Research* 71 (3): 322–329.

Hernandez-Hernandez, O., Sanz, M.L., Kolida, S. et al. (2011). In vitro fermentation by human gut bacteria of proteolytically digested caseinomacropeptide nonenzymatically glycosylated with prebiotic carbohydrates. *Journal of Agricultural and Food Chemistry* 59 (22): 11949–11955.

Hernández-Ledesma, B., García-Nebot, M.J., Fernández-Tomé, S. et al. (2014). Dairy protein hydrolysates: peptides for health benefits. *International Dairy Journal* 38 (2): 82–100.

Hernández-Ledesmaet al (2013). *Food Analysis by HPLC*, 3e (ed. L.M.L. Nollet and F. Toldrá), 69–96. Boca Raton: CRC Press.

Hvas, C.L., Dige, A., Bendix, M. et al. (2016). Casein glycomacropeptide for active distal ulcerative colitis: a randomized pilot study. *European Journal of Clinical Investigation* 46 (6): 555–563.

Idota, T., Kawakami, H., and Nakajima, I. (1994). Growth-promoting effects of N-acetylneuraminic acid-containing substances on bifidobacteria. *Bioscience, Biotechnology, and Biochemistry* 58 (9): 1720–1722.

Inagaki, M., Muranishi, H., Yamada, K. et al. (2014). Bovine kappa-casein inhibits human rotavirus (HRV) infection via direct binding of glycans to HRV. *Journal of Dairy Science* 97 (5): 2653–2661.

Isoda, H., Kawasaki, Y., Tanimoto, M., Dosako, S. and Idota, T. (1999). Use of compounds containing or binding sialic acid to neutralize bacterial toxins. *European Patent* 385112.

Janer, C., Pelaez, C., and Requena, T. (2004). Caseinomacropeptide and whey protein concentrate enhance *Bifidobacterium lactis* growth in milk. *Food Chemistry* 86 (2): 263–267.

Javanmard, M., Lighvani, H., Ghiasi Tarzi, B., and Rashidi, A. (2012). Extraction and purification of glycomacropeptide from whey to be used as a source of protein for phenylketonuria patients. *Iranian Journal of Nutrition Sciences and Food Technology* 7 (2): 9–16.

Jimenez, M., Cervantes-Garcia, D., Munoz, Y.H. et al. (2016). Novel mechanisms underlying the therapeutic effect of Glycomacropeptide on allergy: change in gut microbiota, upregulation of TGF-beta, and inhibition of mast cells. *International Archives of Allergy and Immunology* 171 (3–4): 217–226.

Jimenez, M., Chavez, N.A., and Salinas, E. (2012). Pretreatment with glycomacropeptide reduces allergen sensitization, alleviates immediate cutaneous hypersensitivity and protects from anaphylaxis. *Clinical and Experimental Immunology* 170 (1): 18–27.

Kashket, S. and DePaola, D.P. (2002). Cheese consumption and the development and progression of dental caries. *Nutrition Reviews* 60 (4): 97–103.

Kawasaki, Y., S. Dosako, M. Shimatani, and T. Idota. 1994. Process for producing k-casein glycomacropeptides. Google Patents.

Kawasaki, Y., Isoda, H., Shinmoto, H. et al. (1993a). Inhibition by kappa-casein glycomacropeptide and lactoferrin of influenza virus hemagglutination. *Bioscience, Biotechnology, and Biochemistry* 57 (7): 1214–1215.

Kawasaki, Y., Isoda, H., Shinmoto, H. et al. (1993b). Inhibition by κ-casein glycomacropeptide and lactoferrin of influenza virus hemagglutination. *International Journal of Cancer* 57: 1214–1215.

Kawasaki, Y., Isoda, H., Tanimoto, M. et al. (1992). Inhibition by lactoferrin and kappa-casein glycomacropeptide of binding of cholera toxin to its receptor. *Bioscience, Biotechnology, and Biochemistry* 56 (2): 195–198.

Keogh, J.B., Woonton, B.W., Taylor, C.M. et al. (2010). Effect of glycomacropeptide fractions on cholecystokinin and food intake. *British Journal of Nutrition* 104 (2): 286–290.

Kreuß, M. and Kulozik, U. (2009). Separation of glycosylated caseinomacropeptide at pilot scale using membrane adsorption in direct-capture mode. *Journal of Chromatography A* 1216 (50): 8771–8777.

Krissansen, G.W. (2007). Emerging health properties of whey proteins and their clinical implications. *Journal of the American College of Nutrition* 26 (6): 713S–723S.

LaClair, C.E., Ney, D.M., MacLeod, E.L., and Etzel, M.R. (2009). Purification and use of glycomacropeptide for nutritional management of phenylketonuria. *Journal of Food Science* 74 (4): E199–E206.

Ledoux, N., Mahe, S., Duarry, M. et al. (1999). Intraluminal immunoreactive caseinomacropeptide after milk protein ingestion in humans. *Nahrung* 43 (3): 196–200.

Léonil, J. and Mollé, D. (1991). A method for determination of macropeptide by cation-exchange fast protein liquid chromatography and its use for following the action of chymosin in milk. *Journal of Dairy Research* 58 (3): 321–328.

Li, C., Song, X., Hein, S., and Wang, K. (2010). The separation of GMP from milk whey using the modified chitosan beads. *Adsorption* 16 (1): 85–91.

Li, E. W. Y. and Y. Mine. (2004a). Immunoenhancing effects of bovine glycomacropeptide and its derivatives on the proliferative response and phagocytic activities of human macrophagelike cells, U937. *Journal of Agricultural and Food Chemistry* 52 (9): 2704–2708.

Li, E. W. and Y. Mine. (2004b). Technical note: Comparison of chromatographic profile of glycomacropeptide from cheese whey isolated using different methods. *J Dairy Sci* 87(1):174–177.

Li, E. W. and Y. Mine. (2004c). Comparison of chromatographic profile of glycomacropeptide from cheese whey isolated using different methods. *Journal of dairy science* 87 (1):174–177.

Lieske, B., Konrad, G., and Kleinschmidt, T. (2004). Isolation of caseinomacropeptide from rennet whey by a multi-stage ultrafiltration process II. Influence of pH and heating on the carbohydrate moiety of glycomacropeptide. *Milchwissenschaft* 59 (5–6): 291–294.

Lim, K., van Calcar, S.C., Nelson, K.L. et al. (2007). Acceptable low-phenylalanine foods and beverages can be made with glycomacropeptide from cheese whey for individuals with PKU. *Molecular Genetics and Metabolism* 92 (1–2): 176–178.

Lopez-Posadas, R., Requena, P., Gonzalez, R. et al. (2010). Bovine glycomacropeptide has intestinal antiinflammatory effects in rats with dextran sulfate-induced colitis. *Journal of Nutrition* 140 (11): 2014–2019.

MacLeod, E.L., Gleason, S.T., van Calcar, S.C., and Ney, D.M. (2009). Reassessment of phenylalanine tolerance in adults with phenylketonuria is needed as body mass changes. *Molecular Genetics and Metabolism* 98 (4): 331–337.

Malkoski, M., Dashper, S.G., O'Brien-Simpson, N.M. et al. (2001). Kappacin, a novel antibacterial peptide from bovine milk. *Antimicrobial Agents and Chemotherapy* 45 (8): 2309–2315.

Marshall, K. (2004). Therapeutic applications of whey protein. *Alternative Medicine Review* 9 (2): 136–156.

Martín-Diana, A., Fraga, M., and Fontecha, J. (2002). Isolation and characterisation of caseinmacropeptide from bovine, ovine, and caprine cheese whey. *European Food Research and Technology* 214 (4): 282–286.

Mikkelsen, T.L., Frokiaer, H., Topp, C. et al. (2005). Caseinomacropeptide self-association is dependent on whether the peptide is free or restricted in kappa-casein. *Journal of Dairy Science* 88 (12): 4228–4238.

Molle, D. and Leonil, J. (1995). Heterogeneity of the bovine kappa-casein caseinomacropeptide, resolved by liquid chromatography on-line with electrospray ionization mass spectrometry. *Journal of Chromatography A* 708 (2): 223–230.

Mollé, D. and Léonil, J. (2005). Quantitative determination of bovine κ-casein macropeptide in dairy products by liquid chromatography/electrospray coupled to mass spectrometry (LC-ESI/MS) and liquid chromatography/electrospray coupled to tamdem mass spectrometry (LC-ESI/MS/MS). *International Dairy Journal* 15 (5): 419–428.

Monnai, M., Horimoto, Y., and Otani, H. (1998). Immunomodificatory effect of dietary bovine kappa-caseinoglycopeptide on serum antibody levels and proliferative responses of lymphocytes in mice. *Milchwissenschaft* 53 (3): 129–132.

Monnai, M. and Otani, H. (1997). Effect of bovine kappa-caseinoglycopeptide on secretion of interleukin-1 family cytokines by P388D1 cells, a line derived from mouse monocyte/macrophage. *Milchwissenschaft* 52 (4): 192–196.

Moran, A., Holst, O., Brennan, P., and von Itzstein, M. (2009). Microbial glycobiology. In: *Microbial Glycobiology: Structures, Relevance and Applications* (ed. A. Moran, O. Holst, P. Brennan and M. von Itzstein). London: Elsevier.

Morr, C. and Seo, A. (1988). Fractionation and characterization of glycomacropeptide from caseinate and skim milk hydrolysates. *Journal of Food Science* 53 (1): 80–87.

Moynihan, J.S., d'Mello, F.I., and Howard, C.R. (2000). 48-mer synthetic peptide analogue of the hepatitis B virus "a" determinant induces an anti-HBs antibody response after a single injection. *Journal of Medical Virology* 62 (2): 159–166.

Munoz, F.C., Cervantes, M.M., Cervantes-Garcia, D. et al. (2017). Glycomacropeptide attenuates inflammation, pruritus, and Th2 response associated with atopic dermatitis induced by 2,4-dinitrochlorobenzene in rat. *Journal of Immunology Research* 2017: 6935402.

Nakajima, K., Tamura, N., Kobayashi-Hattori, K. et al. (2005). Prevention of intestinal infection by glycomacropeptide. *Bioscience, Biotechnology, and Biochemistry* 69 (12): 2294–2301.

Nakano, T., Ikawa, N., and Ozimek, L. (2004). Use of epichlorohydrin-treated chitosan resin as an adsorbent to isolate κ-casein glycomacropeptide from sweet whey. *Journal of Agricultural and Food Chemistry* 52 (25): 7555–7560.

Nakano, T., Ikawa, N., and Ozimek, L. (2006a). Separation of glycomacropeptide from sweet whey by using chitosan and a centrifugal filter. *Milchwissenschaft* 61 (2): 191–193.

Nakano, T., Ikawa, N., and Ozimek, L. (2007). Detection of sialylated phosphorylated κ-casein glycomacropeptide electrophoresed on polyacrylamide gels and cellulose acetate strips by the thiobarbituric acid and malachite green dye reactions. *Journal of Agricultural and Food Chemistry* 55 (7): 2714–2726.

Nakano, T., Kusump, S., Ikawa, N., and Ozimek, L. (2006b). Detection of whey in margarine by analysis of sialic acid and κ-casein glycomacropeptide. *Milchwissenschaft* 61 (1): 36–40.

Nakano, T. and Ozimek, L. (1998). Gel chromatography of glycomacropeptide (GMP) from sweet whey on sephacryls-200 at different pH's and on Sephadex G-75 in 6M guanidine hydrochloride. *Milchwissenschaft* 53 (11): 629–633.

Nakano, T. and Ozimek, L. (1999a). Determination of sialic acid by the thiobarbituric acid reaction in sweet whey and its fractions. *Journal of Agricultural and Food Chemistry* 47 (7): 2613–2616.

Nakano, T. and Ozimek, L. (1999b). Purification of glycomacropeptide from non-dialyzable fraction of sweet whey by anion-exchange chromatography. *Biotechnology Techniques* 13 (11): 739–742.

Nakano, T. and Ozimek, L. (2000). Purification of glycomacropeptide from caseinate hydrolysate by gel chromatography and treatment with acidic solution. *Journal of Food Science* 65 (4): 588–590.

Nakano, T. and Ozimek, L. (2002). Isolation of glycomacropeptide from sweet whey by gel chromatography on Sephacryl S-200 at pH 7.0 and 3.5. *Milchwissenschaft* 57 (3): 128–130.

Nakano, T. and Ozimek, L. (2014). A sialic acid assay in isolation and purification of bovine k-casein glycomacropeptide: a review. *Recent Patents on Food, Nutrition and Agriculture* 6 (1): 38–44.

Nakano, T. and Ozimek, L. (2016). Selective removal of phenylalanine impurities from commercial κ-casein glycomacropeptide by anion exchange chromatography. *Journal of Food Processing and Technology* 7 (1).

Nakano, T., Silva-Hernandez, E.R., Ikawa, N., and Ozimek, L. (2002). Purification of kappa-casien glycomacropeptide from sweet whey with undetectable level of phenylalanine. *Biotechnology Progress* 18 (2): 409–412.

Nakano, T., Watanabe, T., Ikawa, N. et al. (2009). Cellulose acetate electrophoresis of whey proteins and glycomacropeptide. *Milchwissenschaft* 64 (3): 291–295.

Neelima, R.S., Rajput, Y.S., and Mann, B. (2013). Chemical and functional properties of glycomacropeptide (GMP) and its role in the detection of cheese whey adulteration in milk: a review. *Dairy Science and Technology* 93 (1): 21–43.

Neeser, J.R., Chambaz, A., del Vedovo, S. et al. (1988). Specific and nonspecific inhibition of adhesion of oral actinomyces and streptococci to erythrocytes and polystyrene by caseinoglycopeptide derivatives. *Infection and Immunity* 56 (12): 3201–3208.

Neeser, J.R., Golliard, M., Woltz, A. et al. (1994). In vitro modulation of oral bacterial adhesion to saliva-coated hydroxyapatite beads by milk casein derivatives. *Oral Microbiology and Immunology* 9 (4): 193–201.

Neeser, J.R., Grafstrom, R.C., Woltz, A. et al. (1995). A 23 kDa membrane glycoprotein bearing NeuNAc alpha 2-3Gal beta 1-3GalNAc O-linked carbohydrate chains acts as a receptor for *Streptococcus sanguis* OMZ 9 on human buccal epithelial cells. *Glycobiology* 5 (1): 97–104.

Ney, D.M., Hull, A.K., van Calcar, S.C. et al. (2008). Dietary glycomacropeptide supports growth and reduces the concentrations of phenylalanine in plasma and brain in a murine model of phenylketonuria. *Journal of Nutrition* 138 (2): 316–322.

Ntemiri, A., Chonchuir, F.N., O'Callaghan, T.F. et al. (2017). Glycomacropeptide sustains microbiota diversity and promotes specific taxa in an artificial colon model of elderly gut microbiota. *Journal of Agricultural and Food Chemistry* 65 (8): 1836–1846.

O'Riordan, N., Kane, M., Joshi, L. and Hickey R. M (2014). Structural and functional characteristics of bovine milk protein glycosylation, *Glycobiology* 24 (3): 220–236.

Oancea, S. (2009). Identification of glycomacropeptide as indicator of milk and dairy drinks adulteration with whey by immunochromatographic assay. *Romanian Biotechnology Letters* 14 (1): 4146–4151.

Ofek, I., Hasty, D.L., and Sharon, N. (2003). Anti-adhesion therapy of bacterial diseases: prospects and problems. *FEMS Immunology and Medical Microbiology* 38 (3): 181–191.

Oliveira, L., Sarubbo, L., Porto, A. et al. (2002). Partition of trypsin in aqueous two-phase systems of poly (ethylene glycol) and cashew-nut tree gum. *Process Biochemistry* 38 (5): 693–699.

Ortega-Gonzalez, M., Capitan-Canadas, F., Requena, P. et al. (2014). Validation of bovine glycomacropeptide as an intestinal anti-inflammatory nutraceutical in the lymphocyte-transfer model of colitis. *British Journal of Nutrition* 111 (7): 1202–1212.

Otani, H., Horimoto, Y., and Monnai, M. (1996). Suppression of interleukin-2 receptor expression on mouse CD4+ T cells by bovine kappa-caseinoglycopeptide. *Bioscience, Biotechnology, and Biochemistry* 60 (6): 1017–1019.

Otani, H. and Monnai, M. (1993). Inhibition of proliferative responses of mouse spleen lymphocytes by bovine milk κ-casein digests. *Food and Agricultural Immunology* 5 (4): 219–229.

Otte, J., Midtgaard, L., and Qvist, K. (1995). Analysis of caseinomacropeptide (s) by free solution capillary electrophoresis. *Milchwissenschaft* 50 (2): 75–79.

Outinen, M., Tossavainen, O., Syväoja, E.-L., and Korhonen, H. (1995). Chromatogrphic isolation of kappa-casein macropeptide from cheese whey with a strong basic anion exchange resin. *Milchwissenschaft* 50, 10: 570–574.

Peterson, G.L. (1983). Determination of total protein. *Methods in Enzymology* 91: 95–119.

Picard, C., Plard, I., Rongdaux-Gaida, D., and Collin, J.-C. (1994). Detection of proteolysis in raw milk stored at low temperature by an inhibition ELISA. *Journal of Dairy Research* 61 (3): 395–404.

Poch, M. and Bezkorovainy, A. (1991). Bovine milk kappa-casein trypsin digest is a growth enhancer for the genus Bifidobacterium. *Journal of Agricultural and Food Chemistry* 39 (1): 73–77.

Rao, P.S., Sharma, R., and Rajput, Y.S. (2012). Direct estimation of sialic acid in milk and milk products by fluorimetry and its application in detection of sweet whey adulteration in milk. *Journal of Dairy Research* 79 (4): 495–501.

Recio, I., García-Risco, M.R., López-Fandiño, R. et al. (2000). Detection of rennet whey solids in UHT milk by capillary electrophoresis. *International Dairy Journal* 10 (5): 333–338.

Recio, I., López-Fandiño, R., Olano, A. et al. (1996). Study of the formation of caseinomacropeptides in stored ultra-high-temperature-treated milk by capillary electrophoresis. *Journal of Agricultural and Food Chemistry* 44 (12): 3845–3848.

Recio, I., Moreno, F., and López-Fandiño, R. (2009). Glycosylated dairy components: their roles in nature and ways to make use of their biofunctionality in dairy products. In: *Dairy-Derived Ingredients: Food and Nutraceutical Uses*, 170–211. Cambridge: Woodhead Publishing.

Requena, P., Daddaoua, A., Guadix, E. et al. (2009). Bovine glycomacropeptide induces cytokine production in human monocytes through the stimulation of the MAPK and the NF-kappaB signal transduction pathways. *British Journal of Pharmacology* 157 (7): 1232–1240.

Requena, P., Daddaoua, A., Martinez-Plata, E. et al. (2008). Bovine glycomacropeptide ameliorates experimental rat ileitis by mechanisms involving downregulation of interleukin 17. *British Journal of Pharmacology* 154 (4): 825–832.

Requena, P., Gonzalez, R., Lopez-Posadas, R. et al. (2010). The intestinal antiinflammatory agent glycomacropeptide has immunomodulatory actions on rat splenocytes. *Biochemical Pharmacology* 79 (12): 1797–1804.

Rhoades, J.R., Gibson, G.R., Formentin, K. et al. (2005). Caseinoglycomacropeptide inhibits adhesion of pathogenic *Escherichia coli* strains to human cells in culture. *Journal of Dairy Science* 88 (10): 3455–3459.

Rigo, J., Boehm, G., Georgi, G. et al. (2001). An infant formula free of glycomacropeptide prevents hyperthreoninemia in formula-fed preterm infants. *Journal of Pediatric Gastroenterology and Nutrition* 32 (2): 127–130.

Robitaille, G. and Champagne, C.P. (2014). Growth-promoting effects of pepsin- and trypsin-treated caseinomacropeptide from bovine milk on probiotics. *Journal of Dairy Research* 81 (3): 319–324.

Robitaille, G., Lapointe, C., Leclerc, D., and Britten, M. (2012). Effect of pepsin-treated bovine and goat caseinomacropeptide on *Escherichia coli* and *Lactobacillus rhamnosus* in acidic conditions. *Journal of Dairy Science* 95 (1): 1–8.

Rojas, E. and Torres, G. (2013). Isolation and recovery of glycomacropeptide from milk whey by means of thermal treatment. *Food Science and Technology (Campinas)* 33 (1): 14–20.

Roldan, N.R., Jimenez, M., Cervantes-Garcia, D. et al. (2016). Glycomacropeptide administration attenuates airway inflammation and remodeling associated to allergic asthma in rat. *Inflammation Research* 65 (4): 273–283.

Rong, Y., Lu, Z., Zhang, H. et al. (2015). Effects of casein glycomacropeptide supplementation on growth performance, intestinal morphology, intestinal barrier permeability and inflammatory responses in *Escherichia coli* K88 challenged piglets. *Animal Nutrition* 1 (2): 54–59.

Royle, P.J., McIntosh, G.H., and Clifton, P.M. (2008). Whey protein isolate and glycomacropeptide decrease weight gain and alter body composition in male Wistar rats. *British Journal of Nutrition* 100 (1): 88–93.

Saito, T. and Itoh, T. (1992). Variations and distributions of O-glycosidically linked sugar chains in bovine kappa-casein. *Journal of Dairy Science* 75 (7): 1768–1774.

Saito, T., Yamaji, A., and Itoh, T. (1991). A new isolation method of caseinoglycopeptide from sweet cheese whey. *Journal of Dairy Science* 74 (9): 2831–2837.

Sandström, O., Lonnerdal, B., Graverholt, G., and Hernell, O. (2008). Effects of alpha-lactalbumin-enriched formula containing different concentrations of glycomacropeptide on infant nutrition. *American Journal of Clinical Nutrition* 87 (4): 921–928.

Sawin, E.A., de Wolfe, T.J., Aktas, B. et al. (2015). Glycomacropeptide is a prebiotic that reduces Desulfovibrio bacteria, increases cecal short-chain fatty acids, and is anti-inflammatory in mice. *American Journal of Physiology. Gastrointestinal and Liver Physiology* 309 (7): G590–G601.

Schüpbach, P., Neeser, J.-R., Golliard, M. et al. (1996). Incorporation of caseinoglycomacropeptide and caseinophosphopeptide into the salivary pellicle inhibits adherence of mutans streptococci. *Journal of Dental Research* 75 (10): 1779–1788.

Sharon, N. and Ofek, I. (2000). Safe as mother's milk: carbohydrates as future anti-adhesion drugs for bacterial diseases. *Glycoconjugate Journal* 17 (7–9): 659–664.

Silva-Hernandez, E.R., Nakano, T., and Ozimek, L. (2002). Isolation and analysis of κ-casein glycomacropeptide from goat sweet whey. *Journal of Agricultural and Food Chemistry* 50 (7): 2034–2038.

Simon, P.M., Goode, P.L., Mobasseri, A., and Zopf, D. (1997). Inhibition of Helicobacter pylori binding to gastrointestinal epithelial cells by sialic acid-containing oligosaccharides. *Infection and Immunity* 65 (2): 750–757.

Singh, D.P.P.a.S. (2015). Chitosan: a multifacet polymer. *International Journal of Current Pharmaceutical Research* 7 (2): 21–28.

Singh, R.H., Rohr, F., Frazier, D. et al. (2014). Recommendations for the nutrition management of phenylalanine hydroxylase deficiency. *Genetics in Medicine* 16 (2): 121–131.

Solverson, P., Murali, S.G., Brinkman, A.S. et al. (2012a). Glycomacropeptide, a low-phenylalanine protein isolated from cheese whey, supports growth and attenuates metabolic stress in the murine model of phenylketonuria. *American Journal of Physiology: Endocrinology and Metabolism* 302 (7): E885–E895.

Solverson, P., Murali, S.G., Litscher, S.J. et al. (2012b). Low bone strength is a manifestation of phenylketonuria in mice and is attenuated by a glycomacropeptide diet. *PLoS One* 7 (9): e45165.

Stan EYa, G.S., Krasil'shchikov, K.B., and Chernikov, M.P. (1983). Effects of κ-casein glycomacropeptide motility in dogs. *Byulleten' Eksperimental'noi Biologii I Metisiney* 96: 10–12.

Strömqvist, M., Falk, P., Hansson, S.B.L. et al. (1995). Human milk K-casein and inhibition of Helicobacter pylori adhesion to human gastric mucosa. *Journal of Pediatric Gastroenterology and Nutrition* 21 (3): 288–296.

Takayama, T., Katsuki, S., Takahashi, Y. et al. (1998). Aberrant crypt foci of the colon as precursors of adenoma and cancer. *New England Journal of Medicine* 339 (18): 1277–1284.

Tanimoto, M., Kawasaki, Y., Dosako, S.I. et al. (1992). Large-scale preparation of κ-casein glycomacropeptide from rennet casein whey. *Bioscience, Biotechnology, and Biochemistry* 56 (1): 140–141.

Taylor, C. and Woonton, B. (2009). Quantity and carbohydrate content of glycomacropeptide fractions isolated from raw and heat-treated milk. *International Dairy Journal* 19 (12): 709–714.

Tek, H.N., Turhan, K.N., and Etzel, M.R. (2005). Effect of conductivity, pH, and elution buffer salinity on glycomacropeptide recovery from whey using anion exchange chromatography. *Journal of Food Science* 70 (4): E295–E300.

Thomä-Worringer, C., Sørensen, J., and López-Fandiño, R. (2006). Health effects and technological features of caseinomacropeptide. *International Dairy Journal* 16 (11): 1324–1333.

Tian, Q., Wang, T., Tang, X. et al. (2014). Developing a potential prebiotic of yogurt: growth of *Bifidobacterium* and yogurt cultures with addition of glycomacropeptide hydrolysate. *International Journal of Food Science and Technology* 50: 120–127.

Tolkach, A. and Kulozik, U. (2005). Fractionation of whey proteins and caseinomacropeptide by means of enzymatic crosslinking and membrane separation techniques. *Journal of Food Engineering* 67 (1): 13–20.

Tullio, L.T., Karkle, E.N.L., and Cândido, L.M.B. (2007). Revisão: isolamento e purificação do glicomacropeptídeo do soro de leite. *Boletim do Centro de Pesquisa de Processamento de Alimentos* 25 (1): 456–457.

van Calcar, S.C., MacLeod, E.L., Gleason, S.T. et al. (2009). Improved nutritional management of phenylketonuria by using a diet containing glycomacropeptide compared with amino acids. *American Journal of Clinical Nutrition* 89 (4): 1068–1077.

Van Riel, J. and Olieman, C. (1995). Determination of caseinomacropeptide with capillary zone electrophoresis and its application to the detection and estimation of rennet whey solids in milk and buttermilk powder. *Electrophoresis* 16 (1): 529–533.

Veldhorst, M.A., Nieuwenhuizen, A.G., Hochstenbach-Waelen, A. et al. (2009). Effects of complete whey-protein breakfasts versus whey without GMP-breakfasts on energy intake and satiety. *Appetite* 52 (2): 388–395.

Vockley, J., Andersson, H.C., Antshel, K.M. et al. (2014). Phenylalanine hydroxylase deficiency: diagnosis and management guideline. *Genetics in Medicine* 16 (2): 188–200.

Walstra, P., J.T.M, W., and Geurts, T.J. (2006). *Dairy Science and Technology*, 2e. Boca Raton: CRC Press.

Wu, J., Wang, Y., and Huang, L. (2012). Separation of glycomacropeptide (a functional protein material) by aqueous two-phase systems. *Advanced Materials Research* 496: doi: 10.4028/www.scientific.net/AMR.496.185.

Xu, Y., Sleigh, R., Hourigan, J., and Johnson, R. (2000). Separation of bovine immunoglobulin G and glycomacropeptide from dairy whey. *Process Biochemistry* 36 (5): 393–399.

Yao, K., Ubuka, T., Masuoka, N. et al. (1989). Direct determination of bound sialic acids in sialoglycoproteins by acidic ninhydrin reaction. *Analytical Biochemistry* 179 (2): 332–335.

Yvon, M., Beucher, S., Guilloteau, P. et al. (1994). Effects of caseinomacropeptide (CMP) on digestion regulation. *Reproduction Nutrition Development* 34 (6): 527–537.

2

Meat Proteins as a Potential Source of Bioactive Ingredients for Food and Pharmaceutical Use

Carlos Álvarez García[1] and Ismael Marcet Manrique[2]

[1] *Teagasc Food Research Centre, Department of Food Quality and Sensory Science, Dublin, Ireland*
[2] *University of Oviedo, Department of Chemical and Environmental Engineering, Oviedo, Spain*

2.1 Introduction

The use of proteins has evolved remarkably in recent decades. Traditionally, proteins were just considered as a source of energy, building blocks for the body and supporters of the health and well-being of human populations (Consultation 2011). However, after the discovery and characterisation of proteolytic enzymes in the first years of the twentieth century (Neurath 1999), there was a growing interest in the 1940s to generate protein hydrolysates by means of such enzymes; the main objective was to predigest food proteins to help patients unable to process whole proteins. These protein hydrolysates are composed mainly of peptides (small amino acid polymers) and free amino acids, which are the basic units that form proteins. Peptides, and free amino acids, are more readily absorbed in the digestive tract when compared to native proteins; in fact, the word 'peptide' is a Greek term meaning 'small digestible' or 'digested'. In the 1950's Mellander discovered the first peptide with biological activity or biopeptide, a peptide derived from milk caseins which enhanced the vitamin D-independent bone calcification (Mellander 1950). Since then, peptides obtained from proteins from all imaginable sources (vegetable, fungi, animal, or even bacteria in origin) have been exhaustively generated and characterised based on their structure, functional properties, and what is more relevant, the type bioactivity they can exhibit.

Bioactive compounds can be defined in several ways: 'food derived components (genuine or generated) that, in addition to their nutritional value exert a physiological effect on the body'(Ryan et al. 2011); or 'Bioactive peptides are specific protein fragments which, above and beyond their nutritional capabilities, have a positive impact on the body's function or condition which may ultimately influence health' (Bhat et al. 2015). The role of food-protein derived peptides as physiological modulators is becoming more relevant, in addition to the inherent nutritional value of the proteins. In recent years many different biological activities have been reported in bioactive peptides, including, for example, antihypertensive, antioxidant, antithrombotic, reducing cholesterol levels, opioid, mineral binding, antimicrobial, antiproliferative, immunomodulatory, cytomodulatory,

Novel Proteins for Food, Pharmaceuticals, and Agriculture: Sources, Applications, and Advances, First Edition. Edited by Maria Hayes.

anti-inflammatory, and gut modulatory, among others (Di Bernardini et al. 2011; Kadam et al. 2015; Lafarga and Hayes 2014; Sharma et al. 2011).

The current trend in biopeptides research is focused on designed peptides, by means of conjugation techniques or by generating synthetic mimics of peptides identified in natural sources (Gabriel et al. 2008; Mojsoska et al. 2015). There is currently huge interest in exploring new sustainable and environmentally friendly sources, such as food wastes or food processing co-products, as well as novel foods such as insects and seaweeds (Mullen et al. 2017).

Meat is considered to be one of the most relevant sources of food proteins, mainly for its high protein content and the excellent nutritive value provided by the high levels of essential amino acids. It is logical, therefore, to think that meat and meat products are excellent candidates to be used as source of bioactive peptides. In this sense, significant efforts are being made to unlock the potential of meat and meat co-products as novel sources of bioactive compounds; either coming from isolated proteins further hydrolysed (Lafarga and Hayes 2014; Mora et al. 2014), peptides naturally occurring in meat (Jayasena et al. 2015a; Jayasena et al. 2015b; Jung et al. 2013), or peptides generated after an ageing/fermentation process (Escudero et al. 2013; Jung et al. 2013; Mora et al. 2008). In addition to these protein-based bioactive compounds, other relevant molecules with biological activity can be recovered from meat sources, the most relevant being conjugated linoleic acid (CLA); which has been reported as a promoter of beneficial health effects (Koba and Yanagita 2014; Lu 2013; Viladomiu et al. 2016); but since this compound is not protein derived, it will not be discussed here.

2.2 Protein-Based Bioactive Compounds

Since protein is the most abundant component in meat and meat by-products, by dry weight, bioactive compounds derived from proteins are the most abundant ones that can be found in meat sources. In fact, the vast majority of research effort focused on generating, characterising, isolating, and identifying new bioactive molecules from meat sources is dedicated to protein-based compounds. Within this group, the protein-based bioactive compounds can be classified into three different categories: peptides generated after an enzymatic hydrolysis process, peptides generated through an aging or fermentation process, and naturally occurring peptides.

2.2.1 Peptides Generated by Enzymatic Hydrolysis

The most common process used to generate bioactive peptides from meat, or from another protein source, is through enzymatic hydrolysis. The production of biopeptides from meat sources using enzymes has been widely studied in recent years. For instance, several papers have reported the use of commercial enzymes to generate anti-hypertensive peptides from diverse meat sources. Jang and Lee (2005), studied the angiotensin converting enzyme (ACE) inhibitory activity of peptides obtained from beef hydrolysates, using three different enzymes (thermolysin, proteinase A, and protease type XIII). After a purification process, involving chromatographic techniques, the authors isolated a small peptide of six amino acids with a strong ACE inhibitory effect. Arihara et al. (2001), hydrolysed porcine skeletal muscle with eight enzymes aiming to generate peptides with strong ACE inhibitory properties. It was found that two peptides generated using thermolysin had the highest activity, they were named as myopentapeptides A and B (Met-Asn-Pro-Pro-Lys and Ile-Thr-Thr-Asn-Pro) since these sequences were found in the myosin heavy chain. The same authors reported that further hydrolysis of such penta-peptides generated ACE inhibitory tri-peptides, whose activity was even higher than the original peptides. Escudero et al. (2010), treated fresh pork skeletal muscle with pancreatin and pepsin, three peptides with antihypertensive activity were obtained. Finally, di Bernardini et al. (2012) studied the hydrolysis of the bovine brisket sarcoplasmic proteins using papain, and the hydrolysates obtained showed both ACE inhibitory activity and antioxidant properties.

The main advantage of the enzymatic treatment is that the peptide generation process can be easily scaled up, allowing the generation of peptides in a predictable way and in sufficient amounts to be used for commercial products in the food, supplement, or even pharmaceutical industries. However, on the other hand, the costs of the enzymes, along with the purification and isolation processes, at industrial scale, are currently prohibitive from an economical point of view (as it will be discussed in the section on challenges). For this reason, in spite of the efficiency in peptide production and the high levels of activity found, the vast majority of the processes developed at lab scale still remain unexploited for commercial uses

The methodology used to obtain biopeptides from meat sources, by means of enzymatic reactions, is described in Figure 2.1. First, the protein source goes through a process of size reduction (mincers are commonly used), which improves the surface volume ratio and enhances the substrate–enzyme interaction. The minced material has to be dispersed in a liquid medium using a mechanical homogeniser. Since the enzymatic activity varies greatly, depending on pH and medium temperature, both parameters are strictly controlled. As the hydrolysis process advances the pH of the medium becomes more acidic (Adler-Nissen 1979), which can lead to an inactivation of the enzyme. For this reason, the reaction medium is usually buffered using phosphate, citric or carbonate salts; or a pH control system which adds an alkali when required to keep the pH constant. Additionally, the temperature of the reactor is controlled using jacketed vessels or heat exchangers in order to keep temperature constant. Once the hydrolysis process has concluded, a complex mixture of peptides and free amino acids is obtained; even non-hydrolysed proteins can be found.

Subsequently, aiming to isolate the most active peptides, further purification steps are performed. Most common techniques are based on molecular size, such as membrane filtration or size exclusion chromatography (Mohammad et al. 2012), their different binding capacity to a chromatographic matrix using different techniques such as reverse phase chromatography or ionic exchange chromatography (Agyei et al. 2016), or differential precipitation by ethanol or salt addition (Bowden et al. 2016). In this phase, the techniques used to isolate the peptides of interest may be combined and performed sequentially; a common approach is to fraction the largest peptides by membrane filtration followed by more specific chromatographic techniques. After each separation or purification step, several fractions are generated. Then, the activity of each one of the fractions is tested, in order to identify the aliquots containing the compounds of interest. The most active ones are selected and purified further, and the new set of fractions generated is

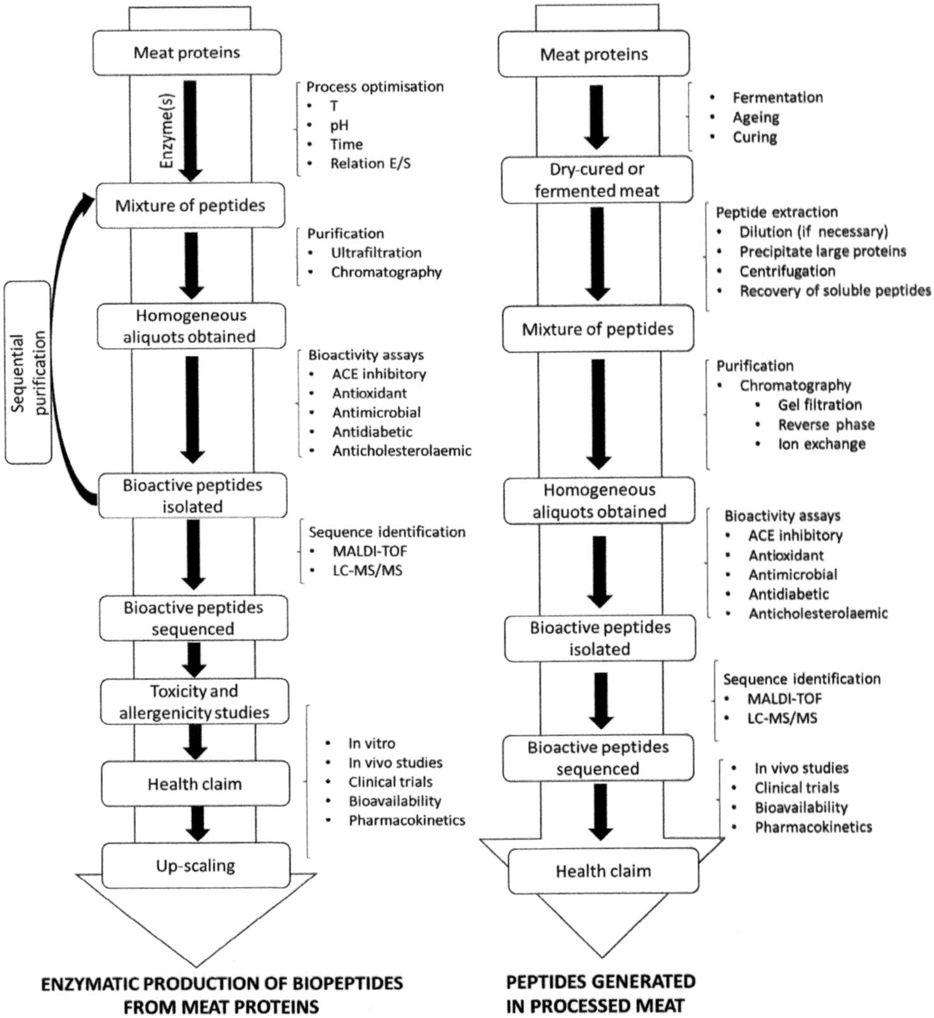

Figure 2.1 Methodologies used to study the bioactive peptides produced by the enzymatic treatment of meat proteins (*left*) or generated after a meat curing or fermentation process (*right*).

tested again. Ideally, as the purification process goes on, activity of the fractions becomes higher, until one single or a few peptides with the highest activity are isolated.

Finally, the sequence of the peptides can be characterised by means of mass spectroscopy techniques, which can determine the amino acid composition and the sequence of the amino acids in the peptide under study. As an additional step, the sequence identified as active can be chemically synthetized and its activity tested, to confirm the relationship between activity and amino acid sequence.

Heme iron polypeptide (HIP) is an example of a currently commercialised peptide generated from a bovine source. The absorption of ingested iron is low, since it can vary from 5% to 35% of the total iron content in the food. This variation is due to the type of iron present in the food (heme or non-heme iron) and other circumstances such as the presence of quelants or absorption inhibitors (McDowell 2003). It has been reported that the iron when forming part of a heme group is absorbed more efficiently by the enterocytes (Hurrell and Egli 2010). In this sense, several authors proposed the hydrolysis of haemoglobin, a rich source of heme iron, to obtain a hydrolysate enriched in heme iron. Lebrun et al. (1998), obtained a peptidic fraction that has the ability to increase the solubilisation of the heme in comparison with the native haemoglobin. In a similar way, it was reported that after the hydrolysis of haemoglobin using commercially available exopeptidase and endopeptidase, a fraction rich in heme iron was isolated (In et al. 2002). It has been reported that in non-dialysis-dependent patients with chronic kidney disease, oral administration of HIP had similar efficacy to injected iron sucrose in maintaining haemoglobin levels, with no significant differences in the negative effects (Nagaraju et al. 2013). However, the product has not yet been evaluated by the FDA and as a disclaimer, the product is not intended to diagnose, treat, cure or prevent any disease.

2.2.2 Peptides Generated in Processed Meat

Despite the promising results obtained using enzymes, in recent years the scientific community has focused not only on the enzymatic production of biopeptides but also on studies regarding the formation and further characterisation of active peptides that have been generated as a consequence of diverse meat processing methods. The promising results obtained in this field have lead to an increased interest in research in the last few years but the philosophy behind these studies differs substantially when compared to the enzymatic production of bioactive peptides (Figure 2.1).

The objective of enzymatic processing is to transform a protein-based source into bioactive peptides. However, when investigating peptides produced as a consequence of traditional curing/ripening or fermentation processes, the final goal is to clarify the effect of the food processing on peptide generation and, eventually, to isolate and characterise such peptide(s), aiming to assess bioactivity and potential benefits to consumer health. Two different processes are used to naturally generate peptides: dry-curing, in which meats are salted and then ripened and dried, and fermentation, in which microorganisms are employed followed by a curing/ageing process (Escudero et al. 2013).

In the production of dry-cured meats, the products are salted (sodium chloride and nitrate/nitrite) to produce a decrease of the food water activity and enhance their colour. The salts are usually applied on the surface of the piece of meat to produce dry-cured hams, shoulders, loins, and so on, or mixed with minced meat to produce dry-fermented sausages, for example (Nip 2008). In these types of meat products, the proteins are mainly degraded by the action of endogenous muscle enzymes (calpains, cathepsins, and peptidases), and the peptides obtained vary according to the animal genetics, the curing time, salt applied, humidity and processing temperature (Mora et al. 2016b). In the case of dry-cured ham, drying treatment can take up to 24 months. This curing time allows the development of new aromas, which arise from proteolysis and lipolysis reactions (Toldrá 1998), as well as an increase in meat tenderness. At advanced stages of ageing, enzymes such as endopeptidases, tripeptidases, and dipeptidases promote the generation of free amino acids and short peptides (Flores and Toldrá 2011).

The fermentation process involves the action of several bacteria, lactic acid bacteria being by far the most relevant, which produce meat acidification due to the production of lactic acid. The low pH reached during the fermentation favours the ripening of meat, the acquisition of curing colours and the inhibition of undesired bacteria that cause food spoilage (Hugas and Monfort 1997). Bacterial metabolism produces the release of low molecular weight peptides and amino acids (Fadda et al. 2010); this phenomenon, added to the activity of the internal proteases, is what generates peptides and free amino acids in the final meat product.

Once the peptides have been generated, by fermentation or curing, they must be isolated and characterised in order to assess their functionality and, if possible, determine their sequence and the protein of origin. One of the main advantages of these kinds of products is that peptides do not need to be included in the formulation and declared as an ingredient, since they are generated by the product manufacturing process, making its commercialisation and acceptance by the consumer easier.

Peptides identified in fermented and dry-cured meats have many different bioactive qualities such as antioxidant or antihypertensive capacity. In Table 2.1, several bioactive peptides isolated from fermented and dry-cured meat products are summarised.

Table 2.1 Bioactive peptides generated after the fermentation or curing process of meat.

Source	Main results	References
Fermented meat sauce	Identification of a tripeptide with antioxidant properties	Ohata et al. (2016)
Dry-fermented camel sausages	Peptides below 3 kDa obtained from meat inoculated with starter bacteria showed antioxidant and antihypertensive capacities	Mejri et al. (2017)
Fermented sausages	Higher or lower ACE inhibitory activity, depending on strain of lactic acid bacteria inoculated into the sausages. The bioactive molecules detected are presumably peptides	Takeda et al. (2017)
Fermented sausages	Identification of small peptides which arose from actin proteolysis during the fermentation process. In this case and in agreement with the cleavage sites identified, the cathepsins showed a relevant role	Berardo et al. (2017)
Pork dry-cured meat products	The peptides produced from hydrolysis of myofibrillar proteins showed better antioxidant properties than those which came from sarcoplasmic proteins. The cured loin meat showed better ability to inhibit fat oxidation	Kęska et al. (2017)
Dry-cured ham	The capacity of three peptides with ACE inhibitory activity to pass through a Caco-2 cell monolayer was studied. In this case, these peptides were degraded during transport but the intact peptides or the small fragments derived showed ACE inhibitory activity	Gallego et al. (2016)
Dry-cured Xuanwei ham	Isolation of peptides with antioxidant activities produced during the ripening period of Xuanwei ham. The sequence Asp-Leu-Glu-Glu was identified as one of the main bioactive peptides	Xing et al. (2016)
Dry-cured hams from Spain, Italy, and Belgium	When the dry-cured hams were compared, differences in pig genetics and the elaboration process led to the production of different bioactive peptides. The ACE inhibitory activity of peptides from the Spanish ham were higher than that from the other hams. However, the antioxidant properties of these peptides were similar for the three hams tested	Mora et al. (2016a)

As a final step, which is common for all peptides generated regardless of the generation method employed, the effect of gastric digestion, absorption through the intestine and how they exert their physiological activity once in the bloodstream need to be studied (Toldrá et al. 2017), as well as whether they have an effect when ingested in regular doses in a balanced diet.

2.2.3 Naturally Occurring Biopeptides

Skeletal muscle contains naturally occurring low molecular weight components which have the potential to be used as ingredients in health-promoting functional foods or even as supplements for sport or active lifestyles. The most relevant, and more studied, among these compounds are anserine, carnosine, L-carnitine, creatine, glutathione, and the amino acid taurine. Their main properties, abundance and bioactivities can be seen in Table 2.2. The biological activities of the low molecular weight molecules make them ideal ingredients for the formulation of functional foods and supplements. For example, L-carnitine and carnosine are included in a currently marketed functional food in Japan and as a drink in the US, as reported by Arihara (2006).

Table 2.2 Naturally occurring biopeptides in meat sources. Chemical formula, molecular weight, abundance in pork (P) and beef (B) muscles, and main biological activities.

Compound	Formula	Beef and pork content *	Biological activity	Reference
Anserine	$C_{10}H_{16}N_4O_3$	P: 17.6 B: 150	Antioxidant, antifatigue, scavenger of hydroxyl radicals	Kikuchi et al. (2004)
Carnosine	$C_9H_{14}N_4O_3$	P: 419 B: 452	Able to scavenge ROS Buffering capacity Aldehyde scavenger	Guiotto et al. (2005)
L-carnitine	$C_7H_{15}NO_3$	P: 21 B: 66	Involved in fatty acid transportation in mitochondrial membrane	Ferrari et al. (2004)
Creatine	$C_4H_9N_3O_2$	P: 373 B: 401	Reservoir of energy, facilitates aqueous two-phase system (ATP) recycling Enhances muscle mass and physical performance	Farshidfar et al. (2017); Twycross-Lewis et al. (2016)
Taurine	$C_2H_7NO_3S$	P: 22.5 B: 38.6	Stimulation of glycolysis and glycogenesis Antioxidant Cell proliferation. Hypotensive Antiarrhythmic	Huxtable (1992)
Glutathione	$C_{10}H_{17}N_3O_6S$	P: 18–23 B: 12–26	Strong antioxidant, cell cycle and caspase modulator	Ballatori et al. (2009)

ROS, reactive oxygen species.

* Values expressed as mg/100 g obtained from Mora et al. (2008), Seline and Johein (2007), Aristoy and Toldrá (1998) and Jones et al. (1992).

Such compounds are naturally ingested with the diet; however, extraction and concentration processes can be performed in order to purify these compounds and subsequently they can be used as ingredients for food formulations or supplements. Common strategies for extraction are those based on heat shock, heat denaturation and further chromatographic purification. However, biosynthesis seems to be a better approach to obtain some of these compounds at industrial scale (Meyer and Robins 2005).

2.3 Potential Applications

The antioxidant and antimicrobial capacity of food systems can be an advantage when intended to improve the stability of food formulations regarding shelf-life and negative bacterial growth. As recently reported by Sohaib et al. (2017), peptides can act as an antioxidant ingredient since they are excellent radical scavengers and metal ion chelators. Such activities are of great benefit to meat and meat-based products, since lipid oxidation is prevented (Jiang and Xiong 2016). Additionally, such peptides can be claimed as clean label ingredients, which will be perceived as positive by consumers. Current research suggests that using meat and meat co-products as a source of bioactive peptides to be included in functional foods could be of interest to meat processors, since this is one of the top trends in the food industry. Therefore, as knowledge about biological activities increases, it can be expected that commercial applications will be found in the near future (Toldrá et al. 2012). Current market trends for functional foods are focused on infant formula (which is the fastest growing functional food), followed by energy drinks and pre- and probiotic dairy products.

Regarding target functionality, those promoting general well-being are the most relevant, followed by weight management, digestive health, energy boosting and endurance (Augustin and Sanguansri 2015). Meat-based biopeptides can play an important role in most of them, especially those targeted for improved sport performance where carnitine, carnosine, and related compounds are appreciated.

Finally, a new market for traditional meat products is emerging; to achieve this goal, the peptides generated during processing need to be properly characterised and the mechanism underpinning their generation needs to be fully understood, in order to control, as strictly as possible, the potential health benefits that can be claimed.

The use of peptides in the pharmaceutical industry offers many advantages due to their high bioactivity and target biospecificity, wide spectrum of therapeutic action, expected low allergenicity and toxicity and, most importantly, a low accumulation level in body tissues (Marx 2005). Recent researchers have reported that peptides derived from food have antiviral properties, improve host defence against pathogens, and even can be used in cancer therapies (Agyei and Danquah 2011). According to these authors, the generation of peptide therapeutics from food sources, including meat, can be an alternative to the current synthetic drugs employed for medical purposes. Such pathways can be exploited by the pharmaceutical industry and they can complement treatments already established. For instance, the number of peptides in new chemical entities (NCEs) that are being selected for clinical studies has increased remarkably in recent decades, reaching 17 per year in the first decade of this century when in the 1970s this figure was only 1.2 (Reichert et al. 2010).

2.4 Challenges

In the last couple of decades, thousands of research papers have been published regarding food-derived peptides and protein hydrolysates with many different biological activities, as it has been shown in previous sections. In spite of the promising results and the myriad of sources and activities tested, very few products can be seen on the market shelves (Li-Chan 2015). According to Udenigwe and Howard (2013), in particular, very few products based on meat-derived bioactive peptides are being commercialised when compared to those from dairy or fish sources. It is even more surprising when considering the quantity and sequence diversity present in meat sources, compared to the most explored dairy and fish proteins. Besides, as new bioactivities are being discovered, it has been forecast that bioactive peptides more suitable for incorporation in marketable functional foods will be found, especially if novel sources, such as meat co-products, are investigated. This lack of products including biopeptides as ingredients can be attributed to the many challenges that this field has to face, and the many questions that need to be answered.

The main challenges can be divided into three main categories: technical aspects, legal framework, and consumer perception. However, they are interconnected and must be considered altogether, in order to develop strategies and processes able to supply active peptides for the growing food, cosmetic, and pharmaceutical industries. Such processes must be feasible from an economic point of view, final products have to comply with the current legal framework and consumer demands have to be fulfilled. In Table 2.3 these challenges are summarised with some of the strategies proposed to overcome them.

2.4.1 Technical Challenges

In order to facilitate the inclusion of meat-derived peptides in commercial products, several technical problems must be overcome; such challenges range from the source material and generation system employed to the final inclusion in food products and their bioavailability once ingested.

The most common process for peptide generation is the use of gastric enzymes (trypsin, pancreatin, or chymotrypsin), vegetable origin enzymes (bromelain or papain), or enyzmes of micro-organism origin (alcalase, thermolysin, or protease) (Bhatet al. 2015). Such processes require strict control of the main parameters and usually the enzymes are expensive. The use of immobilised enzymes could minimise the overall cost of the process since they can be reused (Sousa et al. 2004). A different approach, trying to minimise the cost of processing, is to design continuous processes which are more economically efficient at industrial scale. Novel and emerging technologies such as sub-critical water hydrolysis (Marcet et al. 2016), high hydrostatic pressures (Toldrà et al. 2011) and ultrasound (Kadam et al. 2015) have been tested in order to generate peptides; however, the difficulty of accurately predicting the sequence of peptides generated, along with partial or total degradation of certain amino acids, has not been overcome yet.

In addition to this, meat and meat-derived products have a high value in their own right so the use of these products as source material will increase the total cost of peptide generation. However, meat by-products such as white and red offal, blood, skin,

Table 2.3 Main challenges and possible solutions for industrial exploitation of meat-derived biopeptides.

Technical		Legal		Commercialisation	
Concern	**Solution**	**Concern**	**Main points**	**Concern**	**Solution**
High cost of enzyme	Use of genetically modified proteolytic micro-organisms Enzymatic immobilisation Assisted by emerging technologies Explore the use of novel strains or micro-organisms	Health claims	Strong scientific evidence is required Regulation has to ensure a high level of protection for the consumer	Bitter taste	Encapsulation Novel food formulations Inhibiting ingredients Exopeptidase treatment Taste masking
Upscaling of isolation and purification	Novel monolithic columns Improved membrane filtration systems Electrodialysis with UF membranes (EDUF)	International harmonisation	Different organisations apply different rules (EFSA, USDA, FOSHU products, etc.)	Sustainability of resources	Revalorise protein-rich food co-products and by-products
Quantification and identification	Novel *in silico* strategies and optimised MS/MS analysis. Increase the number of sequenced proteins and identified active peptides in databases	Scientific validation	Pharmacokinetics and pharmacodynamics still elusive for dietary peptides Results of clinical studies are inconsistent Inadequate clinical evidence of bio-efficacy		

Selection of source material and enzyme	*In silico* analysis	Allergenicity and toxicity need to be assessed
Effect of processing	Assessment of processing impact on the structure and function of bioactive peptides	
Matrix effect	Study the bioactivities of the peptides when included in complex food matrixes	
Digestion stability	Encapsulation with triggered release	
	Protection of cleavable bonds	
Bioavailability	Better understanding of peptide transport in the digestive system	
Low solubility	Tailored blend of different peptides each at low concentrations	
	Encapsulation	

EFSA, European Food Safety Authority; FOSHU, foods for specific health uses; USDA, United States Department of Agriculture.

drip loss and exudates may be excellent candidates as biopeptide sources (Lafarga et al. 2017; Toldrá et al. 2016) due to their high protein content and diverse amino acid sequences (Mullen and Álvarez 2016). Additionally, the revalorisation of these products will increase the sustainability of the resources, which will not compromise the growing demand for protein-rich products (Udenigwe 2014).

Reviewing the scientific literature, the greatest handicap when developing industrial processes for bioactive peptides (regardless of the source or mechanism employed) is the further step of purification and concentration (Agyei and Danquah 2011; Agyei et al. 2016; di Bernardini et al. 2011; Ryan et al. 2011). Nowadays, membrane filtration, chromatographic techniques or a combination of both are the methods most often used to purify bioactive peptides from the blend of peptides previously generated, in which most have no bioactivity and therefore might mask the activity of the peptide of interest, so they must be removed. Both techniques have been successfully applied to purify and isolate peptides, as reported by Agyei and Danquah (2011).

Unfortunately, such purification steps, although very effective at laboratory scale, are highly costly when transferred to large-scale operations, this being the main reason for the limited commercial application of peptide-based products. Several alternatives, suitable for industrial-scale applications, have been explored including perfusion chromatography with monolithic columns (Agyei et al. 2016), electromembrane filtration (Bargeman et al. 2002) and electrodialysis with ultrafiltration membranes (Firdaous et al. 2009).

In silico analysis has been a remarkable advance in terms of experimental design, bioactivity prediction and selection of raw materials and enzymes, aiming to generate novel bioactive peptides. The use of these informatic tools have been extensively reviewed and the currently available online software, employed for *in silico* analysis, has been explained and discussed (Lafarga et al. 2014; Udenigwe 2014; Mora et al. 2017). The in silico approach contrasts with the traditional one, in which screening of protein/enzyme combinations occur in order to obtain a bioactive peptides. Based on the current databases of sequenced peptides, it is possible to see the potential for any raw material to be used as a source of peptides with specific bioactivities. The same approach can be used when aiming to detect peptides which are potentially toxic or allergenic (Gupta et al. 2013; Lafarga et al. 2016) although, obviously, *in vivo* and *in vitro* analysis cannot be currently replaced by *in silico* analysis. The competency of *in silico* analysis relies on the current databases, which might not be exhaustive, and hence some bioactive peptides may be missed. Besides, when peptides are generated by fermentation or ageing or by means of non-high-specific enzymes, detecting and sequencing peptides *de novo* are challenging since the classic approach of using tryptic hydrolysis cannot be used (Mora et al. 2017).

It has been reported by Korhonen et al. (1998) that processing can affect proteins and peptides in different ways. For instance, heat treatment can degrade heat-sensitive peptides and promote cross-linkages; extreme pHs lead to destruction of amino acids or oxidation; membrane filtration can modify the amino acid composition; and storage can promote oxidation and amino acid modifications, all of these can lead to a partial or total loss of bioactivity.

The effect of high temperatures, high hydrostatic pressures, ultrasound, irradiation or a combination of all the above on the hydrolysis process has been studied. It has been reported that these physical factors do affect the native protein structure, so the peptide composition of the hydrolysates obtained may be different, since denatured proteins expose more cleavage targets, leading to higher degrees of hydrolysis (Toldrà et al. 2011). For instance, Expósito and Recio (2006) reported that heat-denatured proteins which are further hydrolysed exhibited better antibacterial activity compared to peptides from untreated proteins. On the other hand, biopeptides may become inactive when heat treated (Daliri et al. 2017).

The impact on peptide activity of a single processing factor depends on the peptide itself so it is important to optimise the conditions employed to generate, process, purify, and store each peptide to maintain its activity.

In vitro activity of isolated peptides has been extensively reported, but information about how the peptides behave once ingested is scarce. It has been reported that some of the already identified biopeptides act in the digestive system, inhibiting oxidative stress of the mucosa or fighting against colon cancer or inflammatory bowel diseases (García-Nebot et al. 2014). There is a lack of information about resistance to digestive proteolytic enzymes (*in vivo* stability), availability and accessibility, absorption distribution, metabolism, and excretion (Li-Chan 2015). Some bioactive peptides have very low solubility and low dissolution rates, which lead to very low bioavailability, so these compounds often have to be ingested in higher doses in order to reach therapeutic levels in the bloodstream after ingestion. For example, it was reported that a derived soy peptide, called lunasin, was found in plasma, but its presence was 4.5% of total consumption; indicating that very large amounts of the product have to be consumed to exert an effect (Dia et al. 2009).

Many challenges have to be faced individually (cost of process, upscaling, identification, effect of processing) but some can be solved with a single approach: encapsulation. This process aims to entrap an active component within a secondary material. Encapsulation of bioactive peptides has been reported as an economically and industrially feasible technology which aims to (i) avoid interaction of peptides with the environment and other compounds in the food matrix, (ii) minimise the effect of processing once the peptide has been incorporated into the matrix, (iii) increase the stability and shelf-life, (iv) mask or reduce the bitter taste, (v) protect the peptide against digestive enzymes, (vi) increase solubility and bioavailability, and (vii) allow the design of triggered releasing systems (Augustin and Sanguansri 2015). Two excellent reviews describing the most common materials employed for encapsulation, the most used methodologies, factors affecting the process and how to determine encapsulation efficacy have been recently published (Đorđević et al. 2014; Mohan et al. 2015).

2.4.2 Legal Challenges

From a legal point of view, there is a lack of harmonisation between the main organisations regarding the regulatory environment for the approval of novel functional foods, supplements, and active compounds and how health claims should be processed, approved, and communicated (Hobbs et al. 2014). Some of the main markets for functional foods are regulated by different legislations, and the level of evidence required to make a health claim differs significantly among them. For instance, while in Japan and the United States hundreds of products can be found in the market making health claims, in Europe, in spite of the thousands of claims requested, all have been rejected, with the exception of one dipeptide (Val-Tyr) derived from a sardine hydrolysate, which has been declared as a safe ingredient. The same peptide is broadly used in Japan as an ingredient in functional food for antihypertensive purposes. A similar case took place in relation to the antihypertensive milk peptides isoleucine-proline-proline (IPP) and valine-proline-proline (VPP), which have not been approved to have a health claim by the European Food Safety Authority (EFSA) (EFSA Panel on Dietetic Products and Allergies 2012).

The European organisation regulating health claims is the EFSA, operating under the legal frame of the EU, Regulation EC 1924/2006 (subsequently amended by EC Regulation 1169/2011) for nutrition and health claims made about foods. According to this, all health claims 'shall be based on and substantiated by generally accepted scientific evidence' or 'the food business operator making a health claim shall justify the use of the claim'. In addition, the following information must be provided on the labelling: 'a statement indicating the quantity of the food and pattern of consumption required to obtain the claimed benefit', and a 'statement addressed to persons who should avoid using the food', among others. This legislation ensures that consumers are protected against fraudulent claims and co-ordinates the actions of all Member States. This legal requirement, within Europe, is closely linked to technical challenges regarding bioavailability, identification, dosage, and so on, so it is mandatory to find a solution for all these technical challenges before the legal challenges can be countered satisfactorily.

The Food and Drug Administration (FDA) is the organisation dealing with health claims in the United States. In this country, there is no legal definition for functional food, even though working definitions have been provided by the American Dietetic Association (ADA), for example, 'any food that moves beyond basic nutrition to provide additional health benefits that may reduce disease risk or promote optimal health'. The US regulatory system is more lax, and less evidence is necessary to make a health claim. This flexibility in the legislation, according to Hobbs et al. (2014), encourages more investment on R&D since lower costs are required to substantiate health claims and subsequent commercialization of novel active compounds.

Japan is the leader in the market for functional foods, since a national research programme focused on this type of product was launched in 1984. Japanese authorities, specifically the Japanese Ministry of Health, Labour and Welfare, were the first to formulate a regulatory framework for functional foods. In these regulations, functional food sare divided into two categories: food with nutrient function claims (FNFC) and foods for specified health uses (FOSHU); additionally, in 2005, FOSHU were subdivided into four groups based on the strength of the evidence behind their claims. By 2007, there were over 600 FOSHU products and this number is currently over 1000.

2.4.3 Consumer Challenges

The functional food market is expected to grow to €43 billion in 2017, which is an increase of 25% compared to 2013. A similar trend was observed in food supplements (Gevaert et al. 2016). Such an increment is linked to increased consumer perception that food can play a key role in the prevention of diseases such as obesity, diabetes, gastrointestinal cancer, and blood pressure. So addressing consumer demands, along with legal requirements, is mandatory for the industries involved in order to maintain the current growing trend.

One of the most important challenges from the consumer's point of view is the bitter taste related to peptides. It has been reported that peptides of low molecular weight, precisely those which potentially impart good bioactivity, produce greater bitterness than native proteins. Lafarga and Hayes (2017) reported that consumers have different perceptions of functional foods; for example, in Japan the benefits of these products are more important than the sensory properties, while Western consumers still consider taste and appearance to be among the most important food attributes. However, meat, fish or gelatine hydrolysates are reported to be less bitter than those from other raw materials (Johanna 2007). Several strategies have been suggested to remove, or at least reduce, the bitter taste of active peptides. Some of these, for example encapsulation, use of maskers, bitterness inhibitors or exopeptidases, have been extensively discussed by Goldberg et al. (2017).

One final concern that may affect the consumer's decision when buying functional foods is the sustainability of the production process. The raw materials employed for biopeptide generation are also used as a source of nutritious proteins, demand for which is continuously growing (Mullen et al. 2017). As discussed by Udenigwe (2014), the diversion of large amounts of food proteins to biopeptide generation could lead to an insecure situation relating to the challenge of feeding a growing population. In this regard, efforts have been made in order to utilise meat industry co-products and by-products as novel sources of active biopeptides, since these products are not usually employed for food purposes. There is increasing interest from the scientific community in exploring the potential of meat by- and co-products as sources of bioactive peptides, aiming to improve the sustainability of the meat industry and to make better use of the available resources, and at the same time increase economical profit (Lafarga et al. 2017; Mullen et al. 2015; Toldrá et al. 2016).

2.5 Conclusions

The possibilities that biopeptides from meat offer from a scientific and economic perspective and their benefits to the consumer are potentially enormous. Whether the active molecules are naturally present in the meat, are generated after an ageing or fermentation process, or are produced after an enzymatic process, they will play a very important role in the arena of functional foods and pharmaceutical markets. Food processors, pharmaceutical industries, and research institutes, not without reason, are investing more and more resources in developing novel functional foods, supplements, and drugs including biopeptides as active ingredients. The variety of health benefits discovered in meat-sourced peptides, such as immunomodulatory, opioid, antidiabetic, antioxidant, antihypertensive or antiappetiser effects, to mention just a few, is an incentive to look for solutions to overcome the current challenges. As an additional benefit, the use of meat co-products as raw materials will transform the meat industry into a more sustainable activity, by generating new markets and opportunities.

The use of emerging technologies such as ultrasound, innovative approaches using *in silico* tools and better understanding of the physiological processes involved in digestion, absorption, transport and biological activity, among others, will allow us to establish a solid knowledge base which eventually will help us to maintain future health claims while protecting the interests of consumers.

References

Adler-Nissen, J. (1979). Determination of the degree of hydrolysis of food protein hydrolysates by trinitrobenzenesulfonic acid. *Journal of Agricultural and Food Chemistry* 27 (6): 1256–1262.

Agyei, D. and Danquah, M.K. (2011). Industrial-scale manufacturing of pharmaceutical-grade bioactive peptides. *Biotechnology Advances* 29 (3): 272–277.

Agyei, D., Ongkudon, C.M., Wei, C.Y. et al. (2016). Bioprocess challenges to the isolation and purification of bioactive peptides. *Food and Bioproducts Processing* 98: 244–256.

Arihara, K. (2006). Strategies for designing novel functional meat products. *Meat Science* 74 (1): 219–229.

Arihara, K., Nakashima, Y., Mukai, T. et al. (2001). Peptide inhibitors for angiotensin I-converting enzyme from enzymatic hydrolysates of porcine skeletal muscle proteins. *Meat Science* 57 (3): 319–324.

Aristoy, M. and Toldrá, F. (1998). Concentration of free amino acids and dipeptides in porcine skeletal muscles with different oxidative patterns. *Meat Science* 50 (3): 327–332.

Augustin, M.A. and Sanguansri, L. (2015). Challenges and solutions to incorporation of nutraceuticals in foods. *Annual Review of Food Science and Technology* 6: 463–477.

Bargeman, G., Koops, G.-H., Houwing, J. et al. (2002). The development of electro-membrane filtration for the isolation of bioactive peptides: the effect of membrane selection and operating parameters on the transport rate. *Desalination* 149 (1–3): 369–374.

Berardo, A., Devreese, B., de Maere, H. et al. (2017). Actin proteolysis during ripening of dry fermented sausages at different pH values. *Food Chemistry* 221 (Supplement C): 1322–1332.

Bhat, Z., Kumar, S., and Bhat, H.F. (2015). Bioactive peptides of animal origin: a review. *Journal of Food Science and Technology* 52 (9): 5377–5392.

Bowden, N.A., Kolfschoten, R.C., Sanders, J.P., and Bruins, M.E. (2016). Fractional precipitation of amino acids from agro-industrial residues using ethanol. *Industrial and Engineering Chemistry Research* 55 (27): 7462–7472.

Ballatori, N., Krance, S.M., Notenboom, S. et al. (2009). Glutathione dysregulation and the etiology and progression of human diseases. *Biological Chemistry* 390 (3): 191–214.

Consultation, R. (2011). Dietary protein quality evaluation in human nutrition. *FAO Food and Nutrition Paper* 92.

Di Bernardini, R., Harnedy, P., Bolton, D. et al. (2011). Antioxidant and antimicrobial peptidic hydrolysates from muscle protein sources and by-products. *Food Chemistry* 124 (4): 1296–1307.

Di Bernardini, R., Mullen, A.M., Bolton, D. et al. (2012). Assessment of the angiotensin-I-converting enzyme (ACE-I) inhibitory and antioxidant activities of hydrolysates of bovine brisket sarcoplasmic proteins produced by papain and characterisation of associated bioactive peptidic fractions. *Meat Science* 90 (1): 226–235.

Dia, V.P., Torres, S., de Lumen, B.O. et al. (2009). Presence of lunasin in plasma of men after soy protein consumption. *Journal of Agricultural and Food Chemistry* 57 (4): 1260–1266.

Daliri, E.B.-M., Oh, D.H., and Lee, B.H. (2017). Bioactive peptides. *Foods* 6 (5): 32.

Đorđević, V., Balanč, B., Belščak-Cvitanović, A. et al. (2014). Trends in encapsulation technologies for delivery of food bioactive compounds. *Food Engineering Reviews* 7 (4): 452–490.

EFSA Panel on Dietetic Products, Nutrition, & Allergies (2012). Scientific opinion on the substantiation of health claims related to isoleucine-proline-proline (IPP) and valine-proline-proline (VPP) and maintenance of normal blood pressure (ID 661, 1831, 1832, 2891, further assessment) pursuant to article 13(1) of regulation (EC) no 1924/2006. *EFSA Journal* 10 (6): 2715.

Escudero, E., Sentandreu, M.A., Arihara, K., and Toldrá, F. (2010). Angiotensin I-converting enzyme inhibitory peptides generated from in vitro gastrointestinal digestion of pork meat. *Journal of Agricultural and Food Chemistry* 58 (5): 2895–2901.

Escudero, E., Mora, L., Fraser, P.D. et al. (2013). Purification and identification of antihypertensive peptides in Spanish dry-cured ham. *Journal of Proteomics* 78: 499–507.

Expósito, I.L. and Recio, I. (2006). Antibacterial activity of peptides and folding variants from milk proteins. *International Dairy Journal* 16 (11): 1294–1305.

Fadda, S., López, C., and Vignolo, G. (2010). Role of lactic acid bacteria during meat conditioning and fermentation: peptides generated as sensorial and hygienic biomarkers. *Meat Science* 86 (1): 66–79.

Farshidfar, F., Pinder, M., and Myrie, S. (2017). Creatine supplementation and skeletal muscle metabolism for building muscle mass-review of the potential mechanisms of action. *Current Protein and Peptide Science* 18 (12): 1273–1287.

Ferrari, R., Merli, E., Cicchitelli, G. et al. (2004). Therapeutic effects of l-carnitine and propionyl-l-carnitine on cardiovascular diseases: a review. *Annals of the New York Academy of Sciences* 1033 (1): 79–91.

Firdaous, L., Dhulster, P., Amiot, J. et al. (2009). Concentration and selective separation of bioactive peptides from an alfalfa white protein hydrolysate by electrodialysis with ultrafiltration membranes. *Journal of Membrane Science* 329 (1): 60–67.

Flores, M. and Toldrá, F. (2011). Microbial enzymatic activities for improved fermented meats. *Trends in Food Science and Technology* 22 (2): 81–90.

Gabriel, G.J., Madkour, A.E., Dabkowski, J.M. et al. (2008). Synthetic mimic of antimicrobial peptide with nonmembrane-disrupting antibacterial properties. *Biomacromolecules* 9 (11): 2980–2983.

García-Nebot, M.J., Recio, I., and Hernández-Ledesma, B. (2014). Antioxidant activity and protective effects of peptide lunasin against oxidative stress in intestinal Caco-2 cells. *Food and Chemical Toxicology* 65: 155–161.

Gevaert, B., Veryser, L., Verbeke, F. et al. (2016). Fish hydrolysates: a regulatory perspective of bioactive peptides. *Protein and Peptide Letters* 23 (12): 1052–1060.

Gallego, M., Grootaert, C., Mora, L. et al. (2016). Transepithelial transport of dry-cured ham peptides with ACE inhibitory activity through a Caco-2 cell monolayer. *Journal of Functional Foods* 21: 388–395.

Goldberg, E., Grant, J., Aliani, M., and Eskin, M.N. (2017). 10 methods for removing bitterness in functional foods and nutraceuticals. In: *Bitterness: Perception, Chemistry and Food Processing* (ed. M. Aliani and M.N. Eskin). Chichester: Wiley Blackwell.

Guiotto, A., Calderan, A., Ruzza, P., and Borin, G. (2005). Carnosine and carnosine-related antioxidants: a review. *Current Medicinal Chemistry* 12 (20): 2293–2315.

Gupta, S., Kapoor, P., Chaudhary, K. et al. (2013). In silico approach for predicting toxicity of peptides and proteins. *PLoS One* 8 (9): e73957.

Hobbs, J.E., Malla, S., and Sogah, E.K. (2014). Regulatory frameworks for functional food and supplements. *Canadian Journal of Agricultural Economics/Revue Canadienne d'agroeconomie* 62 (4): 569–594.

Hugas, M. and Monfort, J.M. (1997). Bacterial starter cultures for meat fermentation. *Food Chemistry* 59 (4): 547–554.

Hurrell, R. and Egli, I. (2010). Iron bioavailability and dietary reference values. *American Journal of Clinical Nutrition* 91 (5): 1461S–1467S.

Huxtable, R. (1992). Physiological actions of taurine. *Physiological Reviews* 72 (1): 101–163.

In, M.-J., Jeong Chae, H., and Oh, N.-S. (2002). Process development for heme-enriched peptide by enzymatic hydrolysis of hemoglobin. *Bioresource Technology* 84 (1): 63–68.

Jang, A. and Lee, M. (2005). Purification and identification of angiotensin converting enzyme inhibitory peptides from beef hydrolysates. *Meat Science* 69 (4): 653–661.

Jayasena, D.D., Jung, S., Bae, Y.S. et al. (2015a). Comparison of the amounts of endogenous bioactive compounds in raw and cooked meats from commercial broilers and indigenous chickens. *Journal of Food Composition and Analysis* 37: 20–24.

Jayasena, D.D., Jung, S., Kim, S.H. et al. (2015b). Endogenous functional compounds in Korean native chicken meat are dependent on sex, thermal processing and meat cut. *Journal of the Science of Food and Agriculture* 95 (4): 771–775.

Jiang, J. and Xiong, Y.L. (2016). Natural antioxidants as food and feed additives to promote health benefits and quality of meat products: a review. *Meat Science* 120: 107–117.

Johanna, M. (2007). *Metalloproteases. Industrial Enzymes, Structure, Function and Applications.* Dordrecht: Springer.

Jones, D.P., Coates, R.J., Flagg, E.W. et al. (1992). Glutathione in foods listed in the National Cancer Institute's health habits and history food frequency questionnaire. *Nutrition and Cancer* 17 (1): 57–75.

Jung, S., Bae, Y.S., Kim, H.J. et al. (2013). Carnosine, anserine, creatine, and inosine 5′-monophosphate contents in breast and thigh meats from 5 lines of Korean native chicken. *Poultry Science* 92 (12): 3275–3282.

Kadam, S.U., Tiwari, B.K., Álvarez, C., and O'Donnell, C.P. (2015). Ultrasound applications for the extraction, identification and delivery of food proteins and bioactive peptides. *Trends in Food Science and Technology* 46 (1): 60–67.

Kęska, P., Libera, J., and Stadnik, J. (2017). Comparison of antioxidant activity of protein isolates derived from selected dry-cured meat products. *Journal of Food Processing and Preservation* 41 (3): E12933.

Kikuchi, K., Matahira, Y., and Sakai, K. (2004). Separation and physiological functions of anserine from fish extract. *Developments in Food Science* 42: 97–105.

Koba, K. and Yanagita, T. (2014). Health benefits of conjugated linoleic acid (CLA). *Obesity Research and Clinical Practice* 8 (6): e525–e532.

Korhonen, H., Pihlanto-Leppäla, A., Rantamäki, P., and Tupasela, T. (1998). Impact of processing on bioactive proteins and peptides. *Trends in Food Science and Technology* 9 (8): 307–319.

Lafarga, T. and Hayes, M. (2014). Bioactive peptides from meat muscle and by-products: generation, functionality and application as functional ingredients. *Meat Science* 98 (2): 227–239.

Lafarga, T. and Hayes, M. (2017). Bioactive protein hydrolysates in the functional food ingredient industry: overcoming current challenges. *Food Reviews International* 33 (3): 217–246.

Lafarga, T., O'Connor, P., and Hayes, M. (2014). Identification of novel dipeptidyl peptidase-IV and angiotensin-I-converting enzyme inhibitory peptides from meat proteins using in silico analysis. *Peptides* 59: 53–62.

Lafarga, T., Wilm, M., Wynne, K., and Hayes, M. (2016). Bioactive hydrolysates from bovine blood globulins: generation, characterisation, and in silico prediction of toxicity and allergenicity. *Journal of Functional Foods* 24: 142–155.
Lafarga, T., Álvarez, C., and Hayes, M. (2017). Bioactive peptides derived from bovine and porcine co-products: a review. *Journal of Food Biochemistry* 41 (6): E12418.
Lebrun, F., Bazus, A., Dhulster, P., and Guillochon, D. (1998). Solubility of heme in heme-iron enriched bovine hemoglobin hydrolysates. *Journal of Agricultural and Food Chemistry* 46 (12): 5017–5025.
Li-Chan, E.C.Y. (2015). Bioactive peptides and protein hydrolysates: research trends and challenges for application as nutraceuticals and functional food ingredients. *Current Opinion in Food Science* 1: 28–37.
Lu, S.C. (2013). Glutathione synthesis. *Biochimica et Biophysica Acta-General Subjects* 1830 (5): 3143–3153.
Marcet, I., Alvarez, C., Paredes, B., and Diaz, M. (2016). The use of sub-critical water hydrolysis for the recovery of peptides and free amino acids from food processing wastes. Review of sources and main parameters. *Waste Management* 49: 364–371.
Marx, V. (2005). Watching peptide drugs grow up. *Chemical and Engineering News* 83 (11): 17–24.
McDowell, L.R. (2003). Iron. In: *Minerals in Animal and Human Nutrition*, 2e, 203–233. Amsterdam: Elsevier.
Mejri, L., Vásquez-Villanueva, R., Hassouna, M. et al. (2017). Identification of peptides with antioxidant and antihypertensive capacities by RP-HPLC-Q-TOF-MS in dry fermented camel sausages inoculated with different starter cultures and ripening times. *Food Research International* 100 (Part 1): 708–716.
Meyer, H.-P. and Robins, K.T. (2005). Large scale bioprocess for the production of optically pure L-carnitine. *Monatshefte für Chemie/Chemical Monthly* 136 (8): 1269–1277.
Mellander, O. (1950). The physiological importance of the casein phosphopeptide calcium salts. II. Peroral calcium dosage of infants. Some aspects of the pathogenesis of rickets. *Acta Societatis Botanicorum Poloniae* 55: 247–257.
Mohammad, A.W., Ng, C.Y., Lim, Y.P., and Ng, G.H. (2012). Ultrafiltration in food processing industry: review on application, membrane fouling, and fouling control. *Food and Bioprocess Technology* 5 (4): 1143–1156.
Mohan, A., Rajendran, S.R., He, Q.S. et al. (2015). Encapsulation of food protein hydrolysates and peptides: a review. *RSC Advances* 5 (97): 79270–79278.
Mojsoska, B., Zuckermann, R.N., and Jenssen, H. (2015). Structure-activity relationship study of novel peptoids that mimic the structure of antimicrobial peptides. *Antimicrobial Agents and Chemotherapy* 59 (7): 4112–4120.
Mora, L., Sentandreu, M.Á., and Toldrá, F. (2008). Contents of creatine, creatinine and carnosine in porcine muscles of different metabolic types. *Meat Science* 79 (4): 709–715.
Mora, L., Reig, M., and Toldrá, F. (2014). Bioactive peptides generated from meat industry by-products. *Food Research International* 65: 344–349.
Mora, L., Escudero, E., and Toldrá, F. (2016a). Characterization of the peptide profile in Spanish Teruel, Italian Parma and Belgian dry-cured hams and its potential bioactivity. *Food Research International* 89 (Part 1): 638–646.
Mora, L., Calvo, L., Escudero, E., and Toldrá, F. (2016b). Differences in pig genotypes influence the generation of peptides in dry-cured ham processing. *Food Research International* 86: 74–82.

Mora, L., Gallego, M., Reig, M., and Toldrá, F. (2017). Challenges in the quantitation of naturally generated bioactive peptides in processed meats. *Trends in Food Science and Technology* 69 (B): 306–314.

Mullen, A. and Álvarez, C. (2016). Offal: types and composition. *Encyclopedia of Food and Health* 4: 152–157.

Mullen, A., Álvarez, C., Pojić, M. et al. (2015). Classification and target compounds. In: *Food Waste Recovery: Processing Technologies and Industrial Techniques* (ed. C. Galanakis), 25–57. Amsterdam: Academic Press.

Mullen, A., Álvarez, C., Zeugolis, D.I. et al. (2017). Alternative uses for co-products: harnessing the potential of valuable compounds from meat processing chains. *Meat Science* 132: 90–98.

Nagaraju, S.P., Cohn, A., Akbari, A. et al. (2013). Heme iron polypeptide for the treatment of iron deficiency anemia in non-dialysis chronic kidney disease patients: a randomized controlled trial. *BMC Nephrology* 14 (1): 64.

Neurath, H. (1999). Proteolytic enzymes, past and future. *Proceedings of the National Academy of Sciences* 96 (20): 10962–10963.

Nip, W.-K. (2008). *Dry-Cured Meat Products*. Chichester: Wiley.

Ohata, M., Uchida, S., Zhou, L., and Arihara, K. (2016). Antioxidant activity of fermented meat sauce and isolation of an associated antioxidant peptide. *Food Chemistry* 194 (Supplement C): 1034–1039.

Reichert, J., Pechon, P., Tartar, A., and Dunn, M. (2010). *Development Trends for Peptide Therapeutics*. San Diego: Peptide Therapeutics Foundation.

Ryan, J.T., Ross, R.P., Bolton, D. et al. (2011). Bioactive peptides from muscle sources: meat and fish. *Nutrients* 3 (9): 765–791.

Seline, K.-G. and Johein, H. (2007). The determination of l-carnitine in several food samples. *Food Chemistry* 105 (2): 793–804.

Sharma, S., Singh, R., and Rana, S. (2011). Bioactive peptides: a review. *International Journal Bioautomation* 15 (4): 223–250.

Sohaib, M., Anjum, F.M., Sahar, A. et al. (2017). Antioxidant proteins and peptides to enhance the oxidative stability of meat and meat products: a comprehensive review. *International Journal of Food Properties* 20 (11): 1–13.

Sousa, R. Jr., Lopes, G., Tardioli, P. et al. (2004). Kinetic model for whey protein hydrolysis by alcalase multipoint-immobilized on agarose gel particles. *Brazilian Journal of Chemical Engineering* 21 (2): 147–153.

Takeda, S., Matsufuji, H., Nakade, K. et al. (2017). Investigation of lactic acid bacterial strains for meat fermentation and the product's antioxidant and angiotensin-I-converting-enzyme inhibitory activities. *Animal Science Journal* 88 (3): 507–516.

Toldrá, F. (1998). Proteolysis and lipolysis in flavour development of dry-cured meat products. *Meat Science* 49: S101–S110.

Toldrà, M., Parés, D., Saguer, E., and Carretero, C. (2011). Hemoglobin hydrolysates from porcine blood obtained through enzymatic hydrolysis assisted by high hydrostatic pressure processing. *Innovative Food Science and Emerging Technologies* 12 (4): 435–442.

Toldra, F., Aristoy, M.C., Mora, L., and Reig, M. (2012). Innovations in value-addition of edible meat by-products. *Meat Science* 92 (3): 290–296.

Toldrá, F., Mora, L., and Reig, M. (2016). New insights into meat by-product utilization. *Meat Science* 120: 54–59.

Toldrá, F., Reig, M., Aristoy, M.C., and Mora, L. (2017). Generation of bioactive peptides during food processing. *Food Chemistry* doi: doi.org/10.1016/j.foodchem.2017.06.119.
Twycross-Lewis, R., Kilduff, L.P., Wang, G., and Pitsiladis, Y. (2016). The effects of creatine supplementation on thermoregulation and physical (cognitive) performance: a review and future prospects. *Amino Acids* 48 (8): 1843–1855.
Udenigwe, C.C. (2014). Bioinformatics approaches, prospects and challenges of food bioactive peptide research. *Trends in Food Science and Technology* 36 (2): 137–143.
Udenigwe, C.C. and Howard, A. (2013). Meat proteome as source of functional biopeptides. *Food Research International* 54 (1): 1021–1032.
Viladomiu, M., Hontecillas, R., and Bassaganya-Riera, J. (2016). Modulation of inflammation and immunity by dietary conjugated linoleic acid. *European Journal of Pharmacology* 785: 87–95.
Xing, L.-j., Hu, Y.-y., Hu, H.-y. et al. (2016). Purification and identification of antioxidative peptides from dry-cured Xuanwei ham. *Food Chemistry* 194 (Supplement C): 951–958.

3

Human Gastrointestinal Endogenous Proteins: A Recently Discovered Source of Gut Modulatory Peptides

Lakshmi A. Dave

Riddet Institute, Massey University, Palmerston North, New Zealand

3.1 Introduction

The human gastrointestinal tract (GIT) is essentially a 5–9 m long muscular, open-ended tube that extends from the mouth to the anus (DeSesso and Jacobson 2001; Tortora et al. 2016; Weaver et al. 1991). The epithelial lining of the gut lumen remains in constant contact with the outside environment and hence, anatomically, the gut lumen and its contents are external to the body (DeSesso and Jacobson 2001; Franklin 2015).

The primary function of the GIT is the digestion of ingested food, absorption of nutrients, and propulsion of chyme and digesta along the tract (Johnson 2012), and the GIT secretes a host of proteins to carry out these primary functions (Desnuelle et al. 1986). Examples of such secreted proteins include glycoproteins, digestive enzymes, and other proteins from gastric, pancreatic, biliary, and intestinal cell secretions (Desnuelle et al. 1986). Apart from these secreted proteins that are directly involved in the process of digestion, the GIT also secretes lysozyme (LYS), defensins, mucins, and other proteins involved in the immune response (Chassaing et al. 2014). Further, some body proteins such as serum albumin (SALB) undergo clearance through the gut (Brassinne 1974). In addition, the gut epithelial cells undergo continuous desquamation and turnover, thereby releasing cellular protein in the lumen of the GIT (Williams et al. 2015). The wide variety of adaptive microbiota (Quercia et al. 2014) colonising the upper and lower digestive tract also contribute a significant amount of protein to the gut lumen (Caine et al. 1999). Together, all of the above non-dietary proteinaceous secretions and proteins of epithelial and bacterial origin released into gut lumen are known as gastrointestinal endogenous proteins (GEP).

GEP were first mentioned in light of their contribution to the gut nitrogen pool as early as the 1920s (Mitchell 1924). Table 3.1 gives the major sources of GEP in humans. Table 3.2 gives an overview of the protein nitrogen contributed by different GEP by the end of the terminal ileum of humans given a casein-based diet.

Gut luminal endogenous protein losses or endogenous nitrogen losses (ENL) can be divided into basal and specific losses. Basal losses are non-specific and are defined as

Novel Proteins for Food, Pharmaceuticals, and Agriculture: Sources, Applications, and Advances, First Edition. Edited by Maria Hayes.

Table 3.1 Major sources of gastrointestinal endogenous proteins (GEP) in the GIT of humans.

Type of GEP	Examples
Major secreted GEP	
Salivary glands	Amylase, lipase, mucins, proline-rich proteins, nerve growth factor, epidermal growth factor, renin, peptide hormones, immunoglobulins, LYS
Oesophagus	Mucins
Stomach	Soluble protiens, pepsins, gastric intrinsic factor, lipase, mucin, trefoil factors, hormones
Liver	Plasma proteins and peptides, e.g. SALB and glutathione respectively, hormones, immunoglobulins
Pancreas	Amylase, lipase, phospholipase, zymogens of trypsin, chymotrypsin, elastase, carboxypeptidase
Small intestine	Mucins, LYS, hormones, immunoglobulins
Colon	Mucin, immunoglobulin
Other GEP	Proteins entering the gut for clearance, such as SALB
Non-secretory GEP	
Epithelial cells	Keratin, actin, and aminopeptidase N
Microbiota	Cellular proteins, secreted proteins

Source: Based on data from Deglaire (2008); Desnuelle et al. (1986); Miner-Williams (2012).LYS, lysozyme; SALB, serum albumin.

the obligatory losses associated with the metabolic function and dry matter (DM) intake of the individual. Specific losses are a result of the composition of one's diet and may be induced or modulated by dietary constituents such as fibre, protein content and source, and antinutritional factors in the diet. Apart from dietary factors, other factors influencing GEP losses include individual factors such as age, protein status and overall health (Boisen and Moughan 1996; Ravindran 2016; Rutherfurd et al. 2015a). Figure 3.1 summarises the various factors that influence GEP or ENL. The total gut luminal endogenous protein or GEP losses can be measured using techniques that involve isotope markers, linear regression method, homoarginine technique, protein-free, synthetic amino acid-based diets, protein-free diets with intravenous amino acid infusion, natural proteins devoid of an amino acid, or enzyme-hydrolysed protein (Moughan et al. 1998; Ravindran 2016).

In a healthy individual, a significant part of the crude protein and amino acids in the gut contents (stomach chyme and small intestinal digesta) is composed of GEP. In comparison to dietary proteins, GEP represent a larger and more consistent source of proteins in the GIT. As shown in Figure 3.2, GEP contribute 65–200 g of protein per day, making them quantitatively as significant as dietary proteins. While some GEP are known to be partially indigestible, a majority of GEP undergo simultaneous digestion and reabsorption alongside dietary proteins (Nasset and Ju 1961; Souffrant et al. 1993). Recognising the significance of GEP losses into the gut lumen, the FAO has

Table 3.2 Summary of % total protein nitrogen contributed by major types of GEP by the end of the terminal ileum of humans given a casein-based diet.

Source of protein	Contribution to total GEP in terms of % total protein nitrogen
Bacteria	69.23
Mucin	16.91
Soluble-free protein (albumin, immunoglobulins, digestive enzymes, and low molecular weight bioactive peptides)	7.65
Mucosal cells	6.2

Source: Based on data from Miner-Williams et al. (2012).

recommended the use of the digestible indispensable amino acid score (DIAAS) to assess dietary protein quality. This takes into account the endogenous ileal amino acid flows, and hence provides a more accurate assessment of the quality of dietary proteins (Rutherfurd et al. 2015b).

While the contribution of gut endogenous losses to the total nitrogen pool and specific amino acids in the gut lumen has been the subject of many studies, recent developments have demonstrated, for the first time, that like dietary proteins, GEP can also produce bioactive peptides in the gut. Figure 3.2 gives the daily estimated amounts of bioactive peptides that may be released by dietary protein and GEP. In this context,

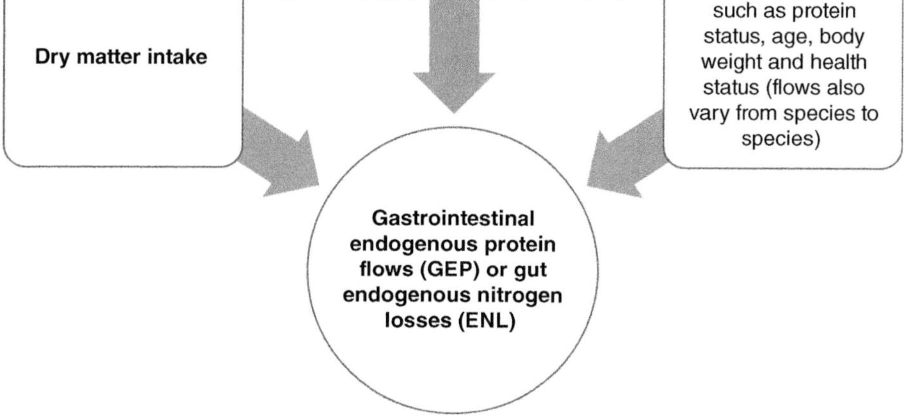

Figure 3.1 Factors impacting GEP protein flows or gut ENL. *Source:* Based on data from Moughan and Rutherfurd (2012); Ravindran (2016); Rutherfurd et al. (2015b).

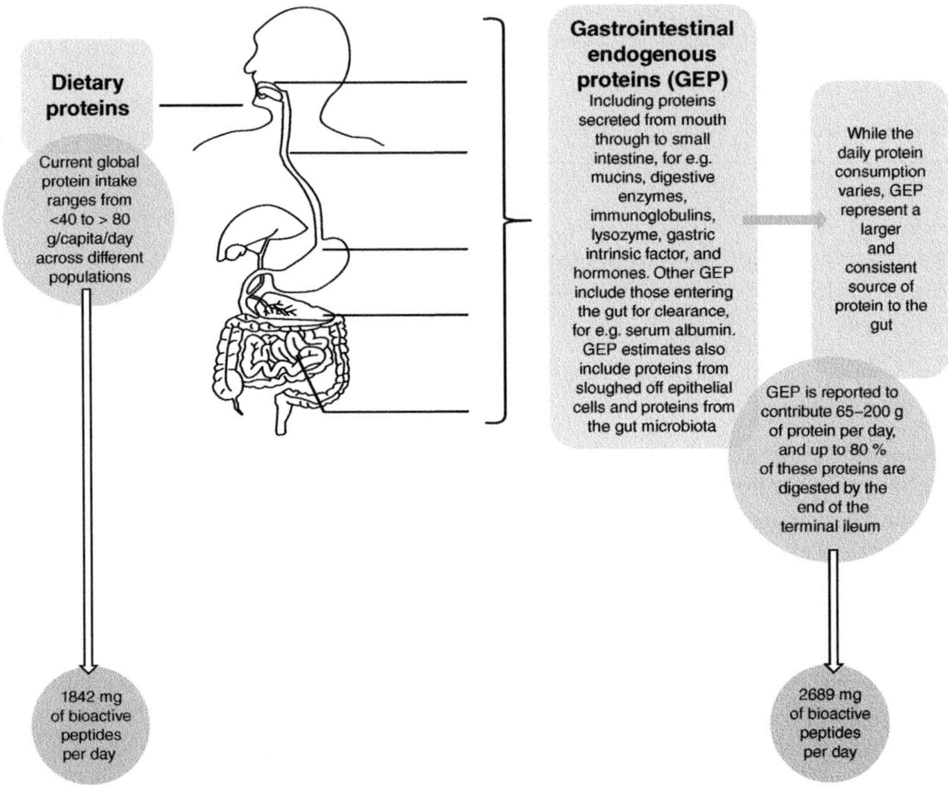

Figure 3.2 The two sources of protein and exogenous bioactive peptides (ExBP) in the gut. Dietary protein and gut endogenous proteins (GEP) are two sources of proteins in the human gut. The amount and quality of dietary protein consumed are known to vary from individual to individual, across populations, across seasons, and from day to day. In comparison, GEP are a consistent source of protein to the gut. This figure depicts the emerging role of GEP as a source of ExBP. Preliminary *in silico, in vitro*, and *in vivo* evidence has shown that, like dietary proteins, GEP can also generate bioactive peptides in the gut. This diagram also shows the daily potential contribution of dietary protein- and GEP-derived ExBP. *Source:* Data collated from various sources (Brunser et al. 1992; Dave et al. 2014, 2015; FAO Expert Consultation 2011; FAOSTAT 2017; Moughan and Rutherfurd 2012; Nasset 1964; Nasset and Ju 1961; UniProt Consortium 2012). Reprinted with permission of Elsevier.

the different physiological regulatory systems at play in the GIT and which may be influenced by bioactive peptides are outlined in Figure 3.3.

It is of note that, because the gut lumen is contiguous with the external environment, both dietary protein and GEP are exogenous sources of protein to the gut. Thus, resultant peptides from these proteins, including those exhibiting bioactivity, are also exogenous (Dave et al. 2015). The classification of bioactive peptides based on the 'site of their generation' is discussed further later in this chapter.

The following sections summarise and re-examine current knowledge about GEP and their potential as a source of bioactive peptides in the gut. They also outline a typical sequential design for the investigation of biologically active peptides from novel sources. Further, the significance of GEP-derived peptides in augmenting the dietary

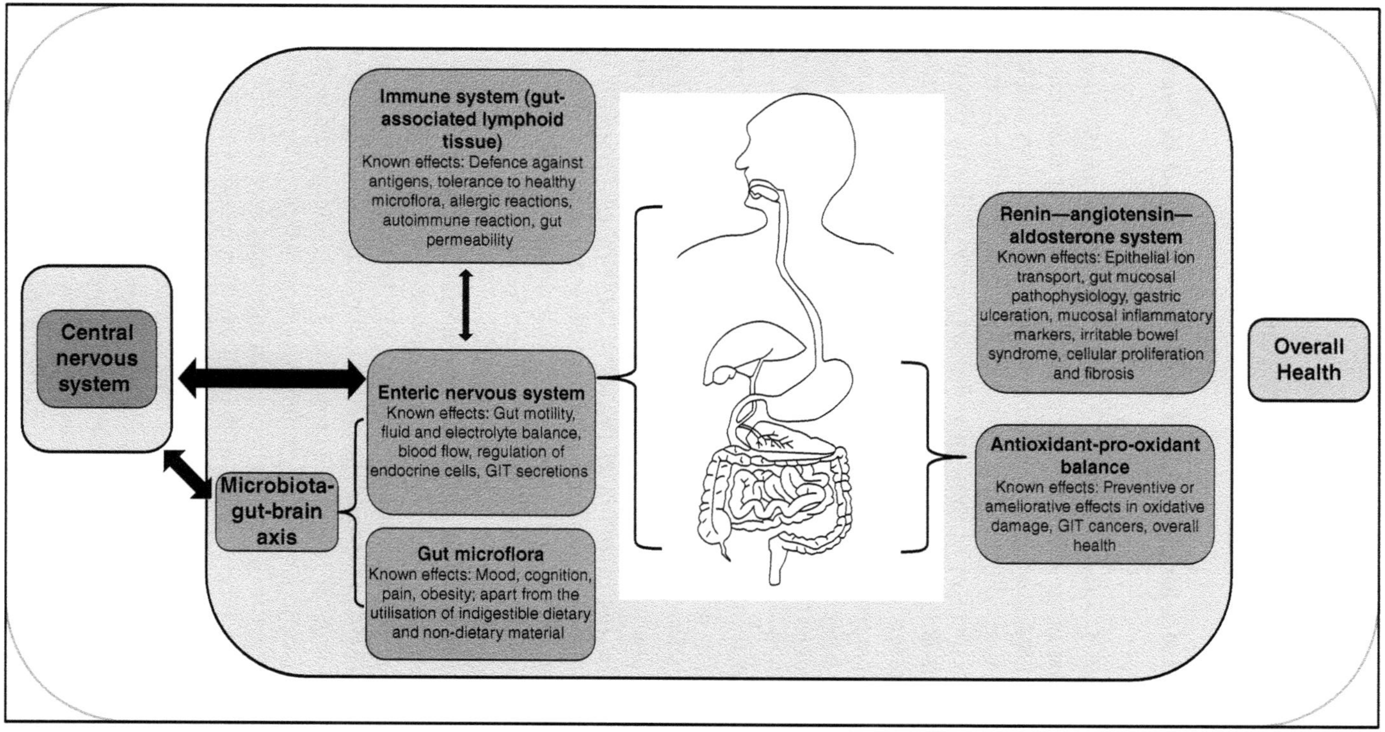

Figure 3.3 The numerous regulatory systems at play in the GIT. *Source:* Reprinted with permission of Elsevier from Dave et al. (2015).

protein-derived bioactive peptides is also discussed. This chapter also dwells on the key research questions raised by the discovery of GEP-derived bioactive peptides.

3.2 A Summary of Current Knowledge Regarding GEP-derived Bioactive Peptides

3.2.1 *In Silico* Evidence for the Presence and Release of Bioactive Peptide Motifs from GEP

In silico models have been used extensively to mine, identify, and document bioactive peptides from a vast range of dietary proteins. A recent study from our group (Dave et al. 2014) utilised the major web-based bioinformatics databases and tools as a first step to explore whether GEP may contain bioactive peptides within their larger protein sequences, and whether any of these peptides are released during the simulated gastric or gastrointestinal digestion of GEP. In the model, the protein sequences of 26 individual GEP and seven known dietary protein sources of bioactive peptides were screened using a bioactive peptide database (BIOPEP) that allowed an estimation of the number of bioactive peptide sequences in each of the proteins. The varied selection of proteins in both groups (i.e. dietary proteins and GEP) had a wide variation in size and their chain lengths ranged from <100 to >5000 amino acid residues. The *in silico* simulated gastrointestinal digestion was carried out using three major known gut proteases – pepsin, trypsin, and chymotrypsin.

This *in silico* study found that the total numbers of bioactive peptide sequences found in intact dietary proteins and intact GEP were comparable. On average, both groups of proteins contained a high number of bioactive peptides encrypted in their primary structure, ranging from ≈60 to 2500 bioactive peptide sequences per molecule. Moreover, the results indicated that GIT digestion of the GEP would probably release numerous peptides, some of which were previously found to have multiple bioactivities.

Interestingly, only a very small fraction (≈1% of both GEP and dietary proteins) of these bioactive peptide sequences was predicted to be released after simulated (*in silico*) gastric digestion. The number of bioactive peptides predicted to be released after the simulated (*in silico*) small intestinal digestion was higher (≈3.3%). Further, *in silico* digestion predicted that GEP released in the small intestine would also give rise to bioactive peptides.

However, the findings of the *in silico* work have to be interpreted with caution as a pure *in silico* approach to simulating protein digestion is limited in several ways. First, only the primary structure of a protein is taken into consideration, and the actions of only some of the proteolytic enzymes of the GIT can be applied. Also, *in silico* digestion models assume that the selected enzymes will break every peptide bond for which they have a known cleavage specificity (i.e. no intermediate peptides would be predicted to be present), but in the *in vivo* situation, this may not necessarily be the case. It is possible, therefore, that not all of the bioactive peptides predicted to be released after simulated *in silico* GIT digestion may be released *in vivo*. Also, the presence of acid, bile, and salts as well as the gut microflora in the GIT may destabilise proteins which in turn may result in the generation of markedly different profiles of bioactive peptides *in vivo*. Regardless, the *in silico* model can be useful in providing a ‘first look’ at what the bioactive peptide profile might look like after GIT digestion.

3.2.2 *In Silico* and *In Vitro* Data on Novel GEP-Derived Bioactive Peptides

In another subsequent study from our group, the *in silico* digestion and *in vitro* bioactivity assays were combined to investigate the potential release of bioactive peptides from GEP in this work (Dave et al. 2016a,b). This study used 11 GEP including SALB, mucin-7, pepsin, salivary amylase, trypsin, cholecystokinin, somatostatin, and LYS based on previous work (Dave. et al. 2014). These proteins were subjected to simulated (*in silico*) gastrointestinal digestion (SIGIT). The other three proteins, consisting of the GEP ghrelin, an important gut hormone, keratin, and CRISPR-associated endoribonuclease Cas2 that represent epithelial and microbial proteins respectively, were also investigated by SIGIT.

Of all the resultant peptides, only those which were predicted to be released after SIGIT, with no previously attributed biological activity, were selected for microwave-assisted chemical synthesis, and were screened for either their dipeptidyl peptidase-IV (DPP-IV) inhibitory and/ or their antioxidant potential using *in vitro* bioactivity assays.

Several novel bioactive peptides such as CK, MIM, FPQW, MPSDR, RHPY, and RPCF were found to inhibit human DPP-IV *in vitro*, with MIM and RPCF being the most potent. Further, SHF, RHPY, RPCF, MSY, CRPK, CCK, DCR, VAW, and QQCP were found to possess diphenyl-1-picrylhydrazyl (DPPH) inhibitory potential, with CCK, RPCF, and CRPK showing inhibition of >90%. In addition, CCK was found to be an effective antioxidant when tested using the ferric reducing antioxidant power (FRAP) bioassay (Dave et al. 2016a,b).

3.2.3 *In Vitro* Release of GEP-Derived Bioactive Peptides

Despite initial *in silico* evidence for the potential of GEP as a source of bioactive peptides, the key question which requires further investigation and corroboration is whether GEP would indeed release the bioactive sequences identified by the *in silico* model. In a follow-up study (Dave et al. 2016a,b), we used the four most common GEP – trypsin, LYS, mucin, and SALB (selected based on predictions from the *in silico* model) – to test this hypothesis.

The proteins were sequentially digested *in vitro* using the INFOGEST method (Minekus et al. 2014) and the resultant gastric and small intestinal digests were analysed for angiotensin-I converting enzyme (ACE-I), renin, platelet-activating factor acetylhydrolase (PAF-AH) and DPP-IV inhibitory activity, and a range of antioxidant bioactivities.

As expected, the GEP were susceptible to cleavage by the gastrointestinal proteases *in vitro*. In addition, the *in vitro* GEP digests and fractions were also found to have bioactivities comparable to dietary protein-derived bioactive peptides. The extent of bioactivity depended on the type of protein and stage of digestion, indicating an influence of progressive digestion of peptides on their respective bioactivities. When select *in vitro* digests were investigated in detail, peptides from the digests of LYS and SALB with molecular weights <10 kDa were found to exhibit the highest levels of antioxidant and renin, PAF-AH, and DPP-IV inhibition. The characterisation of these peptides by mass spectrometry revealed several new previously unidentified sequences with chain lengths ranging from seven to 36 residues.

These findings present an interesting scenario in that the GIT is a dynamic system wherein bioactive peptides are continuously being generated during the hydrolysis of polypeptides by multiple proteases. Further, the hydrolysis of a given bioactive polypeptide may enhance its bioactivity or cause it to manifest an entirely different bioactivity altogether. On the other hand, upon further digestion, bioactive peptides may lose their activity either temporarily or permanently.

3.2.4 Preliminary *In Vivo* Evidence for Bioactive Peptides from Porcine GEP

The presence of GEP-derived bioactive peptides *in vivo* (using a porcine model) was demonstrated successfully by our group (Acharya 2017). Stomach chyme (SC) and jejunal digesta (JD) were collected from experimental pigs fed a protein-free diet. These samples with partially digested GEP (polypeptides) and GEP-derived peptides were tested for a range of bioactivities *in vitro* after extensive purifications.

The protein fragments and peptides isolated from the *in vivo* digestion of GEP possessed measurable bioactivities. The fractions showed significant ACE-I inhibitory activity that was comparable to the positive control captopril. They also showed significant antioxidant potential when assessed using different antioxidant bioassays including FRAP, DPPH inhibition, and MLP inhibition.

This study also identified several potentially bioactive peptide sequences from the most active SC and JD fractions using mass spectrometry. Several of the peptides identified in this study were found to be released by GEP that are known to be consistently present in the GIT; for example, the study identified potentially bioactive sequences from SALB, digestive enzymes, mucin, and salivary proteins. One particular peptide sequence, FAKTCVADESAENCDKS, from the porcine SALB (SC fraction) was found to be the subsequence of a peptide identified from the <10 kDa fraction of the small intestinal *in vitro* digest of human SALB (Dave et al. 2016a,b).

It must be mentioned here that while the general pattern of GEP digestion across samples of a GIT section (i.e. stomach or small intestinal) appeared to be similar between animals, the samples from each animal, perhaps unsurprisingly given variations in time of sampling, showed different bioactivities. This may be attributed in part to the varying conditions in the GIT of individual pigs. Another factor that may have led to variable levels of bioactivity in different pigs is the gut microflora, as there exists a significant difference in individual diversity and abundance of microbes (Heinritz et al. 2016; Lloyd-Price et al. 2016). The upper GIT microbiota may not only have acted as a source of bioactive peptides in the animal gut but also contributed to the release of bioactive peptides by way of bacterial proteases.

3.3 Implications of the Above Findings Regarding GEP as a Source of Bioactive Peptides

3.3.1 Classification of Bioactive Peptides

A considerable amount of literature has been published on biologically active peptides of various kinds and from a wide range of sources, including those in the body derived from classic post-translational modification of precursor proteins to peptides from cryptomic proteins with their own primary functions. However, from the point of view of the gut, it is important to distinguish the bioactive peptides released from different sources.

From the above discussion, it is now clear that the bioactive peptides in the human body can be derived from two different sources: truly endogenous bioactive peptides (TEnBP) and exogenous bioactive peptides (ExBP). While TEnBP are formed biosynthetically, or directly encoded or derived from cryptomic proteins within the body, ExBP are formed outside the body such as during food processing, or via digestion of dietary proteins or GEP in the GIT lumen (Dave et al. 2015). This supports the larger view that every protein, from cellular to organ system level, is a potential cryptomic source of TEnBP or ExBP (Pimenta and Lebrun 2007). The discovery of this vast range of TEnBP and ExBP indicates the significance that the body attaches to the recycling and reuse of macromolecules, especially proteins. Proteolytic activity, a ubiquitous process in the body, is a key step in various physiological processes such as digestion of dietary proteins, proteolytic cleavage of N-terminal signalling sequences, and activation of zymogens. Thus, it follows naturally that there is scope for the body to utilise proteins beyond their primary functions, as cryptomic sources of bioactive peptides relevant to the dynamic local and systemic environments.

3.4 Bioactive Potential of GEP is Comparable to That of Dietary Proteins

While assessing the bioactive potential of GEP using *in silico* models, it may be noted that although the predicted amounts of bioactive peptides released after simulated *in silico* gastric and small intestinal digestion may be low, the numbers of biologically active peptides are significant as it is known that relatively low amounts of bioactive peptides may exert physiological effects (Boutrou et al. 2013).

Using the *in silico* model, we predicted that an average diet for a healthy human adult may give around 1842 mg/day of bioactive peptides, and the daily secretion of GEP may give a comparable amount of 2689 mg/day bioactive peptides (Figure 3.2) (Dave et al. 2014). The findings of this study are important when viewed in light of the fact that while dietary bioactive proteins remain in the GIT for only limited periods of time, several of the GEP are continually present in the GIT. In addition, this study paved the way for further experiments, by helping decide which of the GEP might prove to be the most promising source of bioactive peptides.

Although, among all of the studied GEP, larger proteins such as mucins are expected to contain a large number of bioactive peptide motifs, other proteins showed greater potential as reservoirs of bioactive peptides. When normalised against the chain length, SALB was predicted to be the richest source of bioactive peptides. The different bioactive peptides predicted to be released from SALB included ACE-I, DPP-IV, dipeptidyl aminopeptidase IV, dipeptidyl carboxypeptidase inhibitor, antioxidant, glucose uptake, and vasoactive substance release stimulating peptides (Dave et al. 2014). SALB as a source of bioactive peptides was comparable to the dietary protein beta-casein, considered to be a particularly potent dietary source of bioactive peptides (Shahidi and Zhong 2008).

3.5 The Site of Secretion of GEP Affects its Bioactive Potential

The GIT has a continuous supply of diverse proteins. Further, proteins secreted in different parts of the digestive tract may undergo different digestion patterns. For example, a GEP secreted in the intestinal lumen may only undergo intestinal digestion as against a dietary protein which will undergo both gastric and intestinal digestion. This implies that GEP may give rise to site-specific, distinct functional peptides.

Our study (Dave et al. 2014) attempted to address this query using SALB and somatostatin, which are found in both the stomach and small intestine. It was interesting to note that depending on their site of digestion and the gut proteases they were exposed to, SALB and somatostatin were predicted to have different bioactive peptide profiles that varied in the number of peptides, bioactive peptide sequences, and types of bioactivities.

3.6 Digestion of GEP may Generate Numerous Peptides with Multiple Bioactivities

The simulated gastrointestinal digestion (SGID) of GEP resulted in peptides with a wide range of bioactivities relevant to gut physiology (Dave et al. 2016a,b). We have observed that the SGID of GEP releases peptides with ACE-I, renin, PAF-AH, and DPP-IV inhibition, and antioxidant activity. The *in vitro* SGID and subsequent bioactivity assays of the digests indicated that for a given protein digest, the bioactivity varied with the stage of digestion.

Indeed, the small intestinal digests tended to show a higher quantum of activity than their respective gastric digests. Characterisation studies on the digests indicated that the bioactive peptides were largely <10 kDa while the fraction >10 kDa had comparatively marginal bioactivity. Further, numerous potentially bioactive peptide sequences were identified from the most active fractions (LYS and SALB <10 kDa and <3 kDa fractions). We also noted that the peptide sequences identified were generally longer in chain length (7–36 residues). Similarly, in the subsequent *in vivo* study (Acharya 2017), the peptides identified in the SC and JD fractions of pigs were higher in chain length (10 to >50 amino acid residues).

It is intriguing to enquire why, for a given protein, these peptides found after *in vitro* and *in vivo* digestion were not identified in the *in silico* study. This question can be addressed by considering the evolution of the databases used in the *in silico* study, and the mass spectrometry method used to characterise the peptides here. Most studies and online databases have reported bioactive peptides with short length (2–8 amino acids) since it is known that peptides of these chain lengths may be transported across the intestinal epithelium into the systemic circulation (Aluko 2012; BIOPEP 2015; Miner-Williams et al. 2014b; Minkiewicz et al. 2008; Moller et al. 2008). This may have resulted in larger peptides being overlooked in the *in silico* study.

Low molecular weight peptides are difficult to identify using mass spectrometry (Dave et al. 2013). Indeed, in our own studies on *in vitro* and *in vivo* digests (Acharya 2017; Dave et al. 2016a,b), the low molecular weight data were excluded due to their lower probability scores, which could be explained by their limited ability to be ionised in the mass spectrometer (Dave et al. 2013). It is possible that the small molecular weight peptides were indeed present in the digests in our studies, but they were difficult to identify due to their low probability scores.

3.7 Novel Bioactive Peptides from GEP

Gastrointestinal endogenous protein sequences may contain numerous novel bioactive peptide sequences with bioactivities that are particularly relevant to the GIT. In our study (Dave et al. 2016a,b), we discovered two previously unreported DPP-IV inhibitory peptides (RPCF and MIM), and five potent antioxidant peptides (CCK, RPCF, CRPK, QQCP, and DCR) from the GEP SALB, microbial protein GGCAE, mucin 7, and salivary amylase.

In all of the techniques and experiments (*in silico* methods, chemical synthesis, *in vitro* and *in vivo* work) used to study GEP thus far, SALB was consistently found to be the most potent source of ExBP, and its potential was comparable to the respective dietary proteins used as a control (Acharya 2017; Dave et al. 2014, 2016a,b). LYS was also predicted to be a significant source of ExBP in the gut (Acharya 2017; Dave et al. 2016a,b). These findings are significant also because in their unhydrolysed forms, both SALB and LYS have been classed as bioactive proteins or proteins with bioactive peptide domains (Goncharov et al. 2015; Lonnerdal 2013; Roche et al. 2008; Sitar et al. 2013). While SALB is a known antioxidant (Roche et al. 2008), chicken egg LYS has been reported to possess antibacterial properties (Masschalck and Michiels 2003). Similarly, a 17 amino acid domain of LYS, CNDGRIPGSRNLCNIPC (Li. et al. 1995), is known to prevent acute and chronic oxidant injury by suppressing genes involved in response to exogenous oxidant stress (such as that induced by paraquat and H_2O_2) and reduces the generation of reactive oxygen species induced by advanced glycation end-products (Liu et al. 2006). In the same way, we found that LYS digests and fractions (<10 kDa and <3 KDa) acted as antioxidants, quenching the free radical 2,2-DPPH and 2,2′-azino-bis(3-ethylbenzothiazoline-6-sulphonic acid) (ABTS), and also by way of possessing ferric reducing ability (FRAP) (Dave et al. 2016a,b).

Further, numerous peptide sequences, called albutensins, that are derived from the tryptic digests of bovine SALB have been reported to possess ileum-contracting or agonist activities in porcine ileum preparations (Takahashi et al. 1998). In addition, some of the albutensin sequences correspond to sequences found in porcine and human SALB (Takahashi et al. 1998; Yoshikawa 2015).

3.7.1 Systemic Effects of Bioactive Peptides from GEP

Given that GEP secreted in different regions may undergo digestion to different extents and hence release bioactive peptides of varying lengths, one question that arises is whether relatively larger peptides (7–50 residues long) have any systemic physiological effect, given that they are bound to undergo further digestion in the GIT, brush border membrane and plasma. Indeed, the larger bioactive peptides may not escape GIT digestion and enter the systemic circulation (Miner-Williams et al. 2014b) in small quantities, but the possibility also exists that they exert their effect within the gut lumen. The gut is known to contain numerous receptors, including those involved in the renin-angiotensin-aldosterone system (RAAS) (Fandriks 2011), opioid (Sobczak et al. 2014; Yoshikawa 2015), neurotensin, chemotaxis, anaphylotaxis, anorexia (Yoshikawa 2015), taste recognition (Depoortere 2014), mucosal immune (Mantis et al. 2011), gut hormone (Moran-Ramos et al. 2012), and other functions that depend on enzymes such as

ACE-I and DPP-IV (Dave et al. 2015, 2016a,b), as shown in Figure 3.3. Thus, the larger bioactive peptides may exert physiologically relevant bioactivity before undergoing further digestion. Further, their progressive digestion may result in other peptides which may exhibit the same or different bioactivities, as noted in the *in silico, in vitro* and *in vivo* studies of the current work.

This implies that a given peptide may undergo changes in its bioactivity profile with advancing digestion and in turn regulate gut function accordingly. Nevertheless, there is also some experimental evidence that suggests that some relatively larger peptides ranging between 10 and 51 amino acids in chain length may also be absorbed intact from the intestines to be able to exert an effect elsewhere in the body (Roberts et al. 1999).

3.7.2 Generation of Bioactive Peptides from GEP in *In Vivo* Systems

Significant ACE-I inhibition and antioxidant activities were observed in samples from both the SC and JD fractions collected from pigs fed a protein-free diet. This suggests that a significant amount of ACE-I inhibition and antioxidant activity in the gut lumen may be due to the GEP-derived peptides.

From these findings, it can be assumed that during the normal course of GEP digestion in the GIT of mammals, the generation of bioactive peptides is a natural occurrence. These findings raise further questions regarding the nature and physiological concentration of bioactive peptides that may be generated from GEP on a daily basis, particularly considering that basal levels of GEP are always present in the GIT.

3.8 Important Considerations While Predicting the Behaviour of GEP in the Human GIT: Correlating *In Vitro* and *In Vivo* Studies

Data obtained from *in vivo* samples using any animal model must be interpreted with some caution, as the observed bioactivities may in part result from non-GEP constituents present in the digestive contents. In our study, it was interesting to observe that although the *in vivo* samples showed measurable bioactivities, there was no consistent pattern that emerged with regard to the magnitude of observed bioactivities. This may be explained by the heterogeneity of the peptide composition of extracts resulting from the wide variation in the GIT conditions in each of the animals, as discussed earlier, and by heterogeneity inherent to the sampling procedure. The individual variations could also arise from differences in the degree of hydrolysis by gut proteases, differences in secretion of digestive aids (acid, protease composition, salts, bile, etc.), and interindividual microbiota diversity, and the DM intake. This highlights the complexity involved in correlating *in vitro* data with *in vivo* results.

3.8.1 Gut Microbiota Influences Generation of Bioactive Peptides

The GIT is an ecosystem of a wide variety of bacteria and the balance of these bacteria plays an important role in regulation of important metabolic functions (Oriach et al. 2016; Zhang et al. 2015). The gut microbiota-origin protein makes up over 45% of total

GEP and given their significant numbers in the GIT lumen, bacterial protein is likely to be one of the major constituents of GEP that would undergo digestion. However, identification of such proteins and their peptides upon digestion by gut proteases poses a challenge due to the variability in the type of microflora in the gut and the numerous proteins associated with them, and their minute quantities individually.

Besides, the gut microbes themselves secrete numerous proteases that may utilise dietary proteins and GEP as substrate. Other mechanisms of microbiota-mediated physiological effects in the gut could include the physiological modulation resulting from certain ExBP acting as receptors for bacterial attachment (Lu and Walker 2001). Similarly, certain species of bacteria may secrete gut modulatory peptides and proteins, a phenomenon observed in the probiotic strains of *Bifidobacterium*, *Lactobacillus*, and *Escherichia* spp. (Ewaschuk et al. 2008; O'Shea et al. 2012; Sánchez et al. 2008, 2010). This underlines the complexity in interpretation of data from *in vivo* studies and highlights the challenges in planning such studies.

3.8.2 Epithelial Cells as a Source of Bioactive Peptides in the GIT

Depending on the health of the individual, the mucosal turnover may contribute from 5% to 40% of GEP per day (Dave et al. 2015; Leblond and Walker 1956; Miner-Williams et al. 2014a). Some of the epithelial cell-derived proteins have already been reported to be bioactive in other systems. By way of example, proteolytically derived peptides from the protein ubiquitin (76 residues), that is found in all eukaryotic cells including epithelial cells, have been shown to act as enzyme substrates for peptidylglycine alpha-amidating mono-oxygenase (Chew et al. 2005). Similarly, peptides derived from the actin protein analogue, that forms an integral part of the intestinal microvilli (McConnell et al. 2011), have been found to be bioactive *in vitro* (Ryan et al. 2011). We have also found that the epithelial cell protein keratin may release the precursor (with the amino acid sequence GSR) of an ACE-I inhibitory peptide sequence (GS) upon simulated *in silico* gastrointestinal digestion (Dave et al. 2016a,b).

3.8.3 Structural Aspects of Proteins/Peptides Governing Digestion and Subsequent Bioactivity

An important aspect that is likely to affect the release and action of ExBP is the structural attributes of the parent proteins and the physicochemical interactions occurring during and after their hydrolysis. The structure–activity relationship of bioactive peptides has been discussed in detail in several reviews (Li and Yu 2015; Meisel 1997; Nongonierma and FitzGerald 2016a; Schlimme and Meisel 1995; Udenigwe et al. 2012; Zou et al. 2016). The structure (higher order structure including secondary and tertiary)-mediated bioactivity may assume greater significance in peptides with relatively larger chain lengths. It is likely that the peptides released during the early part of digestion retain a part of their secondary structure and the bioactive patches in the peptide sequences might therefore remain embedded within this structure. These bioactive patches are then released as digestion progresses.

In our *in vivo* study (Acharya 2017), numerous proline-rich peptides and polypeptides were identified. This is in agreement with previous studies that indicate that the proline residues, especially when flanking the amino or carboxyl end of peptides, may have an important role in the bioactivity of the peptides (Manjunatha Kini and Evans 1995). It is also well known that the presence of proline in a protein sequence opposes formation of ordered structure (Rauscher et al. 2006). These two findings are notable since they suggest that the proline residues may either be involved in interactions with receptor sites for bioactive peptides or that their presence might result in an open structure exposing the sequences responsible for the exhibited bioactivity. This presents an interesting area for future work. Furthermore, it has previously been reported that the activity of peptides might be enhanced in the presence of certain amino acids or act synergistically with the bioactivity of amino acids (Davalos et al. 2004). This is not surprising considering the fact that individually, amino acids are also known to have biological activities, such as the antioxidant properties of tyrosine, tryptophan, cysteine, and methionine (del Angel-Meza et al. 2011; Gülçin 2006; Kedare and Singh 2011; Levine et al. 1996).

3.8.4 Effect of Protein–Protein Interactions and Protein Modifications

Since the contents of the gut lumen represent extremely heterogeneous mixtures consisting of numerous endogenous proteins at different stages of digestion, it follows that there is a strong possibility that one or more of the peptides from the same or different proteins may interact with each other. Further, the nature of the interactions or their degree of association may vary with changes in the aqueous conditions prevailing in the GIT. The different protein–protein interactions known to mediate bioactivity of peptides include disulfide bonds (Patil et al. 2015), electrostatic (Wang et al. 2015), and hydrophobic interactions (Mozsolits et al. 1999).

Similarly, the proteins and peptides may also undergo modifications, such as the Maillard reaction, which might alter their ultimate structure and/or bioactive potential (Roscic and Horvat 2006) and also potentially make them less accessible to gut proteases such as trypsin (Tuohy et al. 2006). The latter phenomenon might in turn alter the degree of digestion of glycated proteins and peptides, which will affect the type of bioactive peptides produced from such precursors. With respect to glycation-induced modification of proteins, it is important to note that while dietary protein-derived ExBP might not undergo glycation due to their short half-lives of <2 hours and rapid digestion (Ichikawa et al. 2010; Iwai et al. 2005), certain GEP that are secreted into the gut are susceptible to glycation due to their longer half-lives of 1–2 weeks (Liumbruno et al. 2009). One such example is plasma proteins, including SALB, that enter the GIT for the purpose of clearance (Wetterfors et al. 1960), and have been shown to be a major source of bioactive peptides in the present work.

Thus, the digestion of the cryptome (dietary/GEP) protein and the bioactivity exhibited by the resultant peptides *in vivo* would be the net effect arising from all of these interprotein/interpeptide/inter-amino acid interactions and polypeptide modifications.

3.8.5 Safety and Toxicity of GEP-Derived Bioactive Peptides

Having defined some of the key factors that might have an impact on the ExBP generated in the GIT, it is important to mention that there are a few reports indicating that some of these bioactive peptides might not necessarily have a beneficial effect on the host or consumer. Some bovine beta-casein-derived bioactive peptides from the A1 variant of the protein have been implicated in inflammation, diabetes, and heart disease (Laugesen and Elliott 2003). However, only a few studies in animal models have reported these adverse effects, and more studies are required to substantiate these findings in humans.

Further investigations concerning whether any of the GEP-derived bioactive peptides exhibit adverse effects would therefore pave the way for a greater understanding regarding the safety of GEP-derived bioactive peptides. However, given that GEP are a significant and constant source of protein in the gut, and therefore peptides released from them form a natural part of the gut contents, it is highly unlikely that under normal conditions, GEP-derived peptides have a negative impact on the host. It is important to note here that planning a systematic study for investigating any negative effects of GEP-derived peptides can be hugely challenging.

3.8.6 Some Important Considerations in the Study of GEP- and Dietary Protein-Derived Bioactive Peptides

Unlike dietary proteins, the consumption of which will vary from meal to meal, day to day, and population to population, a certain basal amount of GEP will always be present in the GIT. Some of the factors that affect the secretion of GEP in the human GIT are listed below.

- Age, sex, and overall health of the individual, including the type of lifestyle led (i.e. sedentary, moderate or heavy work) (Duska et al. 2007; Paddon-Jones et al. 2008; Phillips 2004; Tessari 2006). Other conditions that might potentially affect GEP secretion include growth spurts during infancy and adolescence, pregnancy, lactation, and fasting.
- Dietary intake and composition, type of processing applied to the food (Deglaire et al. 2016; Moughan et al. 2005, 2007; Rutherfurd et al. 2015a), and in particular the protein intake (both quantitative and qualitative) of the individual, since it is known that malnutrition can affect protein synthesis (Costarelli and Emery 2009; Millward et al. 2008; Tomé and Bos 2000).
- Medication (Matsui et al. 2011; Roberton et al. 1996; Sostres and Lanas 2011).
- Stress levels (stress is known to affect general health, digestion, gut motility, and gut microbiota) (Bailey et al. 2011; Bercik et al. 2011; Konturek et al. 2011).
- Natural, interindividual variations in gut microflora (Conlon and Bird 2015; Human Microbiome Project Consortium 2012) and dysbiosis (Oriach et al. 2016).

Variations arising from all the above factors may be responsible for differences in GEP-derived ExBP in the gut of different individuals, and may perhaps to some extent make GEP-derived peptides host specific.

3.9 Conclusion

This chapter identifies GEP as an important cryptomic source of exogenously derived bioactive peptides in the gut lumen. The GEP-derived ExBP in turn augment the dietary supply of ExBP. Further, given that basal amounts of GEP are always present in the GIT and that the GIT is a large extracorporeal proteolytic system, GEP-derived

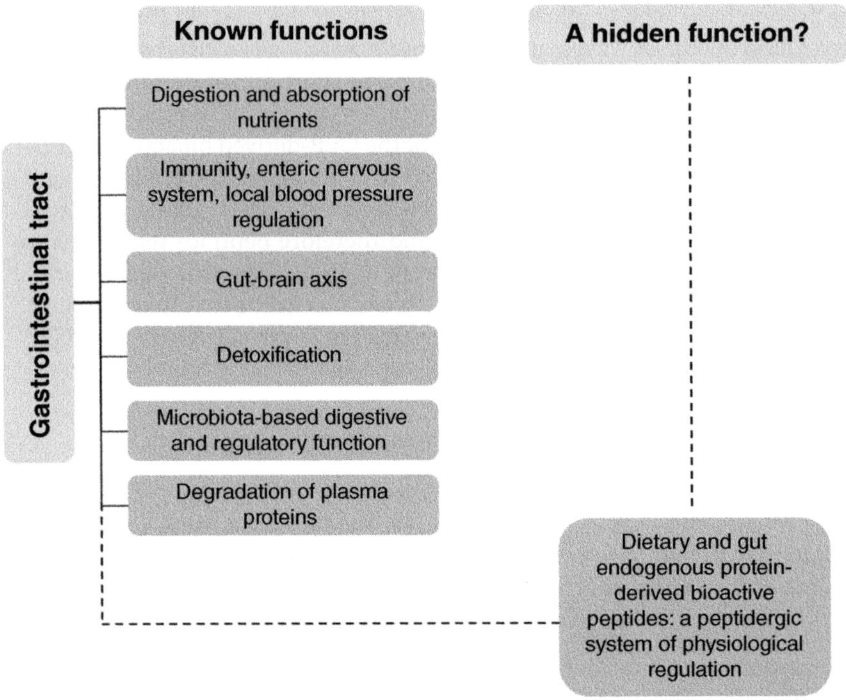

Figure 3.4 The potential role of dietary and gut endogenous protein (GEP)-derived peptides in the GIT. *Source:* Based on data from Acharya (2017); Dave et al. (2015); Moughan et al. (2014). Reprinted with permission of Elsevier.

peptides will be a much more consistent supply of ExBP to the gut than their dietary counterparts. These peptides have a wide range of biological activities that are highly relevant to the GIT. From the above discussion, it is evident that, when studying the role of dietary bioactive peptides, the inherent potential of GEP to generate similar peptide sequences needs to be taken into consideration, as also acknowledged in a recent review (Nongonierma and FitzGerald 2015). Secondly, the diet of an individual may modulate the flows of GEP to obtain desirable GEP-derived bioactive peptides, thus constituting an exquisitely regulated system of homeostatic control (Figure 3.4). In the light of this knowledge, much more research is required to investigate in depth the web of physiological effects that may be mediated through GEP-derived bioactive peptides individually or in conjunction with their dietary counterparts (Figure 3.4).

3.10 Future directions

One of the first known mentions of dietary bioactive peptides can be traced back to the 1950s when casein phosphopeptides were found to have biological activity in improving bone calcification in infants suffering from rickets (Mellander 1950); the eventual discovery of bioactive peptides 'sequences' in milk proteins in the late 1970s gave rise to the concept of dietary proteins as a source of biologically active peptides. In 1978, encrypted bioactive peptides (that were similar in effect to endogenous opioid peptides such as endorphins) were discovered in bovine beta-casein (Teschemacher et al. 1978). Further, Meisel and Schlimme (1990) reported that certain protein fragments of bovine beta-casein possessed an activity that resembled that of enkephalins. Enkephalins, like endorphins, are one of the major classes of endogenously derived (synthesised by the body) opioid peptides (Kastin 2006). Therefore, the analogous diet-derived opioid peptides came to be known as 'exogenous bioactive peptides'.

The knowledge that GEP are a hitherto underexplored and novel source of ExBP points to the need for methods for distinguishing between GEP and dietary protein-derived bioactive peptides (the two natural types of ExBP) in the gut lumen (Dupont 2017; Nongonierma and FitzGerald 2016b). Further research should be carried out to study the impact of the major dietary components (including proteins, peptides, polysaccharides, and non-starch polysaccharides (fibre)) on GEP flows (Butts et al. 1993; Claustre et al. 2002; Cowieson et al. 2004; Deglaire 2008; Deglaire et al. 2006; Han et al. 2008; Miner-Williams et al. 2014a; Moughan et al. 2005, 2007; Moughan and Rutherfurd 1990; Ouellet et al. 2002; Rutherfurd et al. 2015a) and the subsequent impact on the generation of bioactive peptides and profiling of the GEP-derived bioactive peptides of healthy populations.

Significant work has been done towards creating *in vitro* models that closely mimic *in vivo* systems, for example the Human Gastric Simulator (Riddet model) (Ferrua and Singh 2015; Kong and Singh 2010), the TNO Gastro-Intestinal Model (TIM) (Minekus 2015), the semi-dynamic *in vitro* digestion model (Bellmann et al. 2016; Mulet-Cabero et al. 2017), and the ileal fermentation model (Hoogeveen et al. 2017; Tanner et al. 2014; Tran et al. 2016; Venema 2015). However, it is also important to develop an advanced model specifically aimed at understanding the digestion of dietary proteins and GEP. Further progress in the INFOGEST protocol for a semi-dynamic simulation of adult human digestion will also help advance our knowledge of ExBP (INFOGEST 2017). Building on the current INFOGEST *in vitro* digestion model, the new model could include basic bacterial proteases, simulation of post small intestinal digestion by inclusion of brush border proteases, followed by the study of peptide biotransformation under the effect of plasma proteases. Finally, the development of a more robust methodology to characterise smaller peptides is also required.

References

Acharya, P. (2017). *Gastrointestinal Endogenous Proteins as a Source of Bioactive Peptides : Doctor of Philosophy in Nutritional Sciences at Riddet Institute.* Palmerston North, New Zealand: Massey University.

Aluko, R. (2012). *Bioactive Peptides Functional Foods and Nutraceuticals*, 37–61. New York: Springer.

Bailey, M.T., Dowd, S.E., Galley, J.D. et al. (2011). Exposure to a social stressor alters the structure of the intestinal microbiota: implications for stressor-induced immunomodulation. *Brain, Behavior, and Immunity* 25 (3): 397–407.

Bellmann, S., Lelieveld, J., Gorissen, T. et al. (2016). Development of an advanced in vitro model of the stomach and its evaluation versus human gastric physiology. *Food Research International* 88 (Part B): 191–198.

Bercik, P., Park, A.J., Sinclair, D. et al. (2011). The anxiolytic effect of Bifidobacterium longum NCC3001 involves vagal pathways for gut-brain communication. *Neurogastroenterology and Motility* 23 (12): 1132–1139.

BIOPEP. (2015). BIOPEP. Retrieved from: www.uwm.edu.pl/biochemia/index.php/en/biopep

Boisen, S. and Moughan, P.J. (1996). Dietary influences on endogenous ileal protein and amino acid loss in the pig – a review. *Acta Agriculturae Scandinavica, Section A – Animal Science* 46 (3): 154–164.

Boutrou, R., Gaudichon, C., Dupont, D. et al. (2013). Sequential release of milk protein-derived bioactive peptides in the jejunum in healthy humans. *American Journal of Clinical Nutrition* 97 (6): 1314–1323.

Brassinne, A. (1974). Gastric clearance of serum albumin in normal man and in certain gastroduodenal disorders. *Gut* 15 (3): 194–199.

Brunser, O., Espinoza, J., and Araya, M. (1992). Role of the gastrointestinal tract in energy and protein metabolism. In: . Retrieved from: http://archive.unu.edu/unupress/food2/UID07E/UID07E18.HTM.

Butts, C.A., Moughan, P.J., Smith, W.C. et al. (1993). The effect of food dry matter intake on endogenous ileal amino acid excretion determined under peptide alimentation in the 50 kg liveweight pig. *Journal of the Science of Food and Agriculture* 62 (3): 235–243.

Caine, W.R., Tamminga, S., Sauer, W.C. et al. (1999). Bacterial contributions to total and endogenous recoveries of nitrogen and amino acids in ileal digesta of newly weaned piglets fed protease-treated soybean meal. *Livestock Production Science* 57 (2): 147–157.

Chassaing, B., Kumar, M., Baker, M.T. et al. (2014). Mammalian gut immunity. *Biomedical Journal* 37 (5): 246–258.

Chew, G.H., Galloway, L.C., McIntyre, N.R. et al. (2005). Ubiquitin and ubiquitin-derived peptides as substrates for peptidylglycine α-amidating monooxygenase. *FEBS Letters* 579 (21): 4678–4684.

Claustre, J., Toumi, F., Trompette, A. et al. (2002). Effects of peptides derived from dietary proteins on mucus secretion in rat jejunum. *American Journal of Physiology – Gastrointestinal and Liver Physiology* 283 (3 46-3): G521–G528.

Conlon, M.A. and Bird, A.R. (2015). The impact of diet and lifestyle on gut microbiota and human health. *Nutrients* 7 (1): 17–44.

Costarelli, V. and Emery, P.W. (2009). The effect of protein malnutrition on the capacity for protein synthesis during wound healing. *Journal of Nutrition, Health and Aging* 13 (5): 409–412.

Cowieson, A.J., Acamovic, T., and Bedford, M.R. (2004). The effects of phytase and phytic acid on the loss of endogenous amino acids and minerals from broiler chickens. *British Poultry Science* 45 (1): 101–108.

Davalos, A., Miguel, M., Bartolome, B., and Lopez-Fandino, R. (2004). Antioxidant activity of peptides derived from egg white proteins by enzymatic hydrolysis. *Journal of Food Protection* 67 (9): 1939–1944.

Dave, A.C., Loveday, S.M., Anema, S.G. et al. (2013). β-Lactoglobulin self-assembly: structural changes in early stages and disulfide bonding in fibrils. *Journal of Agricultural and Food Chemistry* 61 (32): 7817–7828.

Dave, L.A., Montoya, C.A., Rutherfurd, S.M., and Moughan, P.J. (2014). Gastrointestinal endogenous proteins as a source of bioactive peptides – an in silico study. *PLoS One* 9 (6): e98922.

Dave, L.A., Hayes, M., Montoya, C.A. et al. (2015). Human gut endogenous proteins as a potential source of angiotensin-I-converting enzyme (ACE-I)-, renin inhibitory and antioxidant peptides. *Peptides* 76: 30–44.

Dave, L.A., Hayes, M., Mora, L. et al. (2016a). Gastrointestinal endogenous protein-derived bioactive peptides: an in vitro study of their gut modulatory potential. *International Journal of Molecular Sciences* 17 (4): 482.

Dave, L.A., Hayes, M., Moughan, P.J., and Rutherfurd, S.M. (2016b). Novel dipeptidyl peptidase IV inhibitory and antioxidant peptides derived from human gastrointestinal endogenous proteins. *International Journal of Peptide Research and Therapeutics* 22 (3): 355–369.

Deglaire, A. (2008). Gut endogenous protein flows and postprandial metabolic utilization of dietary amino acids in simple-stomached animals and humans: a thesis presented in partial fulfillment of the requirements for the degree of Doctor of Philosophy in Human Nutrition. Massey University, New Zealand

Deglaire, A., Moughan, P.J., Bos, C., and Tome, D. (2006). Commercial phaseolus vulgaris extract (starch stopper) increases ileal endogenous amino acid and crude protein losses in the growing rat. *Journal of Agricultural and Food Chemistry* 54 (14): 5197–5202.

Deglaire, A., de Oliveira, S., Jardin, J. et al. (2016). Impact of human milk pasteurization on the kinetics of peptide release during in vitro dynamic term newborn digestion. *Electrophoresis* 37 (13): 1839–1850.

Del Angel-Meza, A.R., Dávalos-Marín, A.J., Ontiveros-Martinez, L.L. et al. (2011). Protective effects of tryptophan on neuro-inflammation in rats after administering lipopolysaccharide. *Biomedicine and Pharmacotherapy* 65 (3): 215–219.

Depoortere, I. (2014). Taste receptors of the gut: emerging roles in health and disease. *Gut* 63 (1): 179–190.

DeSesso, J.M. and Jacobson, C.F. (2001). Anatomical and physiological parameters affecting gastrointestinal absorption in humans and rats. *Food and Chemical Toxicology* 39 (3): 209–228.

Desnuelle, P., Sjöström, H., and Norén, O. (1986). *Molecular and Cellular Basis of Digestion*. Amsterdam: Elsevier.

Dupont, D. (2017). Peptidomic as a tool for assessing protein digestion. *Current Opinion in Food Science* 16 (Supplement C): 53–58.

Duska, F., Tuma, P., Mokrejs, P. et al. (2007). Analysis of factors influencing nitrogen balance during acute starvation in obese subject with and without type 2 diabetes. *Clinical Nutrition* 26 (5): 552–558.

Ewaschuk, J.B., Diaz, H., Meddings, L. et al. (2008). Secreted bioactive factors from bifidobacterium infantis enhance epithelial cell barrier function. *American Journal of Physiology – Gastrointestinal and Liver Physiology* 295 (5): G1025–G1034.

Fandriks, L. (2011). The renin-angiotensin system and the gastrointestinal mucosa. *Acta Physiologica* 201 (1): 157–167.

FAO Expert Consultation. (2011). Dietary protein quality evaluation in human nutrition. Retrieved from: www.fao.org/ag/humannutrition/35978-02317b979a686a57aa4593304ffc17f06.pdf

FAOSTAT. (2017). Food and agriculture data. Retrieved from: www.fao.org/faostat/en/#data

Ferrua, M.J. and Singh, R.P. (2015). Human Gastric Simulator (Riddet Model). In: *The Impact of Food Bioactives on Health: In Vitro and Ex Vivo Models* (ed. K. Verhoeckx, P. Cotter, I. López-Expósito, et al.), 61–71. Cham: Springer.

Franklin, M. (2015). *Xenobiotic Absoprtion, Distribution, Metabolism and Excretion*, 3e. NewYork: Wiley.

Goncharov, N.V., Belinskaia, D.A., Razygraev, A.V., and Ukolov, A.I. (2015). On the enzymatic activity of albumin. *BioorganicKhim* 41 (2): 131–144.

Gülçin, İ. (2006). Comparison of in vitro antioxidant and antiradical activities of L-tyrosine and L-Dopa. *Amino Acids* 32 (3): 431–438.

Han, K.S., Deglaire, A., Sengupta, R., and Moughan, P.J. (2008). Hydrolyzed casein influences intestinal mucin gene expression in the rat. *Journal of Agriculture and Food Chemistry* 56 (14): 5572–5576.

Heinritz, S.N., Weiss, E., Eklund, M. et al. (2016). Intestinal microbiota and microbial metabolites are changed in a pig model fed a high-fat/low-fiber or a low-fat/high-Fiber diet. *PLoS One* 11 (4): e0154329.

Hoogeveen, A., Montoya, C. A., & McNabb, W. (2017). Ileal fermentation of dietary fibres using a combined in vivo/in vitro ileal fermentation assay. Paper presented at the Food Structure, Digestion and Health Conference.

Human Microbiome Project Consortium (2012). Structure, function and diversity of the healthy human microbiome. *Nature* 486 (7402): 207–214.

Ichikawa, S., Morifuji, M., Ohara, H. et al. (2010). Hydroxyproline-containing dipeptides and tripeptides quantified at high concentration in human blood after oral administration of gelatin hydrolysate. *International Journal of Food Sciences and Nutrition* 61 (1): 52–60.

INFOGEST. (2017). Food Colloids 2018 INFOGEST Workshop. Retrieved from: www.foodcolloids2018.co.uk/infogest-workshop/

Iwai, K., Hasegawa, T., Taguchi, Y. et al. (2005). Identification of food-derived collagen peptides in human blood after oral ingestion of gelatin hydrolysates. *Journal of Agricultural and Food Chemistry* 53 (16): 6531–6536.

Johnson, L.R. (2012). *Physiology of the Gastrointestinal Tract*. London: Academic Press.

Kastin, A.J. (2006). *Handbook of Biologically Active Peptides*. Boston: Academic Press.

Kedare, S.B. and Singh, R.P. (2011). Genesis and development of DPPH method of antioxidant assay. *Journal of Food Science and Technology* 48 (4): 412–422.

Kong, F. and Singh, R.P. (2010). A human gastric simulator (HGS) to study food digestion in human stomach. *Journal of Food Science* 75 (9): E627–E635.

Konturek, P.C., Brzozowski, T., and Konturek, S.J. (2011). Stress and the gut: pathophysiology, clinical consequences, diagnostic approach and treatment options. *Journal of Physiology and Pharmacology* 62 (6): 591–599.

Laugesen, M. and Elliott, R.B. (2003). Ischaemic heart disease, Type 1 diabetes, and cow milk A1 beta-casein. *New Zealand Medical Journal* 116 (1168): U295.

Leblond, C.P. and Walker, B.E. (1956). Renewal of cell populations. *Physiological Reviews* 36 (2): 255–276.

Levine, R.L., Mosoni, L., Berlett, B.S., and Stadtman, E.R. (1996). Methionine residues as endogenous antioxidants in proteins. *Proceedings of the National Academy of Sciences of the United States of America* 93 (26): 15036–15040.

Li, Y. and Yu, J. (2015). Research progress in structure-activity relationship of bioactive peptides. *Journal of Medicinal Food* 18 (2): 147–156.

Li, Y.M., Tan, A.X., and Vlassara, H. (1995). Antibacterial activity of lysozyme and lactoferrin is inhibited by binding of advanced glycation-modified proteins to a conserved motif. *Nature Medicine* 1 (10): 1057–1061.

Liu, H., Zheng, F., Cao, Q. et al. (2006). Amelioration of oxidant stress by the defensin lysozyme. *American Journal of Physiology: Endocrinology and Metabolism* 290 (5): E824–E832.

Liumbruno, G., Bennardello, F., Lattanzio, A. et al. (2009). Recommendations for the use of albumin and immunoglobulins. *Blood Transfusion* 7 (3): 216–234.

Lloyd-Price, J., Abu-Ali, G., and Huttenhower, C. (2016). The healthy human microbiome. *Genome Medicine* 8: 51.

Lonnerdal, B. (2013). Bioactive proteins in breast milk. *Journal of Paediatrics and Child Health* 49 ((Suppl 1): 1–7.

Lu, L. and Walker, W.A. (2001). Pathologic and physiologic interactions of bacteria with the gastrointestinal epithelium. *American Journal of Clinical Nutrition* 73 (6): 1124S–1130S.

Manjunatha Kini, R. and Evans, H.J. (1995). A novel approach to the design of potent bioactive peptides by incorporation of proline brackets: antiplatelet effects of Arg-Gly-Asp peptides. *FEBS Letters* 375, 1: 15–17.

Mantis, N.J., Rol, N., and Corthesy, B. (2011). Secretory IgA's complex roles in immunity and mucosal homeostasis in the gut. *Mucosal Immunology* 4 (6): 603–611.

Masschalck, B. and Michiels, C.W. (2003). Antimicrobial properties of lysozyme in relation to foodborne vegetative bacteria. *Critical Reviews in Microbiology* 29 (3): 191–214.

Matsui, H., Shimokawa, O., Kaneko, T. et al. (2011). The pathophysiology of non-steroidal anti-inflammatory drug (NSAID)-induced mucosal injuries in stomach and small intestine. *Journal of Clinical Biochemistry and Nutrition* 48 (2): 107–111.

McConnell, R.E., Benesh, A.E., Mao, S. et al. (2011). Proteomic analysis of the enterocyte brush border. *American Journal of Physiology – Gastrointestinal and Liver Physiology* 300 (5): G914–G926.

Meisel, H. (1997). Biochemical properties of regulatory peptides derived from mil proteins. *Peptide Science* 43 (2): 119–128.

Meisel, H. and Schlimme, E. (1990). Milk proteins: precursors of bioactive peptides. *Trends in Food Science and Technology* 1 ((Supplement C)): 41–43.

Mellander, O. (1950). The physiological importance of the casein phosphopeptide calcium salts. II. Peroral calcium dosage of infants. *Acta Societatis Medicorum Upsaliensis* 55 (5–6): 247–255.

Millward, D.J., Layman, D.K., Tomé, D., and Schaafsma, G. (2008). Protein quality assessment: impact of expanding understanding of protein and amino acid needs for optimal health. *American Journal of Clinical Nutrition* 87 (5): 1576S–1581S.

Minekus, M. (2015). The TNO gastro-intestinal model (TIM). In: *The Impact of Food Bioactives on Health: In Vitro and Ex Vivo Models* (ed. K. Verhoeckx, P. Cotter, I. López-Expósito, et al.), 37–46. Cham: Springer.

Minekus, M., Alminger, M., Alvito, P. et al. (2014). A standardised static in vitro digestion method suitable for food – an international consensus. *Food and Function* 5 (6): 1113–1124.

Miner-Williams, W. (2012). The protein composition of endogenous losses in the human gut: a thesis presented in partial fulfillment of the requirements for the degree of Doctor of Philosophy in Human Nutrition at Massey University, New Zealand.

Miner-Williams, W., Deglaire, A., Benamouzig, R. et al. (2012). Endogenous proteins in terminal ileal digesta of adult subjects fed a casein-based diet. *American Journal of Clinical Nutrition* 96 (3): 508–515.

Miner-Williams, W., Deglaire, A., Benamouzig, R. et al. (2014a). Endogenous proteins in the ileal digesta of adult humans given casein-, enzyme-hydrolyzed casein-or crystalline amino-acid-based diets in an acute feeding study. *European Journal of Nutrition* 68 (3): 363–369.

Miner-Williams, W.M., Stevens, B.R., and Moughan, P.J. (2014b). Are intact peptides absorbed from the healthy gut in the adult human? *Nutrition Research Reviews* 27 (2): 308–329.

Minkiewicz, P., Dziuba, J., Iwaniak, A. et al. (2008). BIOPEP database and other programs for processing bioactive peptide sequences. *Journal of AOAC International* 91 (4): 965–980.

Mitchell, H.H. (1924). A method for determining the biological values of protein. *Journal of Biological Chemistry* 58: 873–882.

Moller, N.P., Scholz-Ahrens, K.E., Roos, N., and Schrezenmeir, J. (2008). Bioactive peptides and proteins from foods: indication for health effects. *European Journal of Nutrition* 47 (4): 171–182.

Moran-Ramos, S., Tovar, A.R., and Torres, N. (2012). Diet: friend or foe of Enteroendocrine cells – how it interacts with enteroendocrine cells. *Advances in Nutrition* 3 (1): 8–20.

Moughan, P.J. and Rutherfurd, S.M. (1990). Endogenous flow of total lysine and other amino acids at the distal ileurn of the protein-or peptide-fed rat: the chemical labelling of gelatin protein by transformation of lysine to homoarginine. *Journal of the Science of Food and Agriculture* 52 (2): 179–192.

Moughan, P.J. and Rutherfurd, S.M. (2012). Gut luminal endogenous protein: implications for the determination of ileal amino acid digestibility in humans. *British Journal of Nutrition* 108 ((Suppl 2): S258–S263.

Moughan, P.J., Souffrant, W.B., and Hodgkinson, S.M. (1998). Physiological approaches to determining gut endogenous amino acid flows in the mammal. *Archives of Animal Nutrition* 51 (2–3): 237–252.

Moughan, P.J., Butts, C.A., Rowan, A.M., and Deglaire, A. (2005). Dietary peptides increase endogenous amino acid losses from the gut in adults. *American Journal of Clinical Nutrition* 81 (6): 1359–1365.

Moughan, P.J., Fuller, M.F., Han, K.S. et al. (2007). Food-derived bioactive peptides influence gut function. *International Journal of Sport Nutrition and Exercise Metabolism* 17 ((Suppl): S5–S22.

Moughan, P.J., Rutherfurd, S.M., Montoya, C.A., and Dave, L.A. (2014). Food-derived bioactive peptides – a new paradigm. *Nutrition Research Reviews* 27 (1): 16–20.

Mozsolits, H., Lee, T.H., Wirth, H.J. et al. (1999). The interaction of bioactive peptides with an immobilized phosphatidylcholine monolayer. *Biophysical Journal* 77 (3): 1428–1444.

Mulet-Cabero, A.-I., Rigby, N.M., Brodkorb, A., and Mackie, A.R. (2017). Dairy food structures influence the rates of nutrient digestion through different in vitro gastric behaviour. *Food Hydrocolloids* 67 (Supplement C): 63–73.

Nasset, E.S. (1964). The role of the digestive tract in protein metabolism. *American Journal of Digestive Diseases* 9 (3): 175–190.

Nasset, E.S. and Ju, J.S. (1961). Mixture of endogenous and exogenous protein in the alimentary tract. *Journal of Nutrition* 74: 461–465.
Nongonierma, A.B. and FitzGerald, R.J. (2015). The scientific evidence for the role of milk protein-derived bioactive peptides in humans: a review. *Journal of Functional Foods* 17: 640–656.
Nongonierma, A.B. and FitzGerald, R.J. (2016a). Learnings from quantitative structure-activity relationship (QSAR) studies with respect to food protein-derived bioactive peptides: a review. *RSC Advances* 6 (79): 75400–75413.
Nongonierma, A.B. and FitzGerald, R.J. (2016b). Strategies for the discovery, identification and validation of milk protein-derived bioactive peptides. *Trends in Food Science and Technology* 50 (Supplement C): 26–43.
O'Shea, E.F., Cotter, P.D., Stanton, C. et al. (2012). Production of bioactive substances by intestinal bacteria as a basis for explaining probiotic mechanisms: bacteriocins and conjugated linoleic acid. *International Journal of Food Microbiology* 152 (3): 189–205.
Oriach, C.S., Robertson, R.C., Stanton, C. et al. (2016). Food for thought: the role of nutrition in the microbiota-gut-brain axis. *Clinical Nutrition Experimental* 6: 25–38.
Ouellet, D.R., Demers, M., Zuur, G. et al. (2002). Effect of dietary fiber on endogenous nitrogen flows in lactating dairy cows. *Journal of Dairy Science* 85 (11): 3013–3025.
Paddon-Jones, D., Short, K.R., Campbell, W.W. et al. (2008). Role of dietary protein in the sarcopenia of aging. *American Journal of Clinical Nutrition* 87 (5): 1562S–1566S.
Patil, N., Tailhades, J., Hughes, R. et al. (2015). Cellular disulfide bond formation in bioactive peptides and proteins. *International Journal of Molecular Sciences* 16 (1): 1791.
Phillips, S.M. (2004). Protein requirements and supplementation in strength sports. *Nutrition* 20 (7–8): 689–695.
Pimenta, D.C. and Lebrun, I. (2007). Cryptides: buried secrets in proteins. *Peptides* 28 (12): 2403–2410.
Quercia, S., Candela, M., Giuliani, C. et al. (2014). From lifetime to evolution: timescales of human gut microbiota adaptation. *Frontiers in Microbiology* 5: 587.
Rauscher, S., Baud, S., Miao, M. et al. (2006). Proline and glycine control protein self-organization into elastomeric or amyloid fibrils. *Structure* 14 (11): 1667–1676.
Ravindran, V. (2016). Feed-induced specific ileal endogenous amino acid losses: measurement and significance in the protein nutrition of monogastric animals. *Animal Feed Science and Technology* 221 (Part B): 304–313.
Roberton, A.M., Rabel, B., Stubbs, L. et al. (1996). Aspirin changes the secretion rate and amino acid composition of human small intestinal mucin in subjects with ileal conduits. *Glycoconjugate Journal* 13 (5): 781–789.
Roberts, P.R., Burney, J.D., Black, K.W., and Zaloga, G.P. (1999). Effect of chain length on absorption of biologically active peptides from the gastrointestinal tract. *Digestion* 60 (4): 332–337.
Roche, M., Rondeau, P., Singh, N.R. et al. (2008). The antioxidant properties of serum albumin. *FEBS Letters* 582 (13): 1783–1787.
Roscic, M. and Horvat, S. (2006). Transformations of bioactive peptides in the presence of sugars – characterization and stability studies of the adducts generated via the Maillard reaction. *Bioorganic and Medicinal Chemistry* 14 (14): 4933–4943.
Rutherfurd, S.M., Cui, J., Goroncy, A.K., and Moughan, P.J. (2015a). Dietary protein structure affects endogenous ileal amino acids but not true ileal amino acid digestibility in growing male rats. *Journal of Nutrition* 145 (2): 193–198.

Rutherfurd, S.M., Fanning, A.C., Miller, B.J., and Moughan, P.J. (2015b). Protein digestibility-corrected amino acid scores and digestible indispensable amino acid scores differentially describe protein quality in growing male rats. *Journal of Nutrition* 145 (2): 372–379.

Ryan, J.T., Ross, R.P., Bolton, D. et al. (2011). Bioactive peptides from muscle sources: meat and fish. *Nutrients* 3 (9): 765–791.

Sánchez, B., Bressollier, P., and Urdaci, M.C. (2008). Exported proteins in probiotic bacteria: adhesion to intestinal surfaces, host immunomodulation and molecular cross-talking with the host. *FEMS Immunology and Medical Microbiology* 54 (1): 1–17.

Sanchez, B., Urdaci, M.C., and Margolles, A. (2010). Extracellular proteins secreted by probiotic bacteria as mediators of effects that promote mucosa-bacteria interactions. *Microbiology* 156 (Pt 11): 3232–3242.

Schlimme, E. and Meisel, H. (1995). Bioactive peptides derived from milk proteins. Structural, physiological and analytical aspects. *Nahrung* 39 (1): 1–20.

Shahidi, F. and Zhong, Y. (2008). Bioactive peptides. *Journal of AOAC International* 91 (4): 914–931.

Sitar, M.E., Aydin, S., and Cakatay, U. (2013). Human serum albumin and its relation with oxidative stress. *Clinical Laboratory* 59 (9–10): 945–952.

Sobczak, M., Sałaga, M., Storr, M.A., and Fichna, J. (2014). Physiology, signaling, and pharmacology of opioid receptors and their ligands in the gastrointestinal tract: current concepts and future perspectives. *Journal of Gastroenterology* 49 (1): 24–45.

Sostres, C. and Lanas, A. (2011). Gastrointestinal effects of aspirin. *Nature Reviews: Gastroenterology and Hepatology* 8 (7): 385–394.

Souffrant, W.B., Rerat, A., Laplace, J.P. et al. (1993). Exogenous and endogenous contributions to nitrogen fluxes in the digestive tract of pigs fed a casein diet. III. Recycling of endogenous nitrogen. *Reproduction, Nutrition, Development* 33 (4): 373–382.

Takahashi, M., Moriguchia, S., Minami, T. et al. (1998). Albutensin a, an ileum-contracting peptide derived from serum albumin, acts through both receptors for complements C3a and C5a. *Letters in Peptide Science* 5 (1): 29–35.

Tanner, S.A., Zihler Berner, A., Rigozzi, E. et al. (2014). In vitro continuous fermentation model (PolyFermS) of the swine proximal colon for simultaneous testing on the same gut microbiota. *PLoS One* 9 (4): e94123.

Teschemacher, H., Brantl, V., and Haarmann, I. (1978). *Extraction of Compounds with Opioid Activity from Milk*. Heidelberg: Deutsche Pharmakologische Gesellschaft/ Springer.

Tessari, P. (2006). Nitrogen balance and protein requirements: definition and measurements. In: *Cachexia and Wasting: A Modern Approach* (ed. G. Mantovani, S.D. Anker, A. Inui, et al.), 73–79. Milan: Springer.

Tomé, D. and Bos, C. (2000). Dietary protein and nitrogen utilization. *Journal of Nutrition* 130 (7): 1868S–1873S.

Tortora, G.J., Derrickson, B.H., Burkett, B. et al. (2016). *Principles of Anatomy and Physiology*. Wiley, Milton.

Tran, T.H.T., Boudry, C., Everaert, N. et al. (2016). Adding mucins to an in vitro batch fermentation model of the large intestine induces changes in microbial population isolated from porcine feces depending on the substrate. *FEMS Microbiology Ecology* 92 (2): ii.

Tuohy, K.M., Hinton, D.J.S., Davies, S.J. et al. (2006). Metabolism of Maillard reaction products by the human gut microbiota – implications for health. *Molecular Nutrition and Food Research* 50 (9): 847–857.

Udenigwe, C.C., Li, H., and Aluko, R.E. (2012). Quantitative structure–activity relationship modeling of renin-inhibiting dipeptides. *Amino Acids* 42 (4): 1379–1386.

UniProt Consortium. (2012). UniProtKB. Retrieved from: www.uniprot.org/uniprot/

Venema, K. (2015). The TNO in vitro model of the colon (TIM-2). In: *The Impact of Food Bioactives on Health: In Vitro and Ex Vivo Models* (ed. K. Verhoeckx, P. Cotter, I. López-Expósito, et al.), 293–304. Cham: Springer.

Wang, B., Xie, N., and Li, B. (2015). Charge properties of peptides derived from casein affect their bioavailability and cytoprotection against H2O2-induced oxidative stress. *Journal of Dairy Science* 99 (4): 2468–2479.

Weaver, L.T., Austin, S., and Cole, T.J. (1991). Small intestinal length: a factor essential for gut adaptation. *Gut* 32 (11): 1321.

Wetterfors, J., Gullberg, R., Liljedahl, S.O. et al. (1960). Role of the stomach and small intestine in albumin breakdown1. *Acta Medica Scandinavica* 168 (4): 347–363.

Williams, J.M., Duckworth, C.A., Burkitt, M.D. et al. (2015). Epithelial cell shedding and barrier function: a matter of life and death at the small intestinal villus tip. *Veterinary Pathology* 52 (3): 445–455.

Yoshikawa, M. (2015). Bioactive peptides derived from natural proteins with respect to diversity of their receptors and physiological effects. *Peptides* 72: 208–225.

Zhang, Y.-J., Li, S., Gan, R.-Y. et al. (2015). Impacts of gut bacteria on human health and diseases. *International Journal of Molecular Sciences* 16 (4): 7493–7519.

Zou, T.B., He, T.P., Li, H.B. et al. (2016). The structure-activity relationship of the antioxidant peptides from natural proteins. *Molecules* 21 (1): 72.

4

Cereal Proteins: Potential Health Applications and Allergenicities

Stephen Bleakley

Teagasc Food Research Centre, Food BioSciences Department, Dublin, Ireland

4.1 Introduction

Cereals are classified as belonging to the phylogenetic family Gramineae (grasses). The study of grasses is extremely important due to both their ecological dominance and economic influence. Grasses are found on almost every continent, excluding Antarctica, and cover an estimated 31–43% (depending on the definition of grasslands) of the Earth's surface, including temperate and tropical habitats (Gibson 2009).

The types of cereal that can grow in regions around the world depend on a number of factors, with temperature and water availability being the most relevant. Rice, for example, has a high water demand. This means that it tends to be most common in warm regions where fresh water availability is not an issue, such as areas that have frequent flooding conditions and with high levels of rainfall. These types of environments are typical of many regions in Asia, with China and India therefore being the largest producers of rice (Table 4.1). Maize also has a high fresh water demand, although not to the same extent as rice. Wheat is generally more tolerant of colder conditions and capable of growing in a wider range of temperatures. Sorghum and millet are relatively tolerant to drought stress and therefore more commonly grown in semi-arid conditions, such as Africa and India (Awika 2011). Barley and oats are more tolerant of colder conditions and are therefore mostly produced in Europe, Russia, and Canada (Table 4.1).

4.2 Major Cereal Grains

4.2.1 Maize

Maize (*Zea mays*), also known as corn, is the most widely produced cereal crop, with an estimated 1 037 791 518 tonnes produced worldwide in 2014 (Table 4.2). Maize is believed to have been domesticated from the wild variant teosinte, with molecular analysis indicating that the critical step of liberating the kernels from their previous tough structures (called cupulate fruit-cases) to be controlled by the action of a single

Novel Proteins for Food, Pharmaceuticals, and Agriculture: Sources, Applications, and Advances, First Edition. Edited by Maria Hayes.

Table 4.1 Top five producers of the main cereal grain crops in 2016.

Commodity	Top producing country (1000 Mt)				
Maize	United States (384778)	China (219554)	Brazil (86500)	European Union (60309)	Argentina (36500)
Wheat	European Union (144862)	China (128850)	India (87000)	Russia (72500)	United States (62859)
Rice	China (144850)	India (106500)	Indonesia (36600)	Vietnam (27800)	Bangladesh (34581)
Barley	European Union (59787)	Russia (17500)	Ukraine (9900)	Australia (10600)	Canada (8800)
Millet	India (10750)	Nigeria (4800)	Niger (3200)	China (1900)	Mali (1750)
Sorghum	United States (12199)	Nigeria (6500)	Mexico (5500)	India (5500)	Argentina (3600)
Oats	European Union (7803)	Russia (4750)	Canada (3000)	Australia (1800)	United State (940)
Mixed grain	European Union (15261)	Canada (220)	Turkey (115)	Switzerland (60)	Bangladesh (25)
Rye	European Union (8094)	Russia (2540)	Belarus (800)	Canada (410)	Ukraine (390)

Mt: million tonnes.
Source: Data obtained from USDA (2016).

Table 4.2 Global production and yield of the main cereal grains in 2014, as well as their producer price in the United States.

Commodity	Production (t)	Yield ($t\,ha^{-1}$)	Value ($USD\,t^{-1}$)[a]
Maize	1037791518	5.615	146
Wheat	729012175	3.307	220
Rice	741477711	4.557	295
Barley	144489996	2.923	243
Sorghum	68938587	1.533	159
Millet	28384668	0.903	137
Oats	22721702	2.369	221
Rye	15242551	2.872	305

t: tonnes; ha: hectares.
a) Values given are for the United States of America only.
Source: Figures obtained from FAO (2013).

gene (tga1) belonging to the systolic blood pressure (SBP) domain family of transcriptional regulators (Wang et al. 2005).

Only a very small fraction of maize is used directly for human consumption, with the overwhelming majority used as livestock feed and biomass for bioethanol production to decrease the cost of fuel, lessen reliance on fossil fuels, and reduce pollutants emission. However, there has also been some controversy over the production of maize solely for use in biofuel production, as this takes away arable land that could be used to grow food, thus increasing food prices and exacerbating the global food shortage (Keyzer et al. 2008). Another major use of maize is indirect consumption by humans in the form of corn-starch, sweeteners (such as high fructose corn syrup), and cooking oil.

Maize has been at the forefront of crop genetic modification to express agriculturally desirable traits, such as herbicide and pesticide resistance. This has led to intense public concerns being raised about safety and possible risks (Domingo and Bordonaba 2011).

4.2.2 Wheat

Domesticated wheat (*Triticum aestivum*) is believed to have originated in the Fertile Crescent approximately 9200 years ago (Tanno and Willcox 2006). Wheat is the most important grain in the Western world. Raw wheat, either wholegrain or dehulled, is typically ground into flour which is used to make a wide variety of food and drinks, including bread, pasta, baked goods, breakfast cereals, beer, and vodka. One of the main reasons for wheat's popularity as a food ingredient is the presence of the protein gluten. Gluten provides valuable viscoelastic properties to dough which positively affects its texture and stabilises the shape of the final product. However, gluten is linked to an autoimmune response in individuals with coeliac disease, as well as triggering other diseases such as gluten ataxia, non-coeliac gluten sensitivity, and dermatitis herpetiformis (Ludvigsson et al. 2013).

4.2.3 Rice

Rice (*Oryza sativa*) is the most commonly consumed grain in Asian countries, providing an estimated 35–75% of the required calories for 3 billion Asians (Khush 2005). Rice is one of the oldest domesticated crops, believed to have first been domesticated approximately 8200–13 500 years ago in the Yangtze Valley of China from the wild rice *Oryza rufipogon* (Molina et al. 2011). While rice is well suited to its native lands of Asia due to high rainfall requirements, it can be grown in many other regions around the world with the use of suitable water-controlled terrace systems.

Arsenic is a type-1 carcinogen which can be found in trace amounts in soil and water. Rice has been observed to absorb arsenic more readily than other food crops, largely due to its highly aqueous growing conditions (Rahman et al. 2007). Because of its carcinogenic properties, the Food and Drug Administration (FDA) has therefore imposed a 100 parts per billion (ppb) limit for inorganic arsenic in infant rice cereal, as children are more sensitive to arsenic exposure (FDA 2016). The European Food Safety Authority (EFSA) has also released a Tolerable Weekly Intake limit of $15\,\mu g\,kg^{-1}$ body weight (EFSA 2009).

4.2.4 Barley

Barley (*Hordeum vulgare*) has historically been an important source of energy and human nutrition, with the earliest archaeological remains found in Iran and Syria from approximately 8000 BCE (Newman and Newman 2006). However, poorer mouthfeel and product quality have resulted in barley being widely replaced by wheat and rye. Barley products and production methods have thus seen less development compared to wheat and other cereals. Similarly, there have been fewer attempts at systematically breeding and developing new barley strains for improved production quality and yield (Baik and Ullrich 2008). Barley is currently ranked as the fourth most important cereal worldwide (144 489 996 tonnes) (Table 4.2). However, in the United States only a very small amount of this (~1.5%) is currently directly consumed as food, with the vast majority used as animal feed, and for brewing and malting in alcoholic beverages (Newman and Newman 2006). Nevertheless, barley is still primarily used for direct human nutrition in several regions, such as Morocco, Ethiopia, China, and India.

4.2.5 Oats

Oats (*Avena sativa*) are an annual crop, commonly grown in temperate climates due to their greater tolerance to cold and wet conditions compared to wheat, rye, or barley. Similar to rye, modern oat is a secondary crop, believed to have originated from the Asian wild red oat which grew as a weed in earlier primary cereal crops (Sadiq Butt et al. 2008). Oats are primarily utilised as livestock feed, with 68% of total production in 2013 used for cattle, horses, and pigs, equating to 16 251 946 tonnes (FAO 2013). A lack of gluten makes oats unsuitable for bread making, and they are thus typically consumed rolled or crushed as porridge or breakfast cereals. However, this lack of gluten also makes oats suitable for most individuals with coeliac disease. The dual utilisation of oats as grain feed and fodder adds value to the crop. In addition to the grain used as feed, oat fodder can be used for grazing and oat straw for animal bedding (Suttie and Reynolds 2004). However, due to the presence of tough hulls, wholegrain oats tend to contain high levels of fibre, thus reducing their digestibility and nutritional value without further processing.

4.3 Cereal Proteins

4.3.1 Cereal Storage Proteins

While cereal grains are rich in carbohydrates, they also contain substantial levels of protein (Table 4.3). However, different production systems and fertiliser use can cause significant variations in cereal protein content, in addition to variations due to genetic factors. The majority of cereal grain protein comes from endosperm storage proteins, which provide valuable resources for the seeds during germination (Shewry and Halford 2002). Cereal storage proteins are generally classified based on whether they are soluble in water (albumins), dilute saline (globulins), alcohol/water mixtures (prolamins), and dilute acids/bases (glutelins) (Gangopadhyay et al. 2015).

Prolamins can generally be described as the major storage protein family in cereal grains, including maize, wheat, barley, rye, and sorghum. Prolamins originally got their name from the observation that they were rich in proline and amide nitrogen derived from glutamine, although it is now known that these can be found in lower amounts in some cereals (Shewry and Halford 2002). Prolamins found in Triticaeae can generally be categorised into three groups: high molecular weight, sulfur rich (S-rich), and sulfur poor (S-poor) (Shewry and Halford 2002).

Table 4.3 Nutrient composition of commonly consumed cereals per 100 g.

Nutrient (per 100 g)	Unit	Maize A	Wheat B	Rice C	Barley D	Oats E
Energy	kcal	354	329	361	333	374
Moisture	g	11.3	14.5	10.3	11.4	12.2
Protein	g	9.2	12.4	7.5	11.0	12.6
Carbohydrates	g	65.5	59.5	75	59.5	61.5
Lipids	g	3.8	2.0	2.2	2.1	7.1
Fibre	g	9.2	10.6	4.0	15.0	5.6
Sodium, Na	mg	6.0	10.0	10.0	20.0	10.0
Potassium, K	mg	330.0	0.5	150.0	450.0	350.0
Calcium, Ca	mg	15.0	45.0	20.0	40.0	80
Phosphorus, P	mg	250.0	400.0	320.0	340.0	340.0
Iron, Fe	mg	0.5	3.3	2.6	2.8	5.8
Magnesium, Mg	mg	45.0	140.0	160.0	90.0	—
Zinc, Zn	mg	1.0	4.1	0.8	3.1	4.5
Thiamine (Vitamin B1)	mg	0.5	0.5	0.4	0.3	0.52
Riboflavin (Vitamin B2)	mg	0.15	0.15	0.09	0.18	0.17
Vitamin B6	mg	0.4	0.4	0.65	0.6	1.0
Vitamin B11	mg	30.0	15.0	36.0	50.0	—
Vitamin C	mg	0.0	0.0	0.0	0.0	0.0
Vitamin D	mg	0.0	0.0	0.0	0.0	0.0

A Corn, yellow, raw; B Wheat, whole grain, raw; C Rice, brown, regular, raw; D Barley, raw; E Oats, raw.
Source: Reproduced with permission from de Jong (2017).

Within the prolamin protein family, certain protein classes are more prevalent in different cereal grains. Zeins are the prolamin proteins found within maize, as well as other panicoid cereals including sorghum and millet. The major group of these proteins is alpha-zeins, with other minor groups including beta-, gamma-, and delta-zeins. Zein proteins have great commercial use in the food and drug industries, where they are often used as a confectioner's glaze and to coat pharmaceuticals (Lawton 2002). Gliadins are the major prolamins within wheat and, together with glutenin, make up most of the protein gluten. There are three main gliadin groups: alpha-, gamma-, and omega-gliadin.

Similar to gliadin in wheat, hordein glycoproteins are the main prolamins found in barley, which are intolerable to people with coeliac disease and are the main substance preventing coeliacs from drinking beer. Some of the main barley proteins that have been sequenced and analysed are B hordein, C hordein, D hordein, globulin, and alpha-amylase (Cavazos and Gonzalez de Mejia 2013).

Oats and rice are somewhat unique among cereals as the main storage proteins are globulins (known as avenalins), rather than the prolamin family in other cereal storage proteins (Cunsolo et al. 2012). Avenins are prolamin proteins found within oats to a small degree, although these do not appear to trigger a response in individuals with coeliac disease (Storsrud et al. 2003). Saltwater-soluble globulins are estimated to make up approximately 80% of the total storage proteins in oats, while the dilute alcohol-soluble prolamins only make up approximately 15% of total proteins (Klose and Arendt 2012). This is in contrast to wheat, barley, and rye, in which prolamins constitute 40–50%, 30–50%, and 35–45% of total proteins, respectively. While the sequence and structure of rice proteins are similar to globulins, they are not readily soluble in dilute salt solutions, and are thus classically defined as glutelins.

4.4 Protein Quality

4.4.1 Amino Acid Composition

The quality of protein is largely determined by the amino acid composition and the availability of the amino acids, dictated by the digestibility of the foodstuff containing the protein. The amino acid composition of a single protein source is rarely of significance in most populations due to many individuals getting their protein from a variety of sources. However, in less developed countries many individuals may get the majority of their protein from a single plant protein source. This can lead to deficiencies if the protein source lacks sufficient quantities of certain essential amino acids (EAAs) which the human body is unable to synthesise. The 10 EAAs are histidine, isoleucine, leucine, lysine, methionine, phenylalanine, tyrosine, threonine, tryptophan, and valine (FAO 2011). Cysteine is sometimes included in this list as it is synthesised endogenously from methionine, which is itself an EAA (Nikiforova et al. 2002). Similarly, tyrosine is synthesised in mammals from phenylalanine (Fernstrom and Fernstrom 2007). There are also certain conditional EAAs, including glutamine, glutamic acid, glycine, proline, and taurine. These amino acids may become more important in the diet at certain times, such as for neonates and adults under stress conditions (e.g. burns and infection) (Wu 2013).

The preferred method for determining protein quality is the Protein Digestibility Corrected Amino Acid Score (PDCAAS), as recommended by the FAO/WHO Expert Consultation on Protein Quality Evaluation Committee in 1989 (FAO 1991). PDCAAS is determined by expressing the first limiting EAA in a test protein source as a percentage of the same amino acid content in a reference protein source (casein), based upon the requirements of a preschool-age child (two to three years) (FAO 1991). This reference value is then corrected for remaining faecal nitrogen of the test protein in order to account for digestibility. This corrected value is then referred to as the PDCAAS for the test protein source in question.

Despite the widespread use of PDCAAS for determining protein quality, many have argued that it has several shortcomings. First, the reference EAA requirements were based upon those of very young children, meaning that they may be unsuitable for adults or the elderly (Young and Borgonha 2000). Similarly, PDCAAS does not take into account nitrogen loss in the large intestine due to metabolic activity of microbiota within the hindgut (Rutherfurd et al. 2015). Furthermore, antinutritional factors, such as trypsin inhibitors, glucosinolates, and cellulose, found within many protein sources, can interfere with protein digestion and thus decrease their value to the individual (Sarwar 1997).

The FAO committee therefore recommended a revised score in 2011, called the Digestible Indispensable Amino Acid Score (DIAAS), as a replacement for PDCAAS (FAO 2011). DIAAS is a more accurate method for determining protein quality as it assesses true ileal digestibility of individual amino acids, rather than overall digestibility of a target protein source. The reference EAA intake has also been revised by dividing the requirements into three different age categories: infant (<6 months), child (six months to three years), and adolescent or adult (>6 years). DIASS is found by determining the amount (mg) of each essential (indispensable) amino acid in 1 g of test protein source, divided by the amount (mg) of the same EAA in a reference protein source, typically casein (FAO 2011).

Cereals have a relatively good amino acid profile, generally matching the recommendations of the FAO (Table 4.4). Lysine is the main limiting amino acid found in cereals, although it is found at the highest levels in oats and rice (Biel et al. 2009; Zorovski and Georgieva 2011). Oats are also relatively high in methionine and phenylalanine, while maize has the highest levels of leucine, and barley is a good source of tryptophan (Table 4.4).

Cereals are therefore a good source of most EAAs, although it might also be beneficial to supplement with other plant and animal proteins for an optimum healthy diet.

4.4.2 Digestibility

In addition to amino acid composition, the digestibility of a protein will affect its protein quality and bioavailability. Barley has been reported as having the greatest ileal starch digestibility compared to other cereals such as rye, triticale, and wheat

Table 4.4 Essential amino acid composition of major cereals (g/100 g protein).

	Essential amino acids					
	Barley grain	Oat grain	Wheat grain	Rice milled	Maize	Cornflour
Histidine	2.3	2.2	2.3	2.4	2.7	2.7
Isoleucine	3.7	3.9	3.7	3.8	3.6	3.5
Leucine	7.0	7.4	6.8	8.2	12.5	7.5
Lysine	3.5	4.0	2.8	3.7	2.7	7.3
Cysteine	2.3	1.6	2.3	1.6	1.6	—
Methionine	1.7	2.5	1.2	2.1	1.9	3.5
Phenylalanine	5.2	5.3	4.7	4.8	5.0	7.3
Tyrosine	2.9	3.1	1.7	2.5	3.8	—
Threonine	3.6	3.3	2.9	3.4	3.7	4.2
Tryptophan	1.9	—	1.1	1.3	0.6	1.2
Valine	4.9	5.3	4.4	5.8	5.5	4.9

Source: Values obtained from Shewry (2007) and FAO (2011).

(Rosenfelder-Kuon et al. 2017). The carbohydrate composition and quantity can affect nutrient digestibility, with dehulled barley observed to have greater ileal digestibility (77% ± 1.1%) in weaned piglets compared to hulled barley (64% ± 1.4%) and hulled oats (58% ± 1.5%) (Jha et al. 2010). However, soluble non-starch polysaccharides, such as beta-glucan, were deemed to have little effect on digestibility. Fermentation can also be used to improve the digestibility of cereal protein, with *in vitro* protein digestibility of shambat pearl millet observed to significantly increase from 76.9% to 86.8% following fermentation (Ali et al. 2003).

Protein digestibility can also be hindered by the presence of antinutritional factors, such as enzyme inhibitors, tannins, and phytates (Gilani et al. 2005). In particular, phytate can form complexes in oats and barley, significantly decreasing the bioavailability of protein and minerals (Hübner et al. 2010). One option for overcoming this hurdle involves germination, which has long been used as a means to soften cereal grains, increase nutritional value, and decrease antinutritional factors (Kaukovirta-Norja et al. 2004). Alternatively, microwave treatment has also been shown to effectively reduce antinutritional factors, including trypsin inhibitors, polyphenols, and phytates (Kaur et al. 2012).

4.5 Bioactive Peptides

In addition to providing energy and amino acids essential for growth, it has become increasingly recognised that some dietary proteins contain biologically active peptides that can influence a favourable health response beyond their basic nutritional value. This phenomenon occurs as a result of their hormone-like properties or by acting as antagonistic receptor inhibitors (Korhonen and Pihlanto 2006). Bioactive peptides can trigger a variety of affects, including antihypertensive, anticancer, antioxidant, antidiabetes, and anti-inflammatory activities. Bioactive compounds derived from plants (phytochemicals) have been identified from a number of sources, including soybean (Rho et al. 2009), legumes (Aluko 2008), seaweed (Harnedy and FitzGerald 2011), potato (Pihlanto et al. 2008), and rapeseed (Zhang et al. 2008). Several bioactive peptides have also been identified from cereal grains, although they have yet to be fully exploited (Cavazos and Gonzalez de Mejia 2013).

4.5.1 Antihypertensive Peptides

High blood pressure is the single largest risk factor for deaths worldwide (WHO 2009). Hypertension is responsible for 12.8% of deaths globally, affecting all countries and all income groups (WHO 2009). Furthermore, high systolic blood pressure (SBP) is globally responsible for 51% of stroke cases, 45% of ischaemic heart disease cases, and between 37% (South-East Asia region) and 54% (European countries) of cardiovascular deaths (WHO 2009). Hypertension is therefore a considerable problem in our society, not only placing a great burden on our health but also having substantial impacts on the economy. Direct medical costs of cardiovascular disease (CVD) are estimated to increase from $273 billion in 2010 to $818 billion in 2030 in the United States alone (Heidenreich et al. 2011).

The renin-angiotensin-aldosterone system (RAAS) plays an important role in regulating blood pressure by controlling arteriolar vasoconstriction and intravascular fluid volume. The RAAS itself is largely regulated by the actions of the angiotensin-converting enzyme (ACE; EC 3.4.15.1), which acts to increase blood pressure (Ondetti and Cushman 1982). An alternative approach for treating hypertension is therefore to inhibit the actions of ACE using biologically active peptides as functional foods to avoid the side effects that are typically associated with pharmaceutical ACE inhibitors, which can include a persistent dry cough, anaphylaxis, and renal impairment (Sica and Gehr 2005). Functional foods with antihypertensive properties are also beneficial in individuals that are borderline hypertensive and thus do not warrant the prescription of pharmaceutical medication (Chen et al. 2009).

Several antihypertensive peptides have been isolated from maize proteins. Thermolysin has been observed to effectively generate peptides with ACE inhibition activity, of which Leu-Arg-Pro, Leu-Ser-Pro, and Leu-Gln-Pro had the greatest activity (Miyoshi et al. 1995). These peptides had a similar structure and all contain a proline at the C-terminus of the peptide (Table 4.5). The bioactive dipeptide Ala-Tyr has also been generated from corn gluten meal (CGM), a by-product of maize processing with enriched protein content (Yang et al. 2007; Lin et al. 2011). These results were confirmed *in vivo* with spontaneously hypertensive rats (SHRs), reducing SBP by 9.5 mmHg

Table 4.5 ACE inhibitory peptides derived from cereal grain proteins.

Source	Phytochemical	Protease	IC_{50} (μM)	References
Maize	Leu-Arg-Pro	Thermolysin	0.27	Miyoshi et al. (1995)
	Leu-Arg-Pro		1.7	
	Leu-Gln-Pro		1.9	
	Leu-Asn-Pro		43	
	Leu-Leu-Pro		57	
	Val-Ser-Pro		10	
	Val-Ala-Tyr		16	
	Leu-Ala-Tyr		3.9	
	Phe-Tyr		25	
	Ile-Arg-Ala		6.4	
	Leu-Ala-Ala		13	
	Val-Ala-Ala		13	
	Leu-Gln-Gln		100	
	Ile-Arg-Ala-Gln-Gln		160	
	Ser-Ala-Tyr-Pro-Gly-Gln-Ile-Thr-Ser-Asn	Trypsin	7	Gobbetti et al. (1997)

(Continued)

Table 4.5 (Continued)

Source	Phytochemical	Protease	IC_{50} (μM)	References
	Hydrolysate	Thermolysin	24.5	Yano et al. (1996)
	Ala-Tyr	Alcalase	14.2	Yang et al. (2007)
	Leu-Arg-Pro	Thermolysin	—	Puchalska et al. (2013)
	Leu-Ser-Pro	—		
	Leu-Gln-Pro	—		
Wheat	Thr-Phe	Alkaline protease	17.8	Matsui et al. (1999)
	Leu-Tyr		6.4	
	Tyr-Leu		16.4	
	Ala-Phe		15.2	
	Ile-Tyr		2.1	
	Val-Phe		9.2	
	Ile-Val-Thr		0.48	
	Val-Phe-Pro-Ser		0.46	
	Thr-Val-Pro-Tyr		13.6	
	Thr-Ala-Pro-Tyr		2.0	
	Thr-Val-Val-Pro-Gly		2.2	
	Asp-Ile-Gly-Tyr-Tyr		3.4	
	Asp-Tyr-Val-Gly-Asn		0.72	
	Thr-Tyr-Leu-Gly-Ser		0.86	
	Gly-Gly-Val-Ile-Pro-Asn		0.74	
	Ala-Pro-Gly-Ala-Gly-Val-Tyr		1.7	
	Ile-Ala-Pro	Pepsin-protease M	2.7	Motoi and Kodama (2003)
	Val-Glu-Val	Endogenous proteases	115.2	Yang et al. (2011)
	Asn-Pro-Pro-			
	Ser-Val		40.5	
	Gln-Val		26.8	
	Ala-Met-Tyr		5.8	
Rice	Val-Asn-Pro	Alcalase	6.4	Chen et al. (2013)
	Val-Trp-Pro		4.5	
	Thr-Gln-Val-Tyr Alcalase		18.2	Li et al. (2007)
Barley	Gln-Val-Ser-Leu			
	-Asn-Ser-Gly-Tyr-Tyr	Trypsin	23	Gobbetti et al. (1997)
Oats	Hydrolysate	Thermolysin	—	Cheung et al. (2009)

over two hours following an oral administration dose of 50 mg kg^{-1}. Similarly, CGM hydrolysate at an orally administered dose of 100 mg kg^{-1} reduced SBP of SHRs by 26.57 and 34.45 mmHg in short-term and long-term studies, respectively (Huang et al. 2011).

Hydrolysis of wheat germ with alkaline proteases yielded several peptides with ACE inhibitory activities, the most potent of which was Ile-Val-Tyr with an IC_{50} value of 0.48 μM (Matsui et al. 1999). The same tripeptide (Ile-Val-Tyr) also displayed antihypertensive activity *in vivo*, lowering mean arterial blood pressure in SHRs by 19.2 mmHg following 5 mg kg^{-1} intravenous administration (Matsui et al. 2000). Similarly, wheat gliadin hydrolysates also yielded ACE inhibitory peptides, which also had antihypertensive effects *in vivo* (Table 4.5). Several methods for generating ACE inhibitory peptides have been investigated, including utilising endogenous proteases, combining ultrasound pretreatment and enzymatic hydrolysis, and continuous coupling of enzymatic hydrolysis and membrane separation (Yang et al. 2011; Qu et al. 2013; Huang et al. 2014).

Angiotensin-converting enzyme inhibitory tripeptides (Val-Asn-Pro and Val-Trp-Pro) have been generated from rice protein with potent IC_{50} values of 6.4 and 4.5 μM, respectively (Chen et al. 2013). These competitive ACE inhibitors were observed to be resistant to degradation by gastrointestinal proteases pepsin and chymotrypsin, an important property if the ingested peptide is to reach its site of action. Val-Asn-Pro and Val-Trp-Pro also displayed potent antihypertensive activity *in vivo* in SHRs, lowering SBP by 29 and 38 mmHg following 5 mg kg^{-1} oral administration, respectively. The ACE inhibitory peptide Thr-Gln-Val-Tyr has also been derived from rice protein, which did not display any adverse effects *in vivo* (Li et al. 2007).

In contrast to the other cereal grain proteins described in this chapter, little work has been done to investigate oat and barley protein as a source of bioactive peptides. This is despite *in silico* analysis which revealed that these proteins have the potential to generate ACE inhibitory, renin, and propyl endopeptidase (PEP) inhibitory peptides (Cavazos and Gonzalez de Mejia 2013). Bioactive peptides have traditionally been identified by screening a novel substrate with a selection of proteases based on literature reports and testing bioactivity experimentally. However, this method is time consuming and costly. Online databases and tools, such as bioactive peptide database (BIOPEP) and PeptideCutter (ExPASy), allow for theoretical generation of bioactive peptides of known protein sequences, combined with the known cleavage patterns of commonly used proteases (Lafarga et al. 2014). This enables the optimal substrate and protease to be tested experimentally to achieve the desired bioactive peptides. Prescreening with *in silico* methods can reduce time and costs. As an example, the oat proteins 11S globulin, 12S globulin, and avenin-3 were cleaved with thermolysin, which predicted 162, 158, and 65 previously documented antihypertensive peptides, respectively (Cheung et al. 2009). These results were confirmed *in vitro*, with peptides generated from thermolysin having significantly greater activity than those generated with cleavage from pepsin, subtilisin, and trypsin.

4.5.2 Anticancer Cereal Peptides

After hypertension, cancer is the leading cause of death in the United States (CDC 2015). There are many forms of cancer, with many different causes. However, it is ultimately defined as genetic and cellular changes that lead to uncontrolled cell division and proliferation, thus forming a tumour (Suarez-Jimenez et al. 2012). Traditional chemotherapy produces many adverse effects, including immunosuppression, gastrointestinal distress, anaemia, nausea, and alopecia. Bioactive peptides may exert a preventive cancer effect, including initiation and progression. These peptides are found naturally in food proteins and are considered to be a safer alternative to pharmaceutical therapies (Díaz-Gómez et al. 2017).

Maize peptides generated from hydrolysis with alcalase inhibited hepatoma tumour viability through arresting cell cycle progression and causing apoptosis of Hepg2 cells (Li et al. 2013). Maize peptides also enhanced the host immune system *in vivo* with increased expression of cytokines TNF-alpha, IFN-gamma, and IL-2 in H22-bearing mice, resulting in a prolonged survival time. Similar results were observed by Ortiz-Martinez et al. (2017), where the antiproliferative activity of maize peptides resulted due to reduced expression of antiapoptotic factors, including livin, survivin, cIAP, XIAP, and HIF-1-alpha. Rice bran protein has also displayed potential as a source of anticancer nutraceuticals. The peptide with sequence Glu-Gln-Arg-Pro-Arg at a concentration of 600–700 $\mu g\,mL^{-1}$ inhibited growth by 84% in colon cancer cells (Caco-2, HCT-116), 80% in breast cancer cells (MCF-7, MDA-MB-231), and 84% in liver cancer cells (HepG-2) (Kannan et al. 2010).

Lunasin is a chemopreventive peptide featuring the cell adhesion peptide motif Arg-Gly-Asp, which has been observed to be cytotoxic in established cell lines (Hernandez-Ledesma et al. 2013). Lunasin has been isolated from barley protein, ranging in concentration from 5.9 to 8.7 $ng\,\mu g^{-1}$ protein (Jeong et al. 2002). This is comparable to the concentration of lunasin in soybean (5.5 $ng\,\mu g^{-1}$ protein), in which the chemopreventive peptide was first identified. Purified lunasin from barley supressed colony formation in ras-transfected mouse fibroblast cells at concentrations (100 μM) and activity comparable to those of synthetic lunasin. Lunasin is believed to selectively induce apoptosis in tumour cells by binding to deacetylated histones and inhibiting histone acetylation (Galvez et al. 2001). This was observed *in vivo* using barley lunasin, which significantly decreased histone H3 acetylation in mouse fibroblast NIH3T3 and human cancer MCF-7 cells treated with the histone deacetylase inhibitor sodium butyrate (Jeong et al. 2002). Lunasin has also been reported in wheat, rye, amaranth, triticale, and oats, although at notably lower concentrations to that of barley (Jeong et al. 2007, 2009; Maldonado-Cervantes et al. 2010; Nakurte et al. 2012, 2013).

4.5.3 Antioxidant Peptides

Natural antioxidants have gained increasing popularity in recent years due to their proven health benefits and low adverse effects compared to synthetic antioxidants. Antioxidants reduce the free radical activity of reactive oxygen species within the body, including hydroxyl radicals, superoxide anion radicals, and peroxides (Maritim et al. 2003). These free radicals can cause biochemical damage to cells, often resulting in cell metabolic damage and cell death, promoting the onset of several diseases and increasing the effect of ageing. Furthermore, lipid oxidation can also have a negative effect in the food industry by reducing shelf-life, negatively affecting taste and texture, and decreasing nutritional value.

Maize has been a relatively well-studied source of bioactive peptides with antioxidant properties. This could be explained by the relatively high amounts of hydrophobic amino acids, such as leucine, alanine, and phenylalanine (Zhou et al. 2015). Hydrolysis of CGM with alcalase generated the peptide Gln-Gln-Pro-Gln-Pro-Trp, which was matched to gamma-zein f(50–55) (Wang et al. 2014). This hexapeptide displayed dose-dependent free radical scavenging activity against 2,2-diphenyl-1-picrylhydrazyl (DPPH), 2,2′-azino-bis(3-ethylbenzothiazoline-6-sulfonic acid) (ABTS), and hydroxyl free radicals, with IC_{50} values of 0.95, 0.0112, and 4.43 mg ml^{-1}. Flavourzyme has also yielded antioxidant peptides from CGM, with lower molecular weight having a greater antioxidant activity. The peptides Leu-Pro-Phe, Leu-Leu-Pro-Phe, and Phe-Leu-Pro-Phe were identified using mass spectrometry, with DPPH IC_{50} values of 2.07, 2.22, 1.51 mM respectively, as well as ABTS IC_{50} values of 2.07, 2.11, and 2.83 mM (Zhuang et al. 2013).

Rice bran albumin hydrolysates generated from trypsin have been observed to have antioxidant activity, with an oxygen radical absorbance capacity (ORAC) value of 4.07 μmol of Trolox equivalent (TE)/mg protein (Wattanasiritham et al. 2016). Additionally, several peptides have been identified from defatted rice endosperm protein hydrolysates with neutrase (Zhang et al. 2010). Of these, Phe-Arg-Asp-Glu-His-Lys-Lys was chemically synthesised and effectively inhibited lipid peroxidation in a linoleic emulsion system, suggesting its relevance as a potential natural preservative food ingredient.

Similarly, wheat germ protein hydrolysates also inhibited peroxidation in a linoleic acid emulsion system, with activity comparable to that of alpha-tocopherol (Zhu et al. 2006). Antioxidant peptides have been generated from wheat germ protein using endogenous proteases, with the peptide Val-His-His-His inhibiting hydroxyl radical activity in a dose-dependent manner (Hu et al. 2012). Oat protein hydrolysates generated antioxidant peptides with an ORAC value of 0.67 μM TE μM^{-1} peptide. Oat protein hydrolysates also displayed improved cell viability, attributed to reduced intracellular reactive oxygen species, increased cellular glutathione synthesis, and elevated activity of endogenous antioxidant peptides (Du et al. 2016). The molecular weight of barley-derived bioactive peptides could play a role in the type of antioxidant activity achieved. Barley glutelin hydrolysates generated from alcalase yielded large (greater than 10 kDa) antioxidant peptides which displayed strong DPPH activity, whereas smaller peptides (less than 1 kDa) had increased metal ion chelating and hydroxyl scavenging activity (Xia et al. 2012).

4.5.4 Antidiabetic Peptides

With the estimated number of adults with diabetes expected to increase from 285 million in 2010 to 439 million in 2030, diabetes is predicted to become a global epidemic (Shaw et al. 2010). Glucagon-like peptide 1 (GLP-1) and glucose-dependent insulinotropic peptide (GIP) belong to a family of gut peptides called incretins, responsible for decreasing blood glucose levels after eating by promoting the release of insulin from pancreatic beta-cells. Dipeptidyl peptidase IV (DPP-IV; EC 3.4.14.5) is a serine exopeptidase that cleaves dipeptides from the N-terminus of regulatory peptides with the sequence Xaa-Pro- or, to a lower extent, Xaa-Ala- (where Xaa represents any amino acid) (Mentlein 1999). GLP-1 and GIP are included, resulting in the inactive metabolites GLP-1 (9–36) amide and GIP (3–42), leading to a hyperglycaemic state (Nagashima et al. 2011). Bioactive peptides with DPP-IV inhibitory activity have therefore been suggested as an alternative treatment for the prevention of diabetes.

Rice bran hydrolysed with umamizyme produced dipeptides with the sequence Leu-Pro and Ile-Pro which were observed to inhibit DPP-IV, with IC_{50} values of 0.41 and 2.37 mM (Hatanaka et al. 2012). *In silico* analysis of 72 dietary proteins with DPP-IV inhibitory activity revealed that wheat and oat proteins have some of the highest potentials for potent activity based on their amino acid sequence and composition (Nongonierma and FitzGerald 2014). However, wheat protein-derived peptides have actually been observed to have reduced activity compared to amaranth, with IC_{50} value of 1.1 mg ml^{-1} (Velarde-Salcedo et al. 2013). Alcalase hydrolysis of oat, buckwheat, and flour has also yielded DPP-IV inhibitory peptides, of which the oat glutelin hydrolysate was the most potent with an IC_{50} value of 0.13 mg mL^{-1} (Wang et al. 2015). Furthermore, hydrolysates from maize zein demonstrated dual antihyperglycaemic properties due to increased GLP-1 production and DPP-IV inhibition (Mochida et al. 2010).

4.5.5 Anti-Inflammatory Cereal-Derived Peptides

Hepatitis is inflammation of the liver tissue, which can be caused by a number of factors including virus infection, excessive alcohol abuse, and toxins. The hydrolysate pyroglutamyl leucine (pyroGlu-Leu) from wheat gluten protein showed hepatoprotective activity in D-galactosamine-induced acute rats. The peptide effectively lowered serum aspartate aminotransferase and alanine aminotransferase by 30% and 20% at an oral dose of 20 mg kg^{-1} body weight, respectively, which were near normal levels (Sato et al. 2013). Mass spectrometry analysis also revealed that pyroGlu-Leu is resistant to gastrointestinal proteases, demonstrating its bioavailability *in vivo*. Similarly, CGM hydrolysates have been reported to provide hepatoprotective activity against damage from exposure to lipopolysaccharides, significantly lowering cell necrosis, hepatic lesions, and enzymes associated with hepatotoxicity damage (Guo et al. 2009).

Ulcerative colitis and Crohn's disease are inflammatory bowel diseases resulting in acute inflammation and ulcerative bleeding within the colon. CGM hydrolysates were reported to effectively reduce inflammation, myeloperoxidase activity, and histamine levels in 2,4,5-trinitrobenzene sulfonic acid (TNBS)-induced mice (Mochizuki et al. 2010). Similarly, black rice bran was demonstrated to display anti-inflammation and antiallergenic effects in allergic contact dermatitis inflammatory mouse models (Choi et al. 2010).

4.6 Allergenicity

4.6.1 Immunoglobulin-Mediated

Certain proteins can stimulate an allergenic response in some individuals following ingestion or inhalation. Epitopes are small peptide units which are believed to play a key role in causing an allergic response. B-cell epitopes provide a binding site for immunoglobulin E (IgE), while T-cell epitopes provide a binding site for type-2 lymphocytes, with both reactions capable of causing an allergenic response. In addition to the type of epitope present, other factors, including resistance to proteolysis, glycosylation status, and enzymatic activity, can also affect protein allergenic potential (Huby et al. 2000).

Wheat is associated with two types of allergies: baker's asthma and wheat-dependent exercise-induced anaphylaxis (WDEIA) (Tatham and Shewry 2008). Baker's asthma and rhinitis are allergic responses following the inhalation of wheat and cereal flour and dust. Cereal grain alpha-amylase/trypsin inhibitor proteins have been well documented as being the main proteins responsible for this allergy, which have also been identified in barley and rye. (Salcedo et al. 2004). WDEIA is one of the most common forms of food-dependent exercise-induced anaphylaxis (FDEIA), where neither ingestion of the deleterious food nor exercise alone is sufficient to trigger anaphylaxis (Palosuo et al. 1999). The omega-5-gliadin (Tri a 19) fraction of gluten has been identified as the most potent allergen within wheat (Morita et al. 2003). Cross-reacting allergens with omega-5-gliadin have also been identified as gamma-70 and gamma-36 secalins in rye and gamma-3 hordeins in barley (Palosuo et al. 2001).

A 9 kDa lipid-transfer protein (LTP) has been identified as the major allergen in maize, responsible for causing laryngeal edema and shock in affected patients (Pastorello et al. 2000). This maize LTP has also been demonstrated to survive the cooking process, similar to barley LTP surviving brewing and malting (García-Casado et al. 2001; Pastorello et al. 2003). While heating typically lowers the allergenicity of injurious proteins, it can also increase allergenicity, such as the increase in allergenicity of peanuts following roasting due to the Maillard reaction (Maleki et al. 2000).

Rice allergies are present in 10% of atopic subjects, provoking responses such as dermatitis and eczema (Ogo et al. 2014). Multiple rice proteins are responsible for causing atopic reactions, with major allergenic proteins including alpha-globulin, beta-glyoxalase I, and alpha-amylase/trypsin inhibitor (Zhu et al. 2015). Efforts have therefore been made to silence the production of these injurious proteins through use of transgenic rice, with no reported significant alteration to seed storage proteins and protein folding (Ogo et al. 2014).

4.6.2 Coeliac Disease

While coeliac disease is immunologically mediated, it is not mediated through immunoglobulin antibodies and is therefore not a true allergy. While individuals with allergies to specific proteins can generally consume cereals which do not contain those proteins, coeliacs must avoid all grains containing gluten. Furthermore, symptoms of coeliac disease are generally slow to occur, developing 48–72 hours after consumption of gluten. Alternatively, cereal allergies are generally a rapid reaction, with symptoms sometimes occurring minutes after consumption or inhalation of the injurious protein.

Coeliac disease is therefore a distinct condition that can be defined as both a food intolerance and an autoimmune disorder. It is characterised by a chronic inflammatory state in the proximal small intestine following consumption of foods containing gluten. Innate and adaptive immune responses result in atrophy of the villi within the small intestine and chronic inflammation of intraepithelial compartments (Rostom et al. 2006). This reduces the body's ability to absorb vital nutrients, resulting in malnutrition, diarrhoea, vitamin and mineral deficiencies, as well as possibly anaemia, osteoporosis, intestinal cancer, and infertility (Fasano and Catassi 2001).

Gluten belongs to the alpha-gliadin fraction of prolamins in wheat, and is present in several other cereal prolamins, including barley and rye. However, oats do not contain the injurious gliadin but rather its counterpart avenin (Arentz-Hansen et al. 2004). Moderate amounts of oats (~90 g day^{-1}) have therefore been suggested as an alternative cereal source for patients with coeliac disease following a gluten-free diet to improve nutritional value and provide dietary relief for patients (Storsrud et al. 2003). According to European legislation, the threshold for 'gluten-free' foods is 20 mg kg^{-1} food product, while the limit for 'very low gluten foods' is set to less than 100 mg kg^{-1} food product (Commission Regulation (EC) No 41/2009).

4.7 Non-Protein Health Applications of Cereals

In addition to protein, there are several other constituents of cereals that have great potential in health applications. The effect of the gut microbiome has received increased attention as we learn more about its effect on us in terms of metabolism and health. Consumption of non-digestible fibres can be used as prebiotics to increase levels of good bacteria, typically bifidobacteria and lactic acid bacteria, through selective fermentation (Costabile et al. 2008). These bacteria are important as they confer a benefit to the host, including increased digestion and absorption of nutrients, as well as strengthened immune system. Some common examples of fibres used as prebiotics include resistant starch, pectin, beta-glucans, and oligosaccharides.

Beta-glucan is an unbranched polysaccharide that has been recognised as having therapeutic effects on coronary heart health and lowering cholesterol. In particular, oats have been demonstrated as an excellent source of beta-glucan, to the extent that the FDA has authorised their heart benefits as a health claim (FDA 2002). Daily intake of 3 g beta-glucan can reduce plasma total and low-density lipoprotein (LDL) cholesterol levels by 5–10% (Othman et al. 2011).

4.8 Conclusion

Cereals are an extremely important source of nutrition, forming the foundations of diets in many regions and populations worldwide. Maize is the most produced cereal grain, although only a small fraction of this is used for human consumption, with the majority used for either animal feed or biofuel production. In addition to being a rich source of energy-dense carbohydrates, cereal grains are also high in protein, with a relatively good amino acid composition and digestibility score. Indeed, protein from cereal grains often constitutes the main protein source for many populations in less developed countries where meat may be less common and considered more of a luxury.

While cereals are an excellent source of nutrition to meet caloric and protein needs, cereal proteins have also been demonstrated to display promising functional and bioactive properties beneficial to human health. These primarily occur in the form of bioactive peptides, which can exert a physiological effect beyond their basic nutritional value once released from their parent protein, typically through hydrolysis or fermentation. Bioactive peptides have been identified from all major cereal grains, with antihypertensive, anticancer, antioxidant, antidiabetes, and anti-inflammatory actions. However, despite the work that has been done examining the health benefits of cereal proteins and the promising activity that has been predicted *in silico*, it is dwarfed by what has been done for other protein sources, such as milk and meat. More work is therefore required to fill this knowledge gap, increasing the value of cereal grains and potentially leading to new commercial opportunities in the health food market.

References

Ali, M.A., El Tinay, A.H., and Abdalla, A.H. (2003). Effect of fermentation on the in vitro protein digestibility of pearl millet. *Food Chemistry* 80 (1): 51–54.

Aluko, R.E. (2008). Determination of nutritional and bioactive properties of peptides in enzymatic pea, chickpea, and mung bean protein hydrolysates. *Journal of AOAC International* 91 (4): 947–956.

Arentz-Hansen, H., Fleckenstein, B., Molberg, Ø. et al. (2004). The molecular basis for oat intolerance in patients with celiac disease. *PLoS Medicine* 1 (1): e1.

Awika, J.M. (2011). Major cereal grains production and use around the world. *Advances in Cereal Science: Implications to Food Processing and Health Promotion: ACS Publications* 1–13.

Baik, B.-K. and Ullrich, S.E. (2008). Barley for food: characteristics, improvement, and renewed interest. *Journal of Cereal Science* 48 (2): 233–242.

Biel, W., Bobko, K., and Maciorowski, R. (2009). Chemical composition and nutritive value of husked and naked oats grain. *Journal of Cereal Science* 49 (3): 413–418.

Cavazos, A. and Gonzalez de Mejia, E. (2013). Identification of bioactive peptides from cereal storage proteins and their potential role in prevention of chronic diseases. *Comprehensive Reviews in Food Science and Food Safety* 12 (4): 364–380.

Centers for Disease Control and Prevention (2015) Leading Causes of Death. Retrieved from: www.cdc.gov/nchs/fastats/leading-causes-of-death.htm

Chen, J., Liu, S., Ye, R. et al. (2013). Angiotensin-I converting enzyme (ACE) inhibitory tripeptides from rice protein hydrolysate: purification and characterization. *Journal of Functional Foods* 5 (4): 1684–1692.

Chen, Z.-Y., Peng, C., Jiao, R. et al. (2009). Anti-hypertensive nutraceuticals and functional foods. *Journal of Agricultural and Food Chemistry* 57 (11): 4485–4499.

Cheung, I.W., Nakayama, S., Hsu, M.N. et al. (2009). Angiotensin-I converting enzyme inhibitory activity of hydrolysates from oat (*Avena sativa*) proteins by in silico and in vitro analyses. *Journal of Agricultural and Food Chemistry* 57 (19): 9234–9242.

Choi, S.P., Kim, S.P., Kang, M.Y. et al. (2010). Protective effects of black rice bran against chemically-induced inflammation of mouse skin. *Journal of Agricultural and Food Chemistry* 58 (18): 10007–10015.

Commission Regulation (EC)(2009) No 41/2009 of 20 January 2009 concerning the composition and labelling of foodstuffs for people intolerant to gluten.

Costabile, A., Klinder, A., Fava, F. et al. (2008). Whole-grain wheat breakfast cereal has a prebiotic effect on the human gut microbiota: a double-blind, placebo-controlled, crossover study. *British Journal of Nutrition* 99 (1): 110–120.

Cunsolo, V., Muccilli, V., Saletti, R., and Foti, S. (2012). Mass spectrometry in the proteome analysis of mature cereal kernels. *Mass Spectrometry Reviews* 31 (4): 448–465.

de Jong, F. M. (2017) Nutrition Table. Retrieved from: http://www.foodnutritiontable.com/

Domingo, J.L. and Bordonaba, J.G. (2011). A literature review on the safety assessment of genetically modified plants. *Environment International* 37 (4): 734–742.

Du, Y., Esfandi, R., Willmore, W.G., and Tsopmo, A. (2016). Antioxidant activity of oat proteins derived peptides in stressed hepatic HepG2 cells. *Antioxidants* 5 (4): 39.

Díaz-Gómez, J.L., Castorena-Torres, F., Preciado-Ortiz, R.E., and García-Lara, S. (2017). Anti-cancer activity of maize bioactive peptides. *Frontiers in Chemistry* 5: 44.

EFSA (2009). EFSA Panel on Contaminants in the Food Chain (CONTAM) on the Scientific Opinion on Arsenic in Food. *EFSA Journal* 7 (10): 1351.

FAO (2011) Dietary protein quality evaluation in human nutrition, New Zealand. Food and Agriculture Organisation of the United Nations, Rome.

FAO (2013) Food Balance Sheets. Retrieved from: http://www.fao.org/economic/ess/fbs/en/

FAO, WHO (1991). *Protein Quality Evaluation Report of Joint FAO/WHO Expert Consultation FAO, Food and Nutrient.* Rome: Food and Agriculture Organisation of the United Nations.

Fasano, A. and Catassi, C. (2001). Current approaches to diagnosis and treatment of celiac disease: an evolving spectrum. *Gastroenterology* 120 (3): 636–651.

FDA (2002). Food labeling: health claims; soluble dietary fiber from certain foods and coronary heart disease. Interim final rule. *Federal Register* 67 (191): 61773–61783.

FDA 2016. FDA Statement on Testing and Analysis of Arsenic in Rice and Rice Products. Retrieved from: www.fda.gov/food/foodborneillnesscontaminants/metals/ucm367263.htm

Fernstrom, J.D. and Fernstrom, M.H. (2007). Tyrosine, phenylalanine, and catecholamine synthesis and function in the brain. *Journal of Nutrition* 137 (6): 1539S–1547S.

Galvez, A.F., Chen, N., Macasieb, J., and Ben, O. (2001). Chemopreventive property of a soybean peptide (lunasin) that binds to deacetylated histones and inhibits acetylation. *Cancer Research* 61 (20): 7473–7478.

Gangopadhyay, N., Hossain, M.B., Rai, D.K., and Brunton, N.P. (2015). A review of extraction and analysis of bioactives in oat and barley and scope for use of novel food processing technologies. *Molecules* 20 (6): 10884–10909.

García-Casado, G., Crespo, J.F., Rodríguez, J., and Salcedo, G. (2001). Isolation and characterization of barley lipid transfer protein and protein Z as beer allergens. *Journal of Allergy and Clinical Immunology* 108 (4): 647–649.

Gibson, D.J. (2009). *Grasses and Grassland Ecology*. Oxford: Oxford University Press.

Gilani, G.S., Cockell, K.A., and Sepehr, E. (2005). Effects of antinutritional factors on protein digestibility and amino acid availability in foods. *Journal of AOAC International* 88 (3): 967–987.

Gobbetti, M., Smacchi, E., Corsetti, A., and Bellucci, M. (1997). Inhibition of proteolytic enzymes from *Pseudomonas fluorescens* ATCC 948 and angiotensin I-converting enzyme by peptides from zein, hordein, and gluten hydrolysates. *Journal of Food Protection* 60 (5): 499–504.

Guo, H., Sun, J., He, H. et al. (2009). Antihepatotoxic effect of corn peptides against Bacillus Calmette–Guerin/lipopolysaccharide-induced liver injury in mice. *Food and Chemical Toxicology* 47 (10): 2431–2435.

Harnedy, P.A. and FitzGerald, R.J. (2011). Bioactive proteins, peptides, and amino acids from macroalgae. *Journal of Phycology* 47 (2): 218–232.

Hatanaka, T., Inoue, Y., Arima, J. et al. (2012). Production of dipeptidyl peptidase IV inhibitory peptides from defatted rice bran. *Food Chemistry* 134 (2): 797–802.
Heidenreich, P.A., Trogdon, J.G., Khavjou, O.A. et al. (2011). Forecasting the future of cardiovascular disease in the United States: a policy statement from the American Heart Association. *Circulation* 123 (8): 933–944.
Hernandez-Ledesma, B., Hsieh, C., and de Lumen, B.O. (2013). Chemopreventive properties of peptide Lunasin: a review. *Protein and Peptide Letters* 20 (4): 424–432.
Hu, L., Song, R., and Gu, Z. (2012). An antioxidant peptide produced by autolysis reactions from wheat germ. *African Journal of Biotechnology* 11 (15): 6340–6348.
Huang, L., Liu, B., Ma, H., and Zhang, X. (2014). Combined effect of ultrasound and enzymatic treatments on production of ACE inhibitory peptides from wheat germ protein. *Journal of Food Processing and Preservation* 38 (4): 1632–1640.
Huang, W.-H., Sun, J., He, H. et al. (2011). Antihypertensive effect of corn peptides, produced by a continuous production in enzymatic membrane reactor, in spontaneously hypertensive rats. *Food Chemistry* 128 (4): 968–973.
Huby, R.D., Dearman, R.J., and Kimber, I. (2000). Why are some proteins allergens? *Toxicological Sciences* 55 (2): 235–246.
Hübner, F., O'Neil, T., Cashman, K.D., and Arendt, E.K. (2010). The influence of germination conditions on beta-glucan, dietary fibre and phytate during the germination of oats and barley. *European Food Research and Technology* 231 (1): 27–35.
Jeong, H.J., Jeong, J.B., Kim, D.S. et al. (2007). The cancer preventive peptide lunasin from wheat inhibits core histone acetylation. *Cancer Letters* 255 (1): 42–48.
Jeong, H.J., Lam, Y., and de Lumen, B.O. (2002). Barley lunasin suppresses ras-induced colony formation and inhibits core histone acetylation in mammalian cells. *Journal of Agricultural and Food Chemistry* 50 (21): 5903–5908.
Jeong, H.J., Lee, J.R., Jeong, J.B. et al. (2009). The cancer preventive seed peptide lunasin from rye is bioavailable and bioactive. *Nutrition and Cancer* 61 (5): 680–686.
Jha, R., Rossnagel, B., Pieper, R. et al. (2010). Barley and oat cultivars with diverse carbohydrate composition alter ileal and total tract nutrient digestibility and fermentation metabolites in weaned piglets. *Animal* 4 (5): 724–731.
Kannan, A., Hettiarachchy, N.S., Lay, J.O., and Liyanage, R. (2010). Human cancer cell proliferation inhibition by a pentapeptide isolated and characterized from rice bran. *Peptides* 31 (9): 1629–1634.
Kaukovirta-Norja, A., Wilhelmson, A., and Poutanen, K. (2004). Germination: a means to improve the functionality of oat. *Agricultural and Food Science* 13: 100–112.
Kaur, S., Sharma, S., Dar, B., and Singh, B. (2012). Optimization of process for reduction of antinutritional factors in edible cereal brans. *Revista de Agroquimica y Tecnologia de Alimentos* 18 (5): 445–454.
Keyzer, M.A., Merbis, M.D., and Voortman, R.L. (2008). The biofuel controversy. *Economist* 156 (4): 507–527.
Khush, G.S. (2005). What it will take to feed 5.0 billion rice consumers in 2030. *Plant Molecular Biology* 59 (1): 1–6.
Klose, C. and Arendt, E.K. (2012). Proteins in oats; their synthesis and changes during germination: a review. *Critical Reviews in Food Science and Nutrition* 52 (7): 629–639.
Korhonen, H. and Pihlanto, A. (2006). Bioactive peptides: production and functionality. *International Dairy Journal* 16: 945–960.

Lafarga, T., O'Connor, P., and Hayes, M. (2014). Identification of novel dipeptidyl peptidase-IV and angiotensin-I-converting enzyme inhibitory peptides from meat proteins using in silico analysis. *Peptides* 59: 53–62.

Lawton, J.W. (2002). Zein: a history of processing and use. *Cereal Chemistry* 79 (1): 1–18.

Li, G.H., Qu, M.R., Wan, J.Z., and You, J.M. (2007). Antihypertensive effect of rice protein hydrolysate with in vitro angiotensin I-converting enzyme inhibitory activity in spontaneously hypertensive rats. *Asia Pacific Journal of Clinical Nutrition* 16 (Suppl 1): 275–280.

Li, J.-T., Zhang, J.-L., He, H. et al. (2013). Apoptosis in human hepatoma HepG2 cells induced by corn peptides and its anti-tumor efficacy in H22 tumor bearing mice. *Food and Chemical Toxicology* 51: 297–305.

Lin, F., Chen, L., Liang, R. et al. (2011). Pilot-scale production of low molecular weight peptides from corn wet milling byproducts and the antihypertensive effects in vivo and in vitro. *Food Chemistry* 124 (3): 801–807.

Ludvigsson, J.F., Leffler, D.A. et al. (2013). The Oslo definitions for coeliac disease and related terms. *Gut* 62 (1): 43–52.

Maldonado-Cervantes, E., Jeong, H.J., León-Galván, F. et al. (2010). Amaranth lunasin-like peptide internalizes into the cell nucleus and inhibits chemical carcinogen-induced transformation of NIH-3T3 cells. *Peptides* 31 (9): 1635–1642.

Maleki, S.J., Chung, S.-Y., Champagne, E.T., and Raufman, J.-P. (2000). The effects of roasting on the allergenic properties of peanut proteins. *Journal of Allergy and Clinical Immunology* 106 (4): 763–768.

Maritim, A., Sanders, A., and Watkins, J. III (2003). Diabetes, oxidative stress, and antioxidants: a review. *Journal of Biochemical and Molecular Toxicology* 17 (1): 24–38.

Matsui, T., Li, C.-H., Tanaka, T. et al. (2000). Depressor effect of wheat germ hydrolysate and its novel angiotensin I-converting enzyme inhibitory peptide, Ile-Val-Tyr, and the metabolism in rat and human plasma. *Biological and Pharmaceutical Bulletin* 23 (4): 427–431.

Matsui, T., Li, C.H., and Osajima, Y. (1999). Preparation and characterization of novel bioactive peptides responsible for angiotensin I-converting enzyme inhibition from wheat germ. *Journal of Peptide Science* 5 (7): 289–297.

Mentlein, R. (1999). Dipeptidyl-peptidase IV (CD26)-role in the inactivation of regulatory peptides. *Regulatory Peptides* 85 (1): 9–24.

Miyoshi, S., Kaneko, T., Ishikawa, H. et al. (1995). Production of bioactive peptides from corn endosperm proteins by some proteases. *Annals of the New York Academy of Sciences* 750 (1): 429–431.

Mochida, T., Hira, T., and Hara, H. (2010). The corn protein, zein hydrolysate, administered into the ileum attenuates hyperglycemia via its dual action on glucagon-like peptide-1 secretion and dipeptidyl peptidase-IV activity in rats. *Endocrinology* 151 (7): 3095–3104.

Mochizuki, M., Shigemura, H., and Hasegawa, N. (2010). Anti-inflammatory effect of enzymatic hydrolysate of corn gluten in an experimental model of colitis. *Journal of Pharmacy and Pharmacology* 62 (3): 389–392.

Molina, J., Sikora, M., Garud, N. et al. (2011). Molecular evidence for a single evolutionary origin of domesticated rice. *Proceedings of the National Academy of Sciences* 108 (20): 8351–8356.

Morita, E., Matsuo, H., Mihara, S. et al. (2003). Fast ω-gliadin is a major allergen in wheat-dependent exercise-induced anaphylaxis. *Journal of Dermatological Science* 33 (2): 99–104.
Motoi, H. and Kodama, T. (2003). Isolation and characterization of angiotensin I-converting enzyme inhibitory peptides from wheat gliadin hydrolysate. *Molecular Nutrition and Food Research* 47 (5): 354–358.
Nagashima, M., Watanabe, T., Terasaki, M. et al. (2011). Native incretins prevent the development of atherosclerotic lesions in apolipoprotein E knockout mice. *Diabetologia* 54 (10): 2649–2659.
Nakurte, I., Kirhnere, I., Namniece, J. et al. (2013). Detection of the lunasin peptide in oats (*Avena sativa* L). *Journal of Cereal Science* 57 (3): 319–324.
Nakurte, I., Klavins, K., Kirhnere, I. et al. (2012). Discovery of lunasin peptide in triticale (X Triticosecale Wittmack). *Journal of Cereal Science* 56 (2): 510–514.
Newman, C. and Newman, R. (2006). A brief history of barley foods. *Cereal Foods World* 51 (1): 4–7.
Nikiforova, V., Kempa, S., Zeh, M. et al. (2002). Engineering of cysteine and methionine biosynthesis in potato. *Amino Acids* 22 (3): 259–278.
Nongonierma, A.B. and FitzGerald, R.J. (2014). An in silico model to predict the potential of dietary proteins as sources of dipeptidyl peptidase IV (DPP-IV) inhibitory peptides. *Food Chemistry* 165: 489–498.
Ogo, Y., Wakasa, Y., Hirano, K. et al. (2014). Generation of transgenic rice with reduced content of major and novel high molecular weight allergens. *Rice* 7 (1): 19.
Ondetti, M.A. and Cushman, D.W. (1982). Enzymes of the renin-angiotensin system and their inhibitors. *Annual Review of Biochemistry* 51: 283–308.
Ortiz-Martinez, M., de Mejia, E.G., García-Lara, S. et al. (2017). Antiproliferative effect of peptide fractions isolated from a quality protein maize, a white hybrid maize, and their derived peptides on hepatocarcinoma human HepG2 cells. *Journal of Functional Foods* 34: 36–48.
Othman, R.A., Moghadasian, M.H., and Jones, P.J. (2011). Cholesterol-lowering effects of oat β-glucan. *Nutrition Reviews* 69 (6): 299–309.
Palosuo, K., Alenius, H., Varjonen, E. et al. (2001). Rye γ-70 and γ-35 secalins and barley γ-3 hordein cross-react with ω-5 gliadin, a major allergen in wheat-dependent, exercise-induced anaphylaxis. *Clinical and Experimental Allergy* 31 (3): 466–473.
Palosuo, K., Alenius, H., Varjonen, E. et al. (1999). A novel wheat gliadin as a cause of exercise-induced anaphylaxis. *Journal of Allergy and Clinical Immunology* 103 (5): 912–917.
Pastorello, E.A., Farioli, L., Pravettoni, V. et al. (2000). The maize major allergen, which is responsible for food-induced allergic reactions, is a lipid transfer protein. *Journal of Allergy and Clinical Immunology* 106 (4): 744–751.
Pastorello, E.A., Pompei, C., Pravettoni, V. et al. (2003). Lipid-transfer protein is the major maize allergen maintaining IgE-binding activity after cooking at 100 C, as demonstrated in anaphylactic patients and patients with positive double-blind, placebo-controlled food challenge results. *Journal of Allergy and Clinical Immunology* 112 (4): 775–783.
Pihlanto, A., Akkanen, S., and Korhonen, H.J. (2008). ACE-inhibitory and antioxidant properties of potato (*Solanum tuberosum*). *Food Chemistry* 109 (1): 104–112.
Puchalska, P., Marina, M.L., and García, M.C. (2013). Development of a high-performance liquid chromatography–electrospray ionization-quadrupole-time-of-flight-mass

spectrometry methodology for the determination of three highly antihypertensive peptides in maize crops. *Journal of Chromatography A* 1285: 69–77.

Qu, W., Ma, H., Zhao, W., and Pan, Z. (2013). ACE-inhibitory peptides production from defatted wheat germ protein by continuous coupling of enzymatic hydrolysis and membrane separation: modeling and experimental studies. *Chemical Engineering Journal* 226: 139–145.

Rahman, M.A., Hasegawa, H., Rahman, M.M. et al. (2007). Accumulation of arsenic in tissues of rice plant (*Oryza sativa* L.) and its distribution in fractions of rice grain. *Chemosphere* 69 (6): 942–948.

Rho, S.J., Lee, J.S., Chung, Y.I. et al. (2009). Purification and identification of an angiotensin I-converting enzyme inhibitory peptide from fermented soybean extract. *Process Biochemistry* 44: 490.

Rosenfelder-Kuon, P., Strang, E., Spindler, H. et al. (2017). Ileal starch digestibility of different cereal grains fed to growing pigs. *Journal of Animal Science* 95 (6): 2711–2717.

Rostom, A., Murray, J.A., and Kagnoff, M.F. (2006). American Gastroenterological Association (AGA) Institute technical review on the diagnosis and management of celiac disease. *Gastroenterology* 131 (6): 1981–2002.

Rutherfurd, S.M., Fanning, A.C., Miller, B.J., and Moughan, P.J. (2015). Protein digestibility-corrected amino acid scores and digestible indispensable amino acid scores differentially describe protein quality in growing male rats. *Journal of Nutrition* 145 (2): 372–379.

Sadiq Butt, M., Tahir-Nadeem, M., Khan, M.K. et al. (2008). Oat: unique among the cereals. *European Journal of Nutrition* 47 (2): 68–79.

Salcedo, G., Sanchez-Monge, R., Garcia-Casado, G. et al. (2004). The cereal α-amylase/trypsin inhibitor family associated with baker's asthma and food allergy. In: *Plant Food Allergens* (ed. C. Mills and P.R. Shewry), 70–84. Oxford: Blackwell.

Sarwar, G. (1997). The protein digestibility–corrected amino acid score method overestimates quality of proteins containing antinutritional factors and of poorly digestible proteins supplemented with limiting amino acids in rats. *Journal of Nutrition* 127 (5): 758–764.

Sato, K., Egashira, Y., Ono, S. et al. (2013). Identification of a hepatoprotective peptide in wheat gluten hydrolysate against D-galactosamine-induced acute hepatitis in rats. *Journal of Agricultural and Food Chemistry* 61 (26): 6304–6310.

Shaw, J.E., Sicree, R.A., and Zimmet, P.Z. (2010). Global estimates of the prevalence of diabetes for 2010 and 2030. *Diabetes Research and Clinical Practice* 87 (1): 4–14.

Shewry, P.R. (2007). Improving the protein content and composition of cereal grain. *Journal of Cereal Science* 46 (3): 239–250.

Shewry, P.R. and Halford, N.G. (2002). Cereal seed storage proteins: structures, properties and role in grain utilization. *Journal of Experimental Botany* 53 (370): 947–958.

Sica, D.A. and Gehr, T. (2005). Angiotensin-converting enzyme inhibitors. In: Hypertension. In: , 669–682. Philadelphia: Elsevier-Saunders.

Storsrud, S., Olsson, M., Lenner, R.A., and Nilsson, L.Å. (2003). Adult coeliac patients do tolerate large amounts of oats. *European Journal of Clinical Nutrition* 57 (1): 163.

Suarez-Jimenez, G.-M., Burgos-Hernandez, A., and Ezquerra-Brauer, J.-M. (2012). Bioactive peptides and depsipeptides with anticancer potential: sources from marine animals. *Marine Drugs* 10 (5): 963–986.

Suttie, J. and Reynolds, S. (2004). *Fodder Oats: A World Overview*. Rome: Food and Agriculture Organisation of the United Nations.

Tanno, K.-I. and Willcox, G. (2006). How fast was wild wheat domesticated? *Science* 311 (5769): 1886–1886.
Tatham, A. and Shewry, P. (2008). Allergens to wheat and related cereals. *Clinical and Experimental Allergy* 38 (11): 1712–1726.
USDA (2016) Foreign Agricultural Service, Official USDA Estimates. Retrieved from: www.fas.usda.gov
Velarde-Salcedo, A.J., Barrera-Pacheco, A., Lara-González, S. et al. (2013). In vitro inhibition of dipeptidyl peptidase IV by peptides derived from the hydrolysis of amaranth (*Amaranthus hypochondriacus* L.) proteins. *Food Chemistry* 136 (2): 758–764.
Wang, F., Yu, G., Zhang, Y. et al. (2015). Dipeptidyl peptidase IV inhibitory peptides derived from oat (*Avena sativa* L.), buckwheat (*Fagopyrum esculentum*), and highland barley (Hordeum vulgare trifurcatum (L.) trofim) proteins. *Journal of Agricultural and Food Chemistry* 63 (43): 9543–9549.
Wang, H., Nussbaum-Wagler, T., Li, B. et al. (2005). The origin of the naked grains of maize. *Nature* 436 (7051): 714.
Wang, X.-J., Zheng, X.-Q., Kopparapu, N.-K. et al. (2014). Purification and evaluation of a novel antioxidant peptide from corn protein hydrolysate. *Process Biochemistry* 49 (9): 1562–1569.
Wattanasiritham, L., Theerakulkait, C., Wickramasekara, S. et al. (2016). Isolation and identification of antioxidant peptides from enzymatically hydrolyzed rice bran protein. *Food Chemistry* 192: 156–162.
WHO (2009). *Global Health Risks: Mortality and Burden of Disease Attributable to Selected Major Risks.* Geneva: World Health Organization.
Wu, G. (2013). Functional amino acids in nutrition and health. *Amino Acids* 45 (3): 407–411.
Xia, Y., Bamdad, F., Gänzle, M., and Chen, L. (2012). Fractionation and characterization of antioxidant peptides derived from barley glutelin by enzymatic hydrolysis. *Food Chemistry* 134 (3): 1509–1518.
Yang, R., Zou, Y., Yu, N., and Gu, Z. (2011). Accumulation and identification of angiotensin-converting enzyme inhibitory peptides from wheat germ. *Journal of Agricultural and Food Chemistry* 59 (8): 3598–3605.
Yang, Y., Tao, G., Liu, P., and Liu, J. (2007). Peptide with angiotensin I-converting enzyme inhibitory activity from hydrolyzed corn gluten meal. *Journal of Agricultural and Food Chemistry* 55 (19): 7891–7895.
Yano, S., Suzuki, K., and Funatsu, G. (1996). Isolation from α-zein of thermolysin peptides with angiotensin I-converting enzyme inhibitory activity. *Bioscience, Biotechnology, and Biochemistry* 60 (4): 661–663.
Young, V.R. and Borgonha, S. (2000). Nitrogen and amino acid requirements: the Massachusetts Institute of Technology amino acid requirement pattern. *Journal of Nutrition* 130 (7): 1841S–1849S.
Zhang, J., Zhang, H., Wang, L. et al. (2010). Isolation and identification of antioxidative peptides from rice endosperm protein enzymatic hydrolysate by consecutive chromatography and MALDI-TOF/TOF MS/MS. *Food Chemistry* 119 (1): 226–234.
Zhang, S.B., Wang, Z., and Xu, S.Y. (2008). Antioxidant and antithrombotic activities of rapeseed peptides. *Journal of the American Oil Chemists' Society* 85 (6): 521–527.
Zhou, C., Hu, J., Ma, H. et al. (2015). Antioxidant peptides from corn gluten meal: orthogonal design evaluation. *Food Chemistry* 187: 270–278.

Zhu, F.C., Jia, R.Z., Xu, L. et al. (2015). Genome-wide analysis of potential cross-reactive endogenous allergens in rice (*Oryza sativa* L.). *Toxicology Reports* 2: 1233–1245.

Zhu, K., Zhou, H., and Qian, H. (2006). Antioxidant and free radical-scavenging activities of wheat germ protein hydrolysates (WGPH) prepared with alcalase. *Process Biochemistry* 41 (6): 1296–1302.

Zhuang, H., Tang, N., and Yuan, Y. (2013). Purification and identification of antioxidant peptides from corn gluten meal. *Journal of Functional Foods* 5 (4): 1810–1821.

Zorovski, P. and Georgieva, T. (2011) Essential amino acid contents of winter and spring oat cultivars (*Avena sativa* L.) grown in central South Bulgaria. Climate change: challenges and opportunities in agriculture. AGRISAFE Final Conference, 21–23 March 2011, Budapest, Hungary. Proceedings: Agricultural Research Institute of the Hungarian Academy of Sciences, pp. 481–484.

5

Meat By-Products: New Insights into Potential Technical and Health Applications

Leticia Mora[1], Fidel Toldrá-Reig[2], Milagro Reig[3], and Fidel Toldrá[1]

[1] *Instituto de Agroquímica y Tecnología de Alimentos (CSIC), Valencia, Spain*
[2] *Instituto de Tecnología Química (CSIC-UPV), Valencia, Spain*
[3] *Universidad Politécnica de Valencia, Instituto de Ingeniería de Alimentos para el Desarrollo, Valencia, Spain*

5.1 Introduction

The meat industry generates more than 100 million tonnes per year of protein-containing wastes worldwide. Such large amounts of waste constitute a great problem for the meat industry because of the strict regulations concerning their disposal and high costs of treatment. A large part of the cost can be reduced through appropriate use of these wastes or by-products and by the creation of of added-value products (Toldrá et al. 2012; Mora et al. 2014). There is a wide variety of animal by-products including blood, bones, tendons, meat trimmings, fat or lard, hides, hooves, horns, internal organs and viscera, ears, and feet. Edible by-products can be consumed as a food (Ockerman and Hansen 1988) because of their high nutritional quality, which provides good added value to such by-products (Valta et al. 2015). In view of their high nutritional quality to cost ratio, edible by-products are more usual in countries with low incomes although they can be found in other countries where they form part of highly valued and delicious recipes (Toldrá et al. 2016; Ockerman et al. 2017). They can represent 11.4% and 7.5% of the gross income generated from beef and pork, respectively (Jayathilakan et al. 2012).

More recent applications of animal by-products include the generation of biomolecules of interest with a particular enzyme (Lasekan et al. 2013), a determined functional property (Chernukha et al. 2015) or bioactive peptides (Mora et al. 2014; Martínez-Alvarez et al. 2015).

There are other classic applications of these by-products, such as use as fertilisers, feeds and pet foods, but there are now new and innovative applications in the chemical or pharmaceutical industry or even energy generation in the form of biodiesel produced from waste animal fats (Banković-Ilić et al. 2014; Adewale et al. 2016) or bioderived diesel fuel, also known as bio gas oil (Baladincz and Hancsók 2015).

This chapter reviews the latest developments in the profitability of meat industry by-products (Figure 5.1).

Novel Proteins for Food, Pharmaceuticals, and Agriculture: Sources, Applications, and Advances, First Edition. Edited by Maria Hayes.

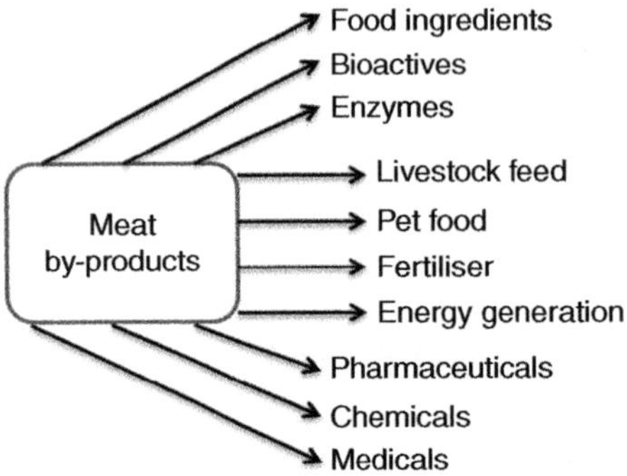

Figure 5.1 Main uses and applications of meat by-products. *Source:* Reproduced from Toldrá, F., Mora, L. & Reig, M. New insights into meat by-products utilization. Meat Science, 2016, 120, 54–59 with permission from Elsevier.

5.2 Meat By-Products

Edible meat by-products such as liver, lung, heart, kidney, brains, spleen, and blood contain nutrients in significant amounts, including essential amino acids (Aristoy and Toldrá 2011), minerals, and vitamins (Honikel 2011; Kim 2011). Such products are highly appreciated in some countries because of their cultural heritage while in other countries, they are considered as wastes with no consideration for use as a food (Nollet and Toldrá 2011). Consumption reduces the amount of waste and their high nutritional value makes them attractive for those populations with low incomes (Toldrá et al. 2016; Ockerman et al. 2017). In addition, some meat by-products are used as ingredients in other foods, such as liver, cheek, and tongue which are used for sausage production; mechanically separated meat which can be used in comminuted meat products and sausages (Pearl 2014); and cooked cracked bones which are used in soups, vegetable dishes, sauces, or gravies for nutritional and flavour purposes (Ockerman and Basu 2014). Edible by-products must be adequately preserved because they are prone to microbial contamination (Ockerman and Basu 2004a, 2014).

5.3 Technical Applications of Meat By-Products

Figure 5.2 shows the main routes of application of animal by-products. Recent strategies are addressed concerning the production of foods, feeds, pet foods, functional ingredients, and processing aids from edible by-products and added-value products from inedible by-products (e.g. plastics, pharmaceuticals, energy) (Ockerman and Basu 2004b).

5.3.1 Use of Meat By-Products as Food Ingredients and Processing Aids

Meat by-products can be used for the generation of new, added-value functional ingredients and processing aids (Zhang et al. 2010; Toldrá and Reig 2011). Such new substances may be used for improved sensory quality (colour, flavour, or texture) by

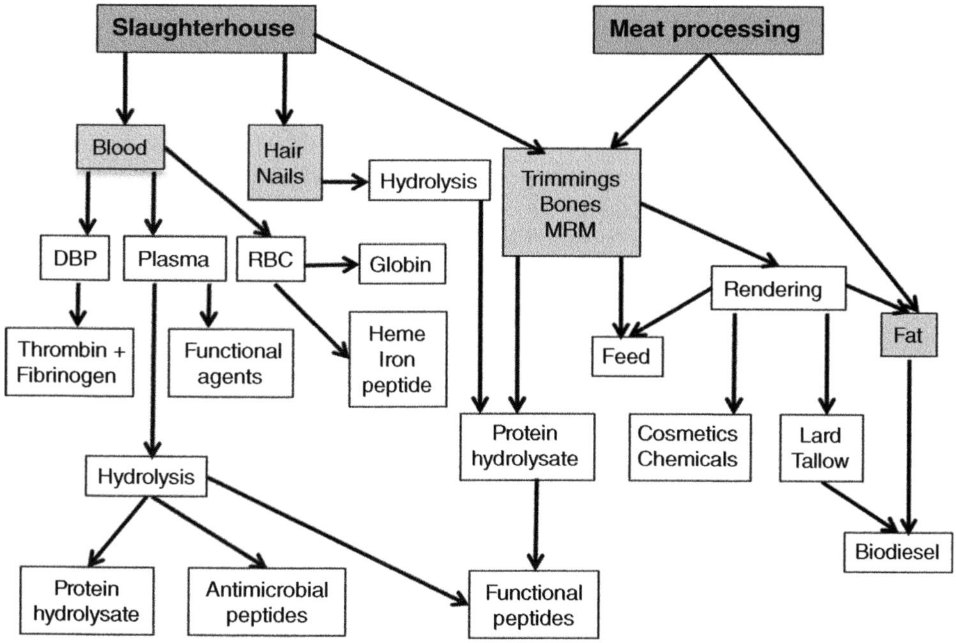

Figure 5.2 Flow diagram of main routes for value-addition to meat by-products. DBP, dried blood protein MRM, mechanically recovered meat; RBC, red blood cells. *Source:* Reproduced from Toldrá, F., Aristoy, M.C., Mora, L. & Reig, M. Innovations in value-addition of edible meat by-products. Meat Science, 2012, 92, 290–296 with permission from Elsevier.

increasing water bonding, protein cross-linking, gelation, foaming, and emulsification, flavour enhancement, and protein enrichment (Toldrá et al. 2012).

Collagen is the major constituent in connective tissues, representing up to 30% of total animal proteins (Ferraro et al. 2016), and is quite abundant in meat by-products, reaching up to 75% of the skin (dw) and 90% of tendons and corneal tissues. There are 29 different types of collagen, types I and III being the most abundant forms (Liu et al. 2015). Collagen can also be obtained from bones, lungs, trachea, tongue, and blood vessels but collagen type I is mainly obtained from bovine and porcine skin and tendon tissues. Gelatine is obtained through the controlled hydrolysis of collagen and has many applications in foods, including gel formation in soups, gravies, and desserts; clarification and stabilisation; formation of protective film for packaging (Djagny et al. 2001; Gómez-Guillén et al. 2011); and protective coating material in bakery, dairy, and meat proteins (Djagny et al. 2001; Ockerman and Basu 2006).

A second group of proteins are those from blood that offer a variety of technological properties of interest for food processing (Hsieh and Ofori 2011). Blood is usually centrifuged to separate plasma from cellular fractions. Some plasma proteins facilitate the cross-linking of proteins (Kang and Lanier 1999), protein enrichment (Yousif et al. 2003), or foaming (Del et al. 2008). Immunoglobulins, fibrinogen, and serum albumin contribute to gelation and emulsification (Cofrades et al. 2000). Red blood cell fractions have been reported to have high antioxidant activity (Bah et al. 2016) while sheep white blood cells have been shown to have antimicrobial activity against *Escherichia coli*,

Staphylococcus aureus, and *Pseudomonas aeruginosa* (Bah et al. 2016). Further, red blood cells, white blood cells, and platelets can be used for colour enhancement although it is not so convenient in practice due to the dark colour of haemoglobin, as well as some negative sensory effects unless haemoglobin is removed (Ofori and Hsieh 2011). Haemoglobin can be hydrolysed and the resulting polypeptides, known as heme iron peptides, can be used for better iron absorption (Nissenson et al. 2003). Another potential application is the partial replacement of fat in meat products (Viana et al. 2005).

Thrombin and fibrinogen are commercially obtained from bovine and porcine blood and used to reconstitute meat steaks or fish pieces through the binding of meat pieces after formation of a three-dimensional fibrin clot network (Lennon et al. 2010; Chen and Lin 2002).

Foams and surfactants may be obtained through rendering (Bressler 2009). Protein hydrolysates can be used as flavourings or as substrates for Maillard reactions, reducing sugars to produce flavourings (Baiano 2014; Toldrá et al. 2016). However, caution must be exercised because the sensory properties depend on the balance and content of small peptides and free amino acids (Maehashi et al. 1999).

5.3.2 Use of Meat By-Products as Feed and Pet Food

Nearly 25 million tonnes per year of animal by-products in the US and 15 million tonnes in the European Union are rendered in order to obtain high-quality fats and proteins (Hamilton 2016) that can be used as ingredients in feeds and pet foods (Murray et al. 1997).

Animal by-products are traditionally used as raw materials for the production of feed and ingredients for pet food because they contain relevant nutrients including essential amino acids, fatty acids, minerals, and trace elements, B-vitamins and some fat-soluble vitamins (Honikel 2011; Nollet and Toldrá 2011). This is the case with blood meal (Alexis and Robert 2004; Pérez-Gálvez et al. 2011), amino acid solutions obtained from blood hydrolysed with specific microbial cultures (Giu and Giu 2010); rendered meat by-products (Murray et al. 1997); bone meal ashes obtained after co-incineration (Coutand et al. 2008); enzymatically hydrolysed heart or lung (Martínez-Alvarez et al. 2015); and meat and bone meal that contains essential amino acids and B-vitamins (Jayathilakan et al. 2012).

Protein hydrolysates are used for feed, pet food, and in aquaculture (Gilbert et al. 2008). Protein hydrolysates obtained from meat by-products contain large amounts of all essential amino acids, within the range of 20–60 $mg\,g^{-1}$ protein (Aristoy and Toldrá 2011), and constitute an interesting alternative to soya meal because of the absence of antinutritional factors or allergenic proteins (Martínez-Alvarez et al. 2015). Hydrolysis increases their utilisation by animals (Piazza and García 2014) but their excessive bitterness impairs sensory acceptability, making it necessary to cleave hydrophobic amino acids from peptides to obtain better palatability (Nchienzia et al. 2010).

Other animal by-products such as hair, nails, feather, and the outer layer of skin are difficult to hydrolyse with the usual peptidases because they contain the fibrous structural protein keratin, requiring hydrolysis with the enzyme keratinase (Deivasigamani and Alagappan 2008; Lasekan et al. 2013). Keratinase is a serine peptidase active in the pH range 6.0–13.0, and under reducing conditions (Brandelli et al. 2015).

5.3.3 Use of Meat By-Products as Fertilisers

Meat and bone meal are typically used to produce ashes for fertilisers with a high phosphorus content. Thermochemical processing is the most common incineration and consists of pyrolysis, combustion, and gasification (Coutand et al. 2008). It contributes to energy production, the heat recovery allows for efficient energy usage (Bujak 2015), and gas emissions comply with international regulations (Coutand et al. 2008; Cascarosa et al. 2012).

5.3.4 Use of Meat By-Products as Plastics and Leather Products

Collagen, gelatine, and glycerine have applications as ingredients for surfactants, paints, varnishes, adhesives, antifreeze, cleaners, polishes, and pharmaceuticals (Pearl 2014). Tallow and lard by-products can generate fatty acids to be used in many chemical processes including rubber, plastic, softeners, and lubricants. Polyhydroxyalkanoate is a biodegradable polymer with plastic-like properties that is produced from rendered fats by a recombinant strain of *Ralstonia eutropha* (Riedel et al. 2015). Collagen and elastin are used as biomaterials for the formation of biodegradable packaging (Ferraro et al. 2016). Shoes, handbags, purses, gloves, leather clothes and belts, and other leather-derived products are obtained from cattle hides (Ockerman and Basu 2004b).

5.3.5 Use of Meat By-Products as an Energy Source

Lard and tallow are used to produce biodiesel that is an alternative to diesel fuel without requiring engine modifications. Fats need to be transesterified with a low molecular weight alcohol in order to obtain a mixture of fatty acid methyl esters and glycerol as a side product (Bhatti et al. 2008; Moreira et al. 2010). Improvements include hydro-oxygenation and hydroisomerisation (Herskowitz 2008), supercritical transesterification (Marulanda et al. 2010), and ultrasound-assisted transesterification (Adewale et al. 2016). Biodiesel has some advantages as it is biodegradable, non-toxic and has a good combustion emission profile. This allows for a significant reduction in carbon dioxide, carbon monoxide, particulate matter, and unburned hydrocarbons (Gerpen 2005; Moreira et al. 2010). However, due to the content of protein and phosphoacylglycerols, a degumming process is required, along with a reduction in water content through vacuum drying and a reduction in the content of saturated fatty acids through a winterization process or the use of additives (Banković-Ilić et al. 2014). Recently, biogas oil was developed using heterogeneous catalytic hydrogenation which converts triacylglycerols into a mixture of iso and normal paraffins (Baladincz and Hancsók 2015).

5.4 Health-Related Applications of Meat By-Products

5.4.1 Bioactive Peptides

Bioactive peptides contribute to improved biological function in human physiological systems. For instance, hypocholesterolaemic, antioxidant, and antithrombotic peptides have been reported to contribute to modulation of the cardiovascular system, mineral-binding peptides contribute to the gastrointestinal system, and immunomodulatory peptides act in the immune system. Bioactive peptides are also able to participate in multiple systems through the action of antimicrobial peptides (Lafarga and Hayes 2014).

The sequences of bioactive peptides usually contain between two and 30 amino acids and are usually located in an inactive form within the parent protein. Bioactive peptides are released from such proteins by peptidases during food processing or gastrointestinal digestion (Mora et al. 2015). In many cases, proteins are hydrolysed by commercial peptidases such as alcalase, neutrase, pronase, pepsin, thermolysin, and trypsin under controlled conditions in order to generate bioactive peptides and control homogeneity between batches (Vercruysse et al. 2005; Toldrá et al. 2016). A flow diagram of enzymatic hydrolysis of edible meat by-products for the generation of bioactive peptides is shown in Figure 5.3. The digestion conditions (temperature, pH, time) and degree of hydrolysis determine the amount and profile of the bioactive peptides obtained. The size and sequence of the obtained peptides are crucial for a given bioactivity (Yu et al. 2006). Examples of commercial peptidases and their main characteristics and origin are given in Table 5.1.

There are many studies focused on bioactive peptides derived from meat by-products, especially from blood and collagen (Ryder et al. 2016) but also from meat trimmings (Ryder et al. 2015). Blood constitutes a rich source of bioactive peptides generated from both the cellular fraction (haemoglobin cells) and plasma fraction proteins (Ofori and Hsieh 2012). Relevant ACE-1 inhibitory (Chang et al. 2007; Adje et al. 2011) and opioid activities (Zhao et al. 1997, 1994; Kapel et al. 2003; Froidevaux et al. 2008) have been reported. Furthermore, many antimicrobial peptides have been obtained from haemoglobin hydrolysates (Nedjar-Arroume et al. 2006, 2008). More specifically, the hydrolysis of bovine haemoglobin with pepsin resulted in the isolation of 67–106, 73–105, 99–105, and 100–105 fragments with antibacterial activity against *Kocuria luteus A270, Listeria innocua, E. coli,* and *S. aureus* respectively, with a minimum inhibitory concentration (MIC) between 187.1 and 35.2 μM. Another fragment, VNFKLLSHSLLVTLASHL, exerted antimicrobial activity against

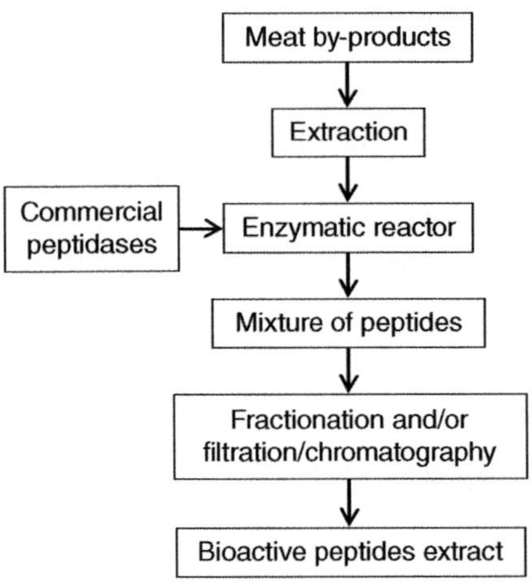

Figure 5.3 Flow diagram of the enzymatic hydrolysis of edible meat by-products for the generation of bioactive peptides. *Source:* Reproduced from Toldrá, F., Aristoy, M.C., Mora, L. & Reig, M. Innovations in value-addition of edible meat by-products. Meat Science, 2012, 92, 290–296 with permission from Elsevier.

Table 5.1 Main characteristics and conditions of use of enzymes typically used for protein hydrolysis of meat by-products.

Enzyme name	Temp[a)]	pH[a)]	Type		Origin
Papain	60–70 °C	pH 6–7	Cysteine protease	Exo- and endopeptidase	Papaya fruit
Bromelain	35–45 °C	pH 7	Sulfhydryl protease	Exo- and endopeptidase	Pineapple fruit
Thermolysin	65–85 °C	pH 5–8.5	Metalloprotease	Exopeptidase	*Bacillus thermoproteolyticus*
Pronase	40–60 °C	pH 7.5	Non-specific protease	Exo- and endopeptidase	*Streptomyces griseus*
Proteinase K	37 to 50–60 °C	pH 4–12	Serine protease	Exo- and endopeptidase	*Engyodontium album*
Neutrase	50 °C	pH 7	Metalloproteinase	Endopeptidase	*Bacillus amyloliquefaciens*
Alcalase	50 °C	pH 8	Serine proteinase	Endo-eptidase	*Bacillus licheniformis*
Crude enzyme extract	40 °C	pH 8	Non-specific proteases	Exo- and endopeptidase	*Raja clavata*

a) Optimum conditions for activity.
Source: Reproduced from Mora, L., Reig, M. & Toldrá, F. Bioactive peptides generated from meat industry by-products. Food Research International, 2014, 65, 344–349 with permission from Elsevier.

Table 5.2 Bioactive peptide sequences isolated and identified after enzyme hydrolysis of haemoglobin or plasma.

Source	Enzyme	Peptide sequences	IC_{50}	Activity	References
Haemoglobin alpha-chain	Pepsin	LDDLPGALSELSDLHAHKLRVDPVNFKLLSHSL	518.29 μM	ACE inhibitor	Adje et al. (2011)
Haemoglobin alpha-chain	Pepsin	KLLSHSL	42.51 μM	ACE inhibitor	Adje et al. (2011)
Haemoglobin alpha-chain	Pepsin	LLSHSL	1095.5 μM	ACE inhibitor	Adje et al. (2011)
Plasma	Trypsin	VYNEGLPAP	3.1 μM	ACE inhibitor	Liu et al. (2010)
Porcine haemoglobin	Pepsin	LGFPTTKTYFPHF	4.92 μM	ACE inhibitor	Yu et al. (2006)
Porcine haemoglobin	Pepsin	VVYPWT	6.02 μM	ACE inhibitor	Yu et al. (2006)
Porcine haemoglobin	Pepsin	WVPSV	$0.37\ mg\ ml^{-1}$	ACE inhibitor	Ren et al. (2011)
Porcine haemoglobin	Pepsin	YTVF	$0.23\ mg\ ml^{-1}$	ACE inhibitor	Ren et al. (2011)
Porcine haemoglobin	Pepsin	VVYPWT	$0.254\ mg\ ml^{-1}$	ACE inhibitor	Ren et al. (2011)
Porcine haemoglobin	Pepsin	QELPG	$0.02\ mg\ ml^{-1}$	ACE inhibitor	Deng et al. (2014)
Haemoglobin	Alcalase	FQKVVA	5.8 μM	ACE inhibitor	Mito et al. (1996)
Haemoglobin	Alcalase	FQKVVAG	7.4 μM	ACE inhibitor	Mito et al. (1996)
Haemoglobin	Alcalase	FQKVVAK	2.1 μM	ACE inhibitor	Mito et al. (1996)
Haemoglobin	Alcalase	GKKVLQ	1.9 μM	ACE inhibitor	Mito et al. (1996)
Plasma	TCA	GVHVV	2 μM	ACE inhibitor	Park et al. (1996)
Plasma	n.d.	LVL	$4.2\ \mu g\ mL^{-1}$	ACE inhibitor	Hazato and Kase (1986)
Haemoglobin alpha-chain	Pepsin	TKAVEHLDDLPGALSELSDLHAHKLRVDPVNFKLLSHSLL	366 μM	ACE inhibitor	Adje et al. (2011)

Haemoglobin	Pepsin	KYR	—	Antimicrobial	Catiau et al. (2011a)
Haemoglobin	Pepsin	RYH	—	Antimicrobial	Catiau et al. (2011b)
Haemoglobin	Pepsin	Alpha-107–136	—	Antibacterial	Daoud et al. (2005)
Haemoglobin alpha-chain	None	VNFKLLSHSLLVTLASHL	—	Antimicrobial	Hu et al. (2011)
Haemoglobin	Pepsin	QADFQKVVAGVANALAHRYH	—	Antimicrobial	Nedjar-Arroume et al. (2006)
Haemoglobin	Pepsin	YPWT	45.2 μM	Opioid	Zhao et al. (1997)
Haemoglobin	Pepsin	TYPWTQ	46.3 μM	Opioid	Zhao et al. (1997)
Haemoglobin	Pepsin	YPWTQR	4.3 μM	Opioid	Zhao et al. (1997)
Haemoglobin	Pepsin	YPWTQRF	2.9 μM	Opioid	Zhao et al. (1997)
Haemoglobin	Pepsin	YPWTQRFF	4.6 μM	Opioid	Zhao et al. (1997)
Haemoglobin	Pepsin	LVVYPWTQ	80.5 μM	Opioid	Zhao et al. (1997)
Haemoglobin	n.d.	LVVYPWTQRF	29.1 μM	Opioid	Glamsta et al. (1992)
Haemoglobin	Pepsin	VVYPWTQ	78.2 μM	Opioid	Zhao et al. (1994)
Haemoglobin	Pepsin	VVYPWTQRF	34.3 μM	Opioid	Zhao et al. (1994)
Plasma	Flavourcyme	VSGVEDVN	0.025 μM	Calcium binding	Lee and Song (2009a)
Plasma	Flavourcyme	DLGEQYFKG	—	Ion binding	Lee and Song (2009b)

ACE, angiotensin-converting enzyme; n.d.: non-determined.
Source: Reproduced from Toldrá, F., Mora, L. & Reig, M. New insights into meat by-products utilization. Meat Science, 2016, 120, 54–59 with permission from Elsevier.

E. coli, *S. aureus*, and *Candida albicans* (Hu et al. 2011). It was concluded that the sequences KYR and RYH were necessary for such antimicrobial activity (Catiau et al. 2011a,b). Such peptides are mostly hydrophobic due to their affinity with the outer membrane of microbials (Lopes et al. 2005).

The hydrolysis of collagen generates large amounts of bioactive peptides (Saiga et al. 2008; Herregods et al. 2011). Most of the published studies have focused on fisheries by-products. Hydrolysis with enzymes like alcalase, trypsin, chymotrypsin, neutrase, flavorenzyme, pepsin, bromelain, and papain generate ACE inhibitory and antioxidant peptides (Saiga et al. 2008; Gómez-Guillén et al. 2011; Herregods et al. 2011; di Bernardini et al. 2012) (Table 5.2).

5.4.2 Biomedical Applications of Meat By-Products

Meat processing by-products such as porcine skin can be applied medically to human skin for protection in cases of burns or skin ulcers (Jayathilakan et al. 2012). Collagen and elastin can be used as biomaterials for a variety of medical applications including skin substitutes, heart valves, elastic cartilage, and vascular grafts (Ferraro et al. 2016). Glands and organs, such as heparin from the liver, insulin from the pancreas, bile from the gall bladder, melatonin from the pineal gland, progesterone and oestrogen from ovaries are consumed for medicinal purposes in China, Japan, and India, or used as a source of pharmaceutical substances (Nollet and Toldrá 2011; Jayathilakan et al. 2012). Collagen also constitutes an excellent substrate for further hydrolysis into added-value bioactive peptides (Jridi et al. 2017). In fact, one of the most popular collagen hydrolysates is used for osteoarthritis in the regeneration of joint cartilages (Bello and Oesser 2006). Osteoarthritis in the joint cartilage may be alleviated with the consumption of collagen hydrolysates, usually commercialised with added magnesium and hyaluronic acid (Bello and Oesser 2006; Gómez-Guillén et al. 2011). Cosmetics applications such as hand and body lotions, creams, and bath products may also be obtained from rendered fat.

5.5 Conclusion

This chapter summarises the most common applications of edible and inedible meat by-products in a variety of fields such as food, feed, pet foods, fertilisers, medicine, pharmaceuticals, chemicals, and energy generation. The creation of added-value products is of primary importance because it allows for a significant reduction in the costs of treatment and disposal of such wastes and also helps to decrease environmental contamination. Research is therefore focused on creating added-value products as well as new industrial applications.

References

Adje, E.Y., Balti, R., Kouach, M. et al. (2011). α 67-106 of bovine hemoglobin: a new family of antimicrobial and angiotensin I-converting enzyme inhibitory peptides. *European Food Research and Technology* 232: 637–646.

Aristoy, M.C. and Toldrá, F. (2011). Essential amino acids. In: *Handbook of Analysis of Edible Animal By-Products* (ed. L.M.L. Nollet and F. Toldrá), 123–135. Boca Raton: CRC Press.

Adewale, P., Dumont, M.-J., and Ngadi, M. (2016). Enzyme-catalyzed synthesis and kinetics of ultra-sonic assisted methanolysis of waste lard for biodiesel production. *Chemical Engineering Journal* 284: 158–165.

Alexis, A. & Robert, J. (2004) Animal byproduct conversion system and method. US patent 7,000,333.

Bah, C.S.F., El-Din, A., Bekhit, A. et al. (2016). Composition and biological activities of slaughterhouse blood from red deer, sheep, pig and cattle. *Journal of the Science of Food and Agriculture* 96: 79–89.

Baiano, A. (2014). Recovery of biomolecules from food wastes – a review. *Molecules* 19: 14821–14842.

Baladincz, P. and Hancsók, J. (2015). Fuel from waste animal fats. *Chemical Engineering Journal* 282: 152–160.

Banković-Ilić, I.B., Stojković, I.J., Stamenković, O.S. et al. (2014). Waste animal fats as feed stocks for biodiesel production. *Renewable and Sustainable Energy Reviews* 32: 238–254.

Bhatti, H.N., Hanif, M.A., Qasim, M., and Rheman, A.U. (2008). Biodiesel production from waste tallow. *Fuel* 87: 2961–2966.

Bello, A.E. and Oesser, S. (2006). Collagen hydrolysate for the treatment of osteoarthritis and other joint disorders: a review of the literature. *Current Medical Research and Opinion* 22: 2221–2232.

Brandelli, A., Sala, L., and Kalil, S.J. (2015). Microbial enzymes for bioconversion of poultry waste into added-value products. *Food Research International* 73: 3–12.

Bressler, D. (2009) Protein based foaming agents and methods of making thereof. World patent WO2009053852.

Bujak, J.W. (2015). New insights into waste management – meat industry. *Renewable Energy* 83: 1174–1186.

Cascarosa, E., Gea, G., and Arauzo, J. (2012). Thermochemical processing of meat and bone meal: a review. *Renewable and Sustainable Energy Reviews* 16: 942–957.

Catiau, L., Traisnel, J., Delval-Dubois, V. et al. (2011a). Minimal antimicrobial peptidic sequence from hemoglobin alpha-chain: KYR. *Peptides* 32: 633–638.

Catiau, L., Traisnel, J., Chihib, N.E. et al. (2011b). RYH: a minimal peptidic sequence obtained from beta-chain hemoglobin exhibiting an antimicrobial activity. *Peptides* 32: 1463–1468.

Chang, C.-Y., Wu, K.-C., and Chiang, S.H. (2007). Antioxidant properties and protein compositions of porcine haemoglobin hydrolysates. *Food Chemistry* 100: 1537–1543.

Chen, M.J. and Lin, C.W. (2002). Factors affecting the water-holding capacity of fibrinogen/plasma protein gels optimized by response surface methodology. *Journal of Food Science* 67: 2579–2582.

Chernukha, I.M., Fedulovaa, L.V., and Kotenkovaa, E.A. (2015). Meat by-product is a source of tissue-specific bioactive proteins and peptides against cardio-vascular diseases. *Procedia Food Science* 5: 50–53.

Cofrades, S., Guerra, N.I.A., Carballo, J. et al. (2000). Plasma protein and soy fiber content effect on bologna sausage properties as influenced by fat levels. *Journal of Food Science* 65: 281–287.

Coutand, M., Cyr, M., Deydier, D. et al. (2008). Characteristics of industrial and laboratory meat and bone meal ashes and their potential applications. *Journal of Hazardous Materials* 150: 522–532.

Del Hoyo, P.H., Rendueles, M., and Díaz, M. (2008). Effect of processing on functional properties of animal blood plasma. *Meat Science* 78: 522–528.

Deivasigamani, B. and Alagappan, K.M. (2008). Industrial application of keratinase and soluble proteins from feather keratine. *Journal of Environmental Biology* 29: 933–936.

Deng, H., Zheng, J., Zhang, F. et al. (2014). Isolation of angiotensin I-converting enzyme inhibitor from pepsin hydrolysate of porcine hemoglobin. *European Food Research and Technology* 239: 933–940.

Di Bernardini, R., Mullen, A.M., Bolton, D. et al. (2012). Assessment of the angiotensin-I-converting enzyme (ACE-I) inhibitory and antioxidant activities of hydrolysates of bovine brisket sarcoplasmic proteins produced by papain and characterisation of associated bioactive peptidic fractions. *Meat Science* 90 (1): 226–235.

Djagny, K.B., Wang, Z., and Xu, S. (2001). Gelatin: a valuable protein for food and pharmaceutical industries, review. *Critical Reviews in Food Science and Nutrition* 41: 481–492.

Ferraro, V., Anton, M., and Santé-Lhoutellier, V. (2016). The "sisters" α-helices of collagen, elastin and keratin recovered from animal by-products: functionality, bioactivity and trends of application. *Trends in Food Science and Technology* 51: 65–75.

Froidevaux, R., Vanhoute, M., Lecouturier, D. et al. (2008). Continuous preparation of two opioid peptides and recycling of organic solvent using liquid/liquid extraction coupled with aluminium oxide column during haemoglobin hydrolysis by immobilized pepsin. *Process Biochemistry* 43: 431–437.

Gerpen, J.V. (2005). Biodiesel processing and production. *Food Processing Technology* 86: 1097–1107.

Gilbert, E.R., Wong, E.A., and Webb, K.E. (2008). Board-invited review: peptide absorption and utilization: implications for animal nutrition and health. *Journal of Animal Science* 86: 2135–2155.

Giu, H.M. & Giu, P.D. (2010) Blood waste treatment system for slaughtered animals, and method for producing high quality amino acid solution using blood waste. Google patent WO2010087660A2.

Glamsta, E.-L., Meyerson, B., Silberring, J. et al. (1992). Isolation of a hemoglobin-derived opioid peptide from cerebrospinal fluid of patients with cerebrovascular bleedings. *Biochemical and Biophysical Research Communications* 184: 1060–1066.

Gómez-Guillén, M.C., Giménez, B., López-Caballero, M.E., and Montero, M.P. (2011). Functional and bioactive properties of collagen and gelatin from alternative sources: a review. *Food Hydrocolloids* 25: 1813–1827.

Hamilton, C.R. (2016) Real and perceived issues involving animal proteins. In: Protein Sources for the Animal Feed Industry. FAO Report. Retrieved from: www.fao.org

Hazato, T. and Kase, R. (1986). Isolation of angiotensin-converting enzyme inhibitor from porcine plasma. *Biochemical and Biophysical Research Communications* 139: 52–55.

Herregods, G., van Camp, J., Morel, N. et al. (2011). Angiotensin I-converting enzyme inhibitory activity of gelatin hydrolysates and identification of bioactive peptides. *Journal of Agricultural and Food Chemistry* 59: 552–558.

Herskowitz, M. (2008) Reaction system for production of diesel fuel from vegetable and animal oils. World patent WO2008035155.

Honikel, K.O. (2011). Composition and calories. In: *Handbook of Analysis of Edible Animal By-Products* (ed. L.M.L. Nollet and F. Toldrá), 105–121. Boca Raton: CRC Press.

Hsieh, Y.H. and Ofori, J.A. (2011). Food-grade proteins from animal by-products. Their usage and detection methods. In: *Handbook of Analysis of Edible Animal By-Products* (ed. L.M.L. Nollet and F. Toldrá), 3–11. Boca Raton: CRC Press.
Hu, J., Xu, M., Hang, B. et al. (2011). Isolation and characterization of an antimicrobial peptide frombovine hemoglobin α-subunit. *World Journal of Microbiology and Biotechnology* 27: 767–771.
Jayathilakan, K., Sultana, K., Radhakrishna, K., and Bawa, A.S. (2012). Utilization of byproducts and waste materials from meat, poultry and fish processing industries: a review. *Journal of Food Science and Technology* 49: 278–293.
Jridi, M., Abdelhedi, O., Mora, L. et al. (2017). Collagenous proteins from black-barred halfbeak skin as a source of food-quality gelatin and bioactive protein hydrol-ysate. *Food Hydrocolloids* 70: 123–133.
Kang, I.S. and Lanier, T.C. (1999). Bovine plasma proteins functions in surimi gelation compared with cysteine protease inhibitors. *Journal of Food Science* 64: 842–846.
Kapel, R., Froidevaux, R., Nedjar-Arroume, N. et al. (2003). Continuous production of a peptidic fraction containing the intermediate opioid peptide LVV-haemorphin-7 (LVVh-7) by peptic hydrolysis of bovine haemoglobin in a continuous membrane reactor. *Biotechnology and Applied Biochemistry* 37: 317–324.
Kim, Y.N. (2011). Vitamins. In: *Handbook of Analysis of Edible Animal By-Products* (ed. L.M.L. Nollet and F. Toldrá), 161–182. Boca Raton: CRC Press.
Lafarga, T. and Hayes, M. (2014). Bioactive peptides from meat muscle and by-products: generation, functionality and application as functional ingredients. *Meat Science* 98 (2): 227–239.
Lasekan, A., Abu Bakar, F., and Hashim, D. (2013). Potential of chicken by-products as sources of useful biological resources. *Waste Management* 33: 552–565.
Lee, S.H. and Song, K.B. (2009a). Isolation of a calcium-binding peptide from hydrolysates of porcine blood plasma protein. *Journal of Korean Society of Applied Biology and Chemistry* 52: 290–294.
Lee, S.H. and Song, K.B. (2009b). Purification of an iron-binding nona-peptide from hydrolysates of porcine blood plasma protein. *Process Biochemistry* 44: 378–381.
Lennon, A.M., McDonald, K., Moon, S.S. et al. (2010). Performance of cold-set binding agents in re-formed beef steaks. *Meat Science* 85: 620–624.
Lopes, S.C., Fedorov, A., and Castanho, M.A. (2005). Lipidic membranes are potential "catalysts" in the ligand activity of multifunctional pentapeptide neokyotorphin. *Chemistry and Biochemistry* 6: 697–702.
Liu, X., Song, C., Chen, R. et al. (2010). Identification of angiotensin I-converting enzyme inhibitors in peptides mixture of hydrolyzed red deer plasma with proteomic approach. *Chinese Journal of Chemistry* 28: 1665–1672.
Liu, D., Nikoo, M., Boran, G. et al. (2015). Collagen and gelatin. *Annual Reviews of Food Science and Technology* 6: 527–557.
Maehashi, K., Matsuzaki, M., Yamamoto, Y., and Udaka, S. (1999). Isolation of peptides from an enzymatic hydrolisate of food proteins and characterization of their taste properties. *Bioscience, Biotechnology and Biochemistry.* 63: 555–559.
Martínez-Alvarez, O., Chamorro, S., and Brenes, A. (2015). Protein hydrolysates from animal processing by-products as a source of bioactive molecules with interest in animal feeding: a review. *Food Research International* 73: 204–212.

Marulanda, V.F., Anitescu, G., and Tavlarides, L.L. (2010). Investigations on supercritical transesterification of chicken fat for biodiesel production from low cost lipid feedstocks. *Journal of Supercritical Fluids* 54: 53–60.

Mito, K., Fujii, M., Kuwahara, M. et al. (1996). Antihypertensive effect of angiotensin I-converting enzyme inhibitory peptides derived from hemoglobin. *European Journal of Pharmacology* 304: 93–98.

Mora, L., Reig, M., and Toldrá, F. (2014). Bioactive peptides generated from meat industry by-products. *Food Research International* 65: 344–349.

Mora, L., Gallego, M., Escudero, E. et al. (2015). Small peptides hydrolysis in dry-cured meats. *International Journal of Food Microbiology* 212: 9–15.

Moreira, A.L., Dias, J.M., Almeida, M.F., and Alvim-Ferraz, C.M. (2010). Biodiesel production through transesterification of poultry fat at 30 °C. *Energy Fuels* 24: 5717–5721.

Murray, S.M., Patil, A.R., Fahey, G.C. Jr. et al. (1997). Raw and rendered animal by-products as ingredients in dog diets. *Journal of Animal Science* 75: 2497–2505.

Nchienzia, H.A., Morawicki, R.O., and Gadang, V.P. (2010). Enzymatic hydrolysis of poultry meal with endo- and exopeptidases. *Poultry Science* 89: 2273–2280.

Nedjar-Arroume, N., Dubois-Delval, V., Miloudi, K. et al. (2006). Isolation and characterization of four antibacterial peptides from bovine hemoglobin. *Peptides* 27: 2082–2089.

Nedjar-Arroume, N., Dubois-Delval, V., Adje, E.Y. et al. (2008). Bovine hemoglobin: an attractive source of antibacterial peptides. *Peptides* 29: 969–977.

Nissenson, A.R., Berns, J.S., Sakiewicz, P. et al. (2003). Clinical evaluation of heme iron polypeptide: sustaining a response to rHuEPO in hemodialysis patients. *American Journal of Kidney Disorders* 42: 325–330.

Nollet, L.M.L. and Toldrá, F. (2011). Introduction – offal meat: definitions, regions, cultures and generalities. In: *Handbook of Analysis of Edible Animal By-Products* (ed. L.M.L. Nollet and F. Toldrá), 3–11. Boca Raton: CRC Press.

Ockerman, H.W. and Basu, L. (2004a). By-products. In: *Encyclopedia of Meat Sciences* (ed. W. Jensen, C. Devine and M. Dikemann), 104–112. London: Elsevier Science.

Ockerman, H.W. and Basu, L. (2004b). Hides and skins. In: *Encyclopedia of Meat Sciences* (ed. W. Jensen, C. Devine and M. Dikemann), 125–138. London: Elsevier Science.

Ockerman, H.W. and Basu, L. (2006). Edible rendering-rendered products for human use. In: *Essential Rendering: All about the Animal by-Products Industry* (ed. D.L. Meeker), 95–110. Arlignton: National Renderers Association.

Ockerman, H.W. and Basu, L. (2014). By-products. In: *Encyclopedia of Meat Sciences*, 2e (ed. C. Devine and M. Dikeman), 104–112. Oxford: Elsevier.

Ockerman, H.W. and Hansen, C.L. (1988). *Animal By-Product Processing.* Ellis Horwood, Chichester.

Ockerman, H.W., Basu, L., and Toldrá, F. (2017). Edible by-products. In: *Lawrie's Meat Science*, 8e (ed. F. Toldrá), 679–696. Cambridge: Woodhead Publishing.

Ofori, J.A. and Hsieh, Y.-H.P. (2011). Blood-derived products for human consumption. *Revelation and Science* 1: 14–21.

Ofori, J.A. and Hsieh, Y.-H.P. (2012). The use of blood and derived products as food additives. In: *Food Additives* (ed. Y. El-Samragy). Retrieved from: www.intechopen.com/books/food-additive/the-use-of-blood-and-derived-products-as-food-additives.

Park, E., Won, M., Lee, H.-H., and Bin Song, K. (1996). Angiotensin converting enzyme inhibitory pentapeptide isolated from supernatant of pig plasma treated by trichloroacetic acid. *Biotechnology Techniques* 10: 479–480.

Pearl, G.G. (2014). Inedible by-products. In: *Encyclopedia of Meat Sciences*, 2e (ed. C. Devine and M. Dikeman), 112–125. Oxford: Elsevier.

Pérez-Gálvez, R., Almecija, M.C., Espejo, F.J. et al. (2011). Biobjective optimisation of the enzymatic hydrolysis of porcine blood protein. *Biochemical Engineering Journal* 53: 305–310.

Piazza, G.J. and García, R.A. (2014). Proteolysis of meat and bone meal to increase utilisation. *Animal Production Science* 54: 200–206.

Ren, Y., Wan, D.-G., Lu, X. et al. (2011). Isolation and characterization of angiotensin I converting enzyme inhibitor peptides derived from porcine hemoglobin. *Scientific Research and Essays* 6: 6262–6269.

Riedel, S.L., Jahnsa, S., Koeniga, S. et al. (2015). Polyhydroxyalkanoates production with Ralstonia eutropha from low quality waste animal fats. *Journal of Biotechnology* 214: 119–127.

Ryder, K., Ha, M., El-Din Bekhit, A., and Carne, A. (2015). Characterisation of novel fungal and bacterial protease preparations and evaluation of their ability to hydrolyse meat myofibrillar and connective tissue proteins. *Food Chemistry* 172: 197–206.

Ryder, K., El-Din Bekhit, A., McConnell, M., and Carne, A. (2016). Towards generation of bioactive peptides from meat industry waste proteins: generation of peptides using commercial microbial proteases. *Food Chemistry* 208: 42–50.

Saiga, A., Iwai, K., Hayakawa, T. et al. (2008). Angiotensin I-converting enzyme-inhibitory peptides obtained from chicken collagen hydrolysate. *Journal of Agricultural and Food Chemistry* 56: 9586–9591.

Toldrá, F. and Reig, M. (2011). Innovations for healthier processed meats. *Trends in Food Science and Technology* 22: 517–522.

Toldrá, F., Aristoy, M.-C., Mora, L., and Reig, M. (2012). Innovations in value-addition of edible meat byproducts. *Meat Science* 92: 290–296.

Toldrá, F., Mora, L., and Reig, M. (2016). New insights into meat by-products utilization. *Meat Science* 120: 54–59.

USDA (2016). USDA Food Composition Databases. Retrieved from: https://ndb.nal.usda.gov/ndb/search/list

Vercruysse, L., van Camp, J., and Smagghe, G. (2005). ACE inhibitory peptides derived from enzymatic hydrolysate of animal protein: a review. *Journal of Agricultural and Food Chemistry* 53: 8106–8115.

Viana, F.R., Silva, V.D.M., Delvivo, F.M. et al. (2005). Quality of ham pâté containing bovine globin and plasma as fat replacers. *Meat Science* 70: 153–160.

Valta, K., Damala, P., Orli, E. et al. (2015). Valorisation opportunities related to wastewater and animal by-products exploitation by the Greek slaughtering industry: current status and future potentials. *Waste Biomass Valor* 6: 927–945.

Yousif, A.M., Cranston, P., and Deeth, H.C. (2003). Incorporation of bovine dry blood plasma into biscuit flour for the production of pasta. *LWT Food Science and Technology* 36: 295–302.

Yu, Y., Hu, J., Miyaguchi, Y. et al. (2006). Isolation and characterization of angiotensin I-converting enzyme inhibitory peptides derived from porcine hemoglobin. *Peptides* 27: 2950–2956.

Zhang, W., Xiao, S., Samaraweera, H. et al. (2010). Improving functional value of meat products. *Meat Science* 86: 15–31.

Zhao, Q., Sannier, F., Garreau, I. et al. (1994). Inhibition and inhibition kinetics of angiotensin converting enzyme activity by hemorphins isolated from peptic bovine hydrolysate. *Biochemical and Biophysical Research Communications* 204: 216–223.

Zhao, Q., Garreau, I., Sanier, F., and Piot, J.M. (1997). Opioid peptides derived from hemoglobin: hemorphins. *Biopolymers* 43: 75–98.

6

Potential Applications of Plant-Derived Proteins in the Food Industry

Tomas Lafarga

Institut de Recerca i Tecnologia Agroalimentàries (IRTA), Postharvest Programme, Processed Fruits and Vegetables, Lleida, Spain

6.1 Introduction

The human population is predicted to grow by two to four billion by 2050 (Cohen 2003). This expanded population is expected to consume twice as much protein than currently consumed today and meat production and consumption are increasing in several countries, including the US (Daniel et al. 2011) and in most emerging economies (Delgado 2003). However, raising animals for human food is an inefficient process and the observed increased appetite for foods containing animal protein renders the food system unsustainable (Sabaté and Soret 2014). For this reason, together with concerns for animal welfare and personal health, vegetarian nutrition has gained a growing international following (Ruby et al. 2013). Indeed, meat consumption has decreased by approximately 10% in several countries such as Germany and other Western societies (Leitzmann 2014). In order to produce sufficient protein for everyone, a number of changes will need to be made in the food industry and these include the use of novel protein sources for both human and animal nutrition and the use of plant-based substitutes for animal protein (Boland et al. 2013).

Proteins derived from fruits and vegetables can be used not only as a nutritious alternative to animal-derived proteins but also as raw materials for the generation of novel valuable ingredients . For example, several plant-derived proteins have been used as resources for the generation of bioactive hydrolysates and peptides with varied bioactivities. Bioactive peptides are short sequences of approximately 2–30 amino acids in length, that are inactive within the sequence of the parent protein but have positive impacts on systems of the body, beyond their inherent nutritional value, once released (Korhonen and Pihlanto 2006). Bioactive peptides such as antimicrobial peptides are naturally found in plants (Tam et al. 2015) or generated *in vitro* by microbial fermentation or enzymatic hydrolysis. Bioactive hydrolysates and peptides generated from plant-derived proteins include peptides with antioxidant (Sun et al. 2004; Segura Campos et al. 2010), antihypertensive (Segura Campos et al. 2010 ; Lau et al. 2014), antidiabetic (Velarde-Salcedo et al. 2013), or anticarcinogenic (Kim et al. 1999) properties. Moreover, because of their unique and varied structures and interactions with other food

Novel Proteins for Food, Pharmaceuticals, and Agriculture: Sources, Applications, and Advances,
First Edition. Edited by Maria Hayes.

ingredients, proteins exhibit a wide variety of functional properties. These have a profound impact on food quality and include, but are not limited to, gelling, foaming, and water-binding properties.

This chapter provides an introduction to the potential applications of plant-derived proteins in the food industry beyond basic nutrition. These include their use as techno-functional ingredients with, for example, foaming or gelling properties. Furthermore, plant-derived proteins can also be used as novel resources for the generation of bioactive hydrolysates and peptides with varied bioactivities including antioxidant, antimicrobial, and antihypertensive properties. Such hydrolysates and peptides could be used as novel health-promoting ingredients in the development of functional foods, one of the top trends in the food industry.

6.2 Plant-Derived Proteins: Sources and Composition

Proteins are large and complex nitrogen-containing molecules that are formed by amino acids and are required for the structure, function, and regulation of the body's tissues and organs. Proteins are used by the human body as, for example, antibodies, which bind to specific foreign particles to help protect the body, as enzymes, which carry out almost all the chemical reactions that take place in our cells, or as structural components in muscle and other tissues.

Plant-derived proteins can be divided into four major groups: albumins, globulins, prolamins, and glutelins (Day 2013). Albumins are water soluble and coagulate by heat, whereas globulins are soluble in saline solutions and insoluble in water. Albumins are compact globular proteins consisting of two polypeptide chains with molecular weightsof 4 and 9 kDa (Shewry et al. 1995). Prolamins are insoluble in both water and saline solutions but can be extracted in concentrated aqueous alcohol solutions (60–70% v/v). In addition, glutelins, which are the major storage protein in rice, can be extracted in dilute aqueous acid or alkali solutions (Day 2013).

Amino acids play key roles both as building blocks of proteins and as intermediates in metabolism (Wu 2009). Twenty natural amino acids are found within proteins, and their chemical properties determine the biological activity of the protein. On the basis of dietary needs for nitrogen balance or growth, amino acids are classified as nutritionally essential or non-essential. Non-essential amino acids can be synthesised *de novo* by the human body to meet optimal requirements. Essential amino acids must be provided from the diet to meet optimal requirements and include histidine, isoleucine, leucine, lysine, methionine, phenylalanine, threonine, tryptophan, valine, and arginine (mainly for young mammals) (Wu 2009).

Animal proteins, such as those found in eggs, milk, fish, or meat, provide the highest quality rating of food sources, mainly because of the high occurrence of essential amino acids within their sequences (Hoffman and Falvo 2004). However, consumption of these proteins is generally linked with high intakes of saturated fats and cholesterol. For this reason, consumption of vegetable proteins, combined to contain all of the essential amino acids, provides an excellent alternative to meat consumption. Table 6.1 lists the protein content and concentration of several amino acids in a number of seeds, legumes, fruits, and vegetables. Seeds, legumes, nuts, fruits, and vegetables also provide numerous health-promoting nutrients such as fibre and phytochemicals that are important in

Table 6.1 Protein and amino acid content of common cereals, seeds, legumes, fruits, and vegetables. Data obtained from the Food and Agriculture Organisation of the United Nations (FAO 1981).

Food	Botanical name	Protein content (%)	Isoleucine (mg per 100 g food)	Leucine (mg per 100 g food)	Lysine (mg per 100 g food)	Methionine (mg per 100 g food)	Phenylalanine (mg per 100 g food)	Threonine (mg per 100 g food)	Valine (mg per 100 g food)	Arginine (mg per 100 g food)	Histidine (mg per 100 g food)
Apple	*Malus sylvestris*	0.4	13	23	22	3	10	14	15	10	7
Apricot	*Prunus armeniaca*	0.8	14	23	23	4	13	16	19	10	13
Asparagus	*Asparagus officinalis*	2.1	55	96	96	28	55	60	79	83	34
Avocado	*Persea americana*	1.4	47	76	59	29	48	40	63	47	25
Barley	*Hordeum vulgare*	11.0	421	784	406	196	603	389	592	555	248
Bean	*Phaseolus vulgaris*	22.1	927	1685	1593	234	1154	878	1016	1257	627
Broccoli	*Brassica oleracea italica*	4.3	186	236	218	61	177	161	210	242	78
Brown rice	*Oryza sativa*	7.5	300	648	299	183	406	307	433	650	197
Buckwheat	*Fagopyrum sagittatum*	12.2	415	720	464	183	464	439	817	1195	255
Cabbage	*Brassica oleracea capitata*	1.6	50	86	50	17	49	61	68	135	41

(Continued)

Table 6.1 (Continued)

Food	Botanical name	Protein content (%)	Isoleucine (mg per 100 g food)	Leucine (mg per 100 g food)	Lysine (mg per 100 g food)	Methionine (mg per 100 g food)	Phenylalanine (mg per 100 g food)	Threonine (mg per 100 g food)	Valine (mg per 100 g food)	Arginine (mg per 100 g food)	Histidine (mg per 100 g food)
Cauliflower	*Brassica oleracea v. botrytis*	1.1	33	50	44	14	31	32	50	50	16
Celery	*Apium graveolens*	1.1	44	76	27	25	51	38	54	45	17
Chickpea	*Cicer arietinum*	20.1	891	1505	1376	209	1151	756	913	1891	531
Coconut	*Cocos nucifera*	6.6	305	524	275	150	354	265	424	1028	160
Cucumber	*Cucumis sativus*	0.8	25	25	35	8	19	21	28	51	11
Eggplant	*Solanum melongena*	1.2	52	72	63	13	49	44	61	59	25
Fig	*Ficus carica*	1.2	36	51	48	10	28	38	46	27	17
Peanut	*Arachis hypogaea*	25.6	990	1876	1036	338	1459	764	1224	3269	694
Lettuce	*Lactuca sativa*	1.3	50	83	50	24	67	54	71	59	21
Oats	*Avena sativa*	12.0	526	1012	517	234	698	462	711	876	292
Onion	*Allium cepa*	1.4	20	37	63	16	38	20	30	256	14
Orange	*Citrus sinensis*	0.8	23	22	43	12	30	12	31	12	12
Peach	*Prunus persica*	0.8	13	29	30	31	18	27	40	17	17
Potato	*Solanum tuberosum*	2.0	76	121	96	26	80	75	93	100	30

Pumpkin	*Cucurbita pepo*	1.0	37	52	43	9	33	27	48	36	16
Quinoa	*Chenopodium quinoa*	12.0	432	720	672	240	492	219	281	438	150
Radish	*Taphanus sativus*	1.1	54	75	48	10	48	42	126	29	29
Soybean	*Glycine max*	38.0	1889	3232	2653	525	2055	1603	1995	3006	1051
Spinach	*Spinacia oleracea*	2.2	106	208	159	46	133	244	133	139	56
Sweet potato	*Ipomoea batatas*	1.3	48	71	45	22	51	50	59	64	18
Tomato	*Solanum lycopersicum*	1.1	20	30	32	7	20	25	24	24	17
Turnip	*Brassica rapa*	0.9	22	36	17	10	18	25	14	7	41
Wheat	*Triticum aestivum*	12.2	426	871	374	196	589	382	577	602	299

healthy diets (Hoffman and Falvo 2004). Indeed, a study carried out by Clarys et al. (2014) compared the nutritional quality of vegan, vegetarian, semi-vegetarian, and omnivorous diets and concluded, using indexing systems, that vegan diets were the most healthy. Popular sources of vegetable proteins include legumes, nuts, seeds, peas, and soy.

Peas (*Pisum sativum*) and lentils (*Lens culinaris*), which are produced mainly in Asia and North America (Figure 6.1), are used as protein sources for flours, concentrates, and isolates. Earth Protein (Bulk Nutrients, Tasmania, Australia) and LENTEIN™ (Parabel USA Inc., Florida, USA) are examples of commercialised products containing pea and lentil proteins. Soy (*Glycine max*) is a species of legume native to East Asia, widely known for its edible bean. Figure 6.1 shows the annual production of soybean per country. It is one of the most important agricultural commodities because of its high protein content (Table 6.1). Soy protein is a major component of animal feed and is increasingly important in the human diet (Friedman and Brandon 2001). Other applications of soybean include ethanol or biodiesel production (Koc et al. 2011). As mentioned previously, plant-derived proteins are generally deficient in essential amino acids, and soy protein, deficient in methionine, is a clear example (Friedman and Brandon 2001).

Common flax (*Linum usitatissimum*) has been cultivated since ancient times and is an attractive crop because of its high content of linolenic acid, fibre, and high-quality proteins (Rabetafika et al. 2011). The protein content of flaxseed varies from 20% to 30%, constituting 80% globulins and 20% glutelin (Hall et al. 2006). As with soybean, flaxseed proteins are not considered to be complete due to their low lysine concentration (Goyal et al. 2014).

Rice (*Oryza sativa*) protein is also common in several countries. Rice is considered one of the most appropriate cereals for producing gluten-free bread because of its high digestibility and hypoallergenic properties (Kim et al. 2014). Although rice protein has low concentrations of essential amino acids, a study carried out in 2015 concluded that rice bran protein showed superior nutritional quality in comparison to other vegetable proteins (including soy) but its biological availability was lower than that of animal proteins, including casein and proteins derived from whey (Han et al. 2015).

Some nuts and seeds commonly consumed worldwide are rich in protein. For example, almonds (*Prunus dulcis*), walnuts (*Juglans regia*), cashews (*Anacardium occidentale*), peanuts (*Arachis hypogaea*), and pistachios (*Pistacia vera*) contain approximately 23.8%, 14.4%, 20.2%, 14.9%, and 24.6% protein, respectively (Chung et al. 2013). Nuts and seeds are rich in essential amino acids. For example, peanuts contain a high concentration of leucine and phenylalanine and pistachios are rich in lysine and leucine (Chung et al. 2013). It is important to highlight that the essential amino acid content of a protein is only one of the factors determining its nutritional quality. Other factors, such as digestibility and bioavailability of the protein, are of key importance and can also affect the potential utilisation of proteins by humans.

6.3 Bioactive Peptides Generated from Fruits and Vegetables

Several proteins found in food exert physiological effects on humans after ingestion, either directly or upon degradation, releasing bioactive peptides (Korhonen and Pihlanto 2006). As mentioned previously, bioactive peptides are short sequences of

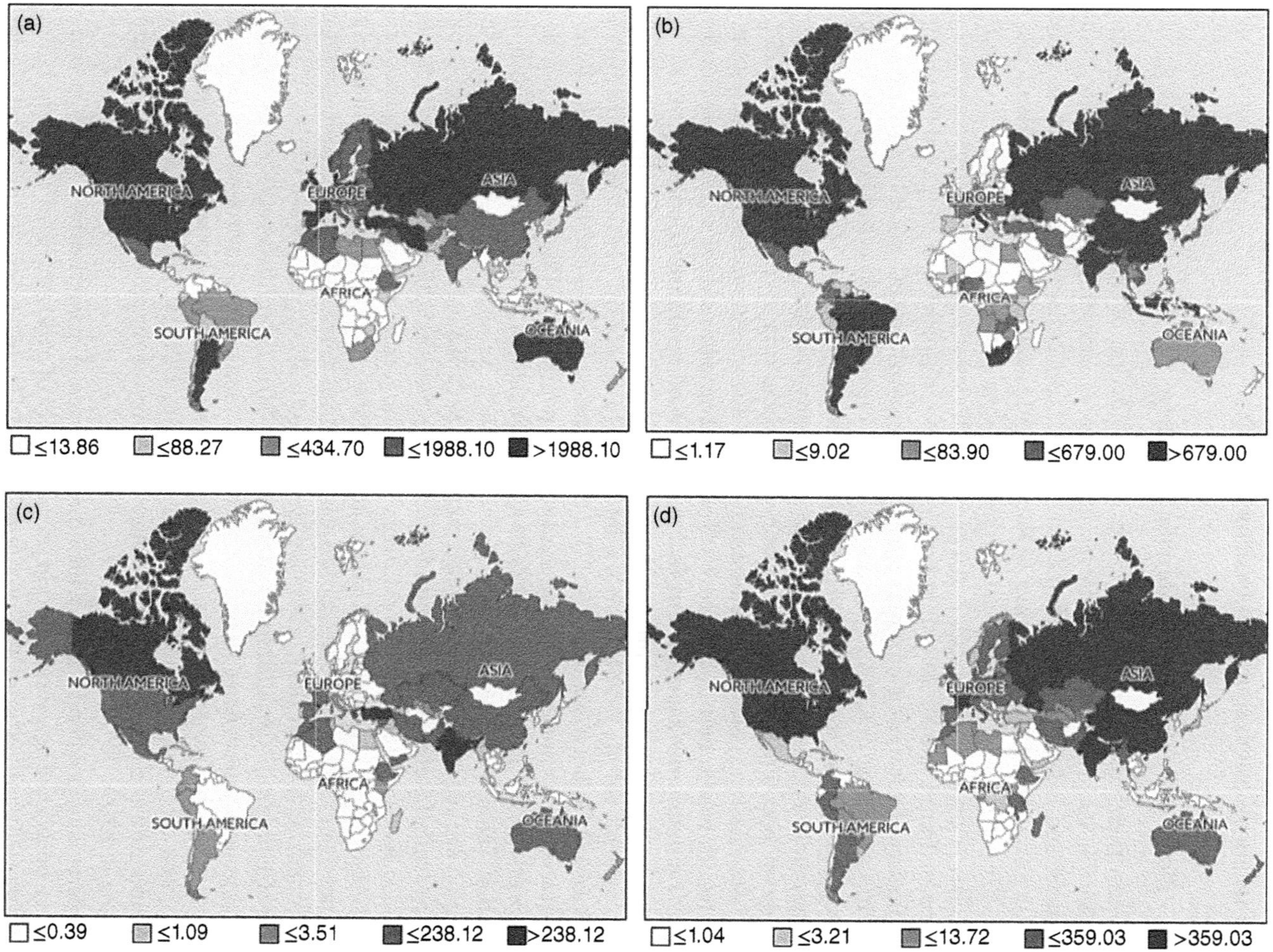

Figure 6.1 Approximate production quantities of barley (a), soybean (b), lentils (c), and dry peas (d) by country per year. Data are expressed in thousands of tonnes and were obtained from the Food and Agriculture Organization Corporate Statistical Database (FAOSTAT) available at www.fao.org/faostat/en/.

amino acids that have positive impacts on the body, including the cardiovascular, digestive, immune, or nervous systems (Kitts and Weiler 2003). Bioactive peptides have been generated from a wide variety of foods and food co-products such as meat (Lafarga and Hayes 2014), milk (Marcone et al. 2017), fish (Jo et al. 2017), and seaweed (Ovando et al. 2016). Even human gastrointestinal endogenous proteins have been used for the generation of bioactive peptides (Dave et al. 2014). Previous studies suggested that although the protein content of fruits and vegetables is lower compared to that of milk or meat, these are also excellent sources of bioactive peptides.

6.3.1 Bioactive Peptides and Metabolic Syndrome

The term metabolic syndrome describes a combination of medical disorders which increase the risk of developing cardiovascular diseases, including diabetes, obesity, lipid disorders, and hypertension (Fulop et al. 2006). Type 2 diabetes is characterised by excessive accumulation of free glucose in blood and is defined as a fasting plasma glucose $\geq 126\,mg\,dL^{-1}$, a two-hour oral glucose tolerance test value $\geq 200\,mg\,dL^{-1}$, or random plasma glucose $\geq 200\,mg\,dL^{-1}$ in a patient with classic symptoms of hyperglycemia (Rosendorff et al. 2015). Retarding lipid or glucose absorption through inhibition of carbohydrate-hydrolysing enzymes such as alpha-amylase (EC 3.2.1.1), maltase (EC 3.2.1.20), and lipase (EC 3.1.1.3) is one of the therapeutic approaches currently being investigated to decrease diabetes (Hamden et al. 2012).

In addition, a promising enzyme related to obesity and diabetes treatment and prevention is beta-site amyloid precursor protein-cleaving enzyme-1 (BACE-1) (EC 3.4.23.46). Meakin et al. (2012) demonstrated that downregulation of the genes responsible for its production decreased body weight and adiposity in mice. Inhibition of BACE-1 also improved glucose disposal and increased insulin sensitivity.

However, one of the most promising pharmacological approaches for the prevention and treatment of type 2 diabetes is the inhibition of dipeptidyl peptidase-IV (DPP-IV) (EC 3.4.14.5) (Juillerat-Jeanneret 2013). This enzyme degrades and inactivates a number of incretin hormones, including glucagon-like peptide-1 (GPL-1) and gastric inhibitory peptide (GIP), which contribute to the enhancement of glucose-induced insulin secretion (Drucker 2006). DPP-IV inhibitors such as sitagliptin, vildagliptin, and alogliptin are approved for the management of type 2 diabetes in the European Union and the United States (Lacroix and Li-Chan 2012). Several DPP-IV inhibitory peptides have also been generated from food sources, including meat co-products (Lafarga et al. 2016). Plant-derived proteins can also be used for the generation of DPP-IV inhibitory peptides. For example, Velarde-Salcedo et al. (2013) obtained an amaranth (*Amaranthus hypochondriacus* L.) hydrolysate using gastrointestinal enzymes with a DPP-IV IC_{50} value of $1.1\,mg\,mL^{-1}$. In addition, Hatanaka et al. (2012) identified two DPP-IV inhibitory peptides sequenced as IP and LP within a umamizyme G hydrolysate of defatted rice bran. This hydrolysate had an IC_{50} value of $2.3\,mg\,mL^{-1}$. Enzymatic hydrolysates of hemp, pea, rice, and soy protein could also inhibit DPP-IV, with different potencies depending on the combination of protein and enzyme employed (Nongonierma and FitzGerald 2015).

Hypertension is a condition in which the blood vessels have constant raised pressure and can be defined as a systolic blood pressure (SBP) of ≥ 140 mmHg or a diastolic blood pressure (DBP) of ≥ 90 mmHg (Rosendorff et al. 2015). According to the World Health

Organization (WHO), hypertension accounts for 9.4 million deaths worldwide every year (WHO 2013). Estimated deaths caused by cardiovascular diseases by sex and cause are represented in Figure 6.2.

Arterial blood pressure is regulated by diverse metabolic pathways, including the renin-angiotensin-aldosterone system (RAAS). Inhibition of the RAAS can lower blood

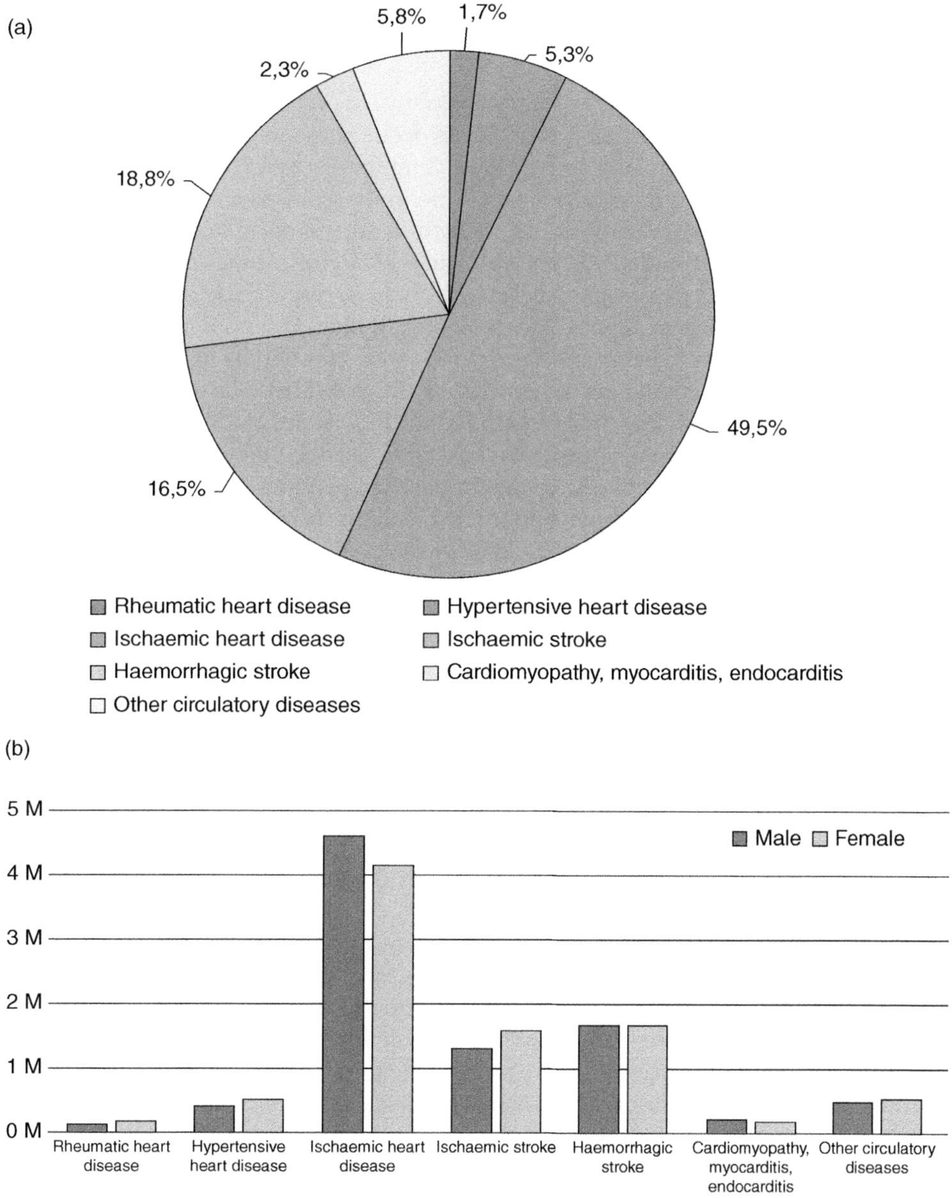

Figure 6.2 Estimated deaths caused by cardiovascular diseases by cause (a) and sex (b). Data obtained from the Global Health Observatory (GHO) data repository, available at www.who.int/gho/database/en/.

pressure and prevent atherosclerosis development and the harmful effects of diabetes (Steckelings et al. 2009). The main enzymes in the RAAS are renin (EC 3.4.23.15) and angiotensin-1-converting enzyme (ACE-1) (EC 3.4.15.1). There are over 10 ACE-1 inhibitors currently available on the market, including the first such compound, captopril or D-3-mercapto-2-methylpropanoyl-L-proline. Side effects of captopril, including dry cough, which appears in 5–20% of patients, and angio-oedema, which affects between 0.1% and 0.5% of patients (Guang et al. 2012), resulted in the development of novel ACE-1 inhibitors and the promotion and development of bioactive peptides with renin and ACE-1 inhibitory properties.

Few studies have looked at the side effects of bioactive peptides generated from natural sources. However, it is thought that if side effects occur, these would be milder (Norris and Fitzgerald 2013). Several plant-derived ACE-1 inhibitory hydrolysates and peptides have been studied, including Alcalase, Flavourzyme, and pepsin-pancreatin hydrolysates of cowpea (*Vigna unguiculata*) protein. These hydrolysates showed IC_{50} values ranging between 24 and 123, 0.04 and 170.6, and 44.7 and 112 µg of protein per millilitre for alcalase, flavourzyme, and pepsin-pancreatin hydrolysates, respectively (Segura Campos et al. 2010). In addition, Lau et al. (2014) identified eight highly functional ACE-1 inhibitory peptides from button mushroom (*Agaricus bisporus*) protein by high-performance liquid chromatography tandem mass spectrometry (LC-MS/MS). The most potent peptides were AHEPVK, RIGLF, and PSSNK with ACE-1 IC_{50} values of 63, 116, and 129 µM, respectively. In addition, vegetable proteins have been used for the generation of renin inhibitory peptides. For example, the patatin-derived peptides PRY and WG were recently investigated for their renin and ACE-1 inhibitory activities using kinetics, intrinsic fluorescence, and molecular docking by Fu et al. (2017). The authors of this study concluded that the peptide PRY was more potent at inhibiting renin and ACE-1, and it inhibited the enzymes as a mixed-type and a competitive inhibitor, respectively.

6.3.2 Fruit- and Vegetable-Derived Peptides with Antioxidant Properties

An antioxidant is a substance which inhibits the oxidation of a molecule. Oxidative processes are one of the main causes of food deterioration (Calligaris et al. 2016) and diets rich in antioxidants have been repeatedly associated with better health and functional longevity (Zhang and Tsao 2016). Several antioxidant hydrolysates and peptides have been generated from varied food sources such as meat (Lafarga and Hayes 2017) and milk (Pihlanto 2006).

Fruits and vegetables are rich sources of antioxidant compounds. Although the influence of phenolic flavonoids, lycopene, carotenoids, and glucosinolates on health has been widely studied (Kaur and Kapoor 2001), little attention has been paid to antioxidant peptides generated from fruits and vegetables. Whole hydrolysates containing antioxidant peptides, in sufficient quantities, are more likely to be industry relevant than purified antioxidant peptides, due to cost considerations. For this reason, most studies focused on antioxidant activities were conducted on enzymatic hydrolysates and not on purified peptides. For example, Alcalase, Flavourzyme, and pepsin-pancreatin hydrolysates of cowpea (*V. unguiculata*) protein showed a trolox equivalent antioxidant coefficient (TEAC) of 303–1457, 357–10211, and 267–2830 $mmol\,L^{-1}\cdot mg$ of protein, respectively (Segura Campos et al. 2010). Some studies have also assessed the antioxidant potential of purified peptides. Sun et al. (2004) studied the antioxidant potential of a peptide-containing fraction isolated from fermented mushroom (*Ganoderma lucidum*) obtained using ultrafiltration followed by gel chromatography via different oxidation systems. Results demonstrated that soybean lipoxygenase activity was blocked by the peptide in a dose-dependent manner with an IC_{50} value of 27.1 μg mL^{-1} and the peptide showed scavenging activity toward hydroxyl radicals produced in a deoxyribose system with an IC_{50} value of 25 μg mL^{-1}.

6.3.3 Other Bioactivities

Natural production of antibiotic peptides has emerged as an important mechanism of immunity in plants and animals, including humans (Ganz 2003). In addition, the growing problem of resistance to conventional antibiotics has led to an increased scientific interest in the development of antimicrobial peptides as human therapeutics. Antimicrobial proteins and peptides are especially interesting because they have different mechanisms of action and show broad-spectrum antimicrobial activity (Zasloff 2002; Brogden 2005). Such peptides can be naturally present in animals or plants and can also be generated *in vitro* by fermentation or enzymatic hydrolysis of proteins. Antimicrobial peptides have been generated from a wide variety of food proteins including milk (Benkerroum 2010), meat (Lafarga and Hayes 2014), egg (Mine et al. 2004), and marine sources such as seaweed or fish (Aneiros and Garateix 2004).

Although no studies have been published using fruit and vegetable proteins for the *in vitro* generation of antibiotic peptides, fruits and vegetables are naturally rich in antimicrobial proteins and peptides. Indeed, antimicrobial peptides represent a defensive mechanism that has been used by plants in combating pathogens for thousands of years (Tam et al. 2015). Ruiz-Ruiz et al. (2017) recently studied the antimicrobial properties of proteins extracted from *Bromelia pinguin* L. The proteins were fractionated by gel filtration chromatography, analysed by gel electrophoresis, and assessed for antimicrobial bioactivity against *Escherichia coli* ATCC 25922 and *Staphylococcus aureus* ATCC 25923. The extracted proteins exhibited antimicrobial activity and a minimum inhibitory concentration (MIC) against *E. coli* and *S. aureus* of 0.34 and 0.68 mg mL^{-1}, respectively. The use of plant-derived proteins obtained from co-products for the generation of antibiotic peptides shows potential for increasing their value and reducing the amount of food discarded as waste or used for low-value purposes such as animal feed.

Cancer is one of the leading causes of death and disability worldwide. Bioactive peptides generated from plant proteins have been reported to have antitumour or anticancer activities. It has been suggested that bioactive peptides with anticarcinogenic activities act by preventing the different stages of cancer including initiation, promotion, and progression (de Mejia and Dia 2010). Peptides derived from animal sources are most often studied for anticarcinogenic properties (Hernández-Ledesma and Hsieh 2017) but studies have demonstrated the anticarcinogenic and anticancer properties of bioactive hydrolysates and peptides derived from plants. For example, peptides derived from hydrolysed soybean exhibited cytotoxic effects on different cell lines (Kim et al. 1999).

6.4 Technofunctional Properties

In order to facilitate access to low-cost and tasty foods, the food industry has relied on the use of fats, sugars, chemical processing aids, and plastics (Lamppa et al. 2014). Over the last decade, consumers have become more aware of the relationship between diet and health, resulting in increased interest in natural ingredients, especially those derived from plants.

Plant-derived proteins have technofunctional properties that affect their behaviour in food matrices through interactions with other ingredients. Their behaviour depends on several factors including their hydration mechanisms for solubility, water/oil retention capacity, structure, viscosity, elasticity, gelation, and their interfacial properties for forming foams or emulsions (Rabetafika et al. 2011). The method used for the extraction or purification of plant-derived proteins can affect their functionality. Joshi et al. (2011) generated a lentil (*Lens culinaris* M.) protein isolate which was converted into powder using freeze drying, spray drying, or vacuum drying. The authors of this study found differences in the proteins' particle size, distribution, subunit composition, colour, solubility, water absorption, and gelation ability. Functional properties of plant-derived proteins can also be affected by the use of novel technologies such as ultrasound, which was able to improve the solubility of previously poorly soluble proteins and their emulsifying performance (O'Sullivan et al. 2016).

Temperature and pH can also significantly affect a protein's functionality (Ogunwolu et al. 2009). For this reason, although most scientific studies generally report valuable quantitative measurements of the functional properties of food protein, the definite potential of a protein's functional property should be assessed after inclusion in an accepted model system or in the final product (Foegeding and Davis 2011).

6.4.1 Solubility of Plant-Derived Proteins

Solubility is the most important technofunctional property as it has a strong influence on other technofunctional properties such as emulsification and foaming (McCarthy et al. 2013). Proteins are soluble at their isoelectric point, and factors affecting protein solubility include pH, ionic strength, solvent, and temperature. Most plant-derived proteins contain low contents of charged amino acid residues and their solubility at neutral pH is generally extremely low (Day 2013). However, some plant-derived proteins, such as those derived from cashews, showed excellent water solubility at their natural pH. Ogunwolu et al. (2009) did not find any significant difference between the solubility in water of proteins extracted from cashew nuts and defatted cashew nuts and found that it was adequate for use in most food formulations without need for adjusting the pH. Thermal processing can be also used as a pre-treatment to alter the solubility of plant-derived proteins. Indeed, Neto et al. (2001) studied the solubility of raw and heat-treated cashew nut kernel proteins and although the raw proteins exhibited higher solubility at pH above the isoelectric point (pH 5.0), both isolates had a comparable solubility at lower pH values. Similar results were obtained for proteins derived from pea (Megha and Grant 1986), soy (Sosulski and McCurdy 1987), and beans (Okezie and Bello 1988).

6.4.2 Gelling Properties

The ability of food proteins to form gels is of key importance in food processing and formulation. Common foods such as tofu and bread doughs are examples of plant-derived protein forming gels or protein networks in foods.

Several plant-derived proteins have been shown to present excellent gelling properties. For example, Boye et al. (2010) studied the gelling capacity of different plant-derived proteins including those derived from pea, chickpeas, and red and green lentils using the least gelling concentration method. A low least gelling concentration indicates better capacity for the protein to form gels. In this study, proteins were extracted using two different methods, isoelectric precipitation and ultrafiltration. A strong gel was formed using proteins extracted from green lentil by ultrafiltration at a concentration of 6% (w/v) while no gels were formed using a concentration of 2% or 4% (w/v). Results also suggested that proteins extracted using ultrafiltration had better gelling properties (lower least gelling concentration). Ogunwolu et al. (2009) determined the least gelation capacity of a cashew nut protein isolate, a cashew nut protein concentrate, and a defatted cashew nut powder as 13.5%, 10.0%, and 6.5%, respectively. Higher protein concentration is usually required for the gelation of globular proteins. Gels formed by plant-derived proteins such as soy or peas are rather similar, suggesting a common pattern of protein denaturation and gel formation (Day 2013).

6.4.3 Emulsifying Properties

Emulsions form a major part of many processed food formulations. The potential of a protein to be used as a food emulsifier is governed by its amino acid sequence, structure, and properties at colloidal interfaces. Animal-derived proteins, especially those derived from eggs and milk, are commonly used to stabilise emulsions and foams (Lynch et al. 2017). The interfacial structures and properties of plant-derived proteins have not been studied in detail but, in general, plant proteins form a relatively thicker interfacial layer at oil/water interfaces caused by their lower molecular size and structural constraint by disulfide cross-links (Wong et al. 2012). This, together with the weak protein interactions once they are adsorbed at interfaces, contributes to a superior stability of emulsions stabilised by a number of plant-derived proteins when compared with, for example, dairy proteins (Day 2013).

Soy proteins have been studied more than other plant-derived proteins for their emulsification properties (Day 2013). Several plant-derived proteins, including those derived from peanut (Yu et al. 2007), rice (Chandi and Sogi 2007), lentil (Joshi et al. 2012), potato (Cheng et al. 2010), and pea (Gharsallaoui et al. 2012), have been assessed for emulsifying properties. For example, Guo and Mu (2011) studied the effect of different protein concentrations ranging between 0.1% and 2.0% and different oil volume fractions ranging between 5% and 45% on the properties of stabilised emulsions of proteins derived from sweet potato. Both factors, protein concentration and oil volume fraction, affected the emulsifying properties of the emulsions generated using sweet potato proteins and the authors suggested the potential use of proteins derived from sweet potato as emulsifiers in the food industry. In addition, Cheng et al. (2010) investigated the efficacy of an Alcalase hydrolysate of potato protein for the stabilisation of an oil droplets in soybean oil in water emulsion. Although results suggested that the studied hydrolysate was a weak emulsifier, the observed partitioning of the hydrolysate at the emulsion interface allowed an excellent hydrolysate-Tween® 20 emulsion that was stable both physically and chemically.

Legume- and seed-derived proteins have also demonstrated strong emulsifying properties. Akasha et al. (2016) determined the emulsifying index and emulsion stability index of proteins extracted from date palm (*Phoenix dactylifera* L.). In this study, although date seed proteins showed lower emulsifying activity compared to that of whey proteins, the stability of the emulsions produced was very similar and the authors suggested that date seed proteins could be used as functional food ingredients. Joshi et al. (2012) studied the dynamic interfacial tension at the oil–water interface and emulsifying properties, including emulsion activity indices, emulsion stability indices, and droplet sizes, of a lentil protein isolate compared to sodium caseinate, bovine serum albumin, lysozyme, and a whey protein hydrolysate. The emulsion activity index ranged between 90.3 and 123.3 $m^2\,g^{-1}$ and was 93.3 $m^2\,g^{-1}$ in the lentil protein hydrolysate emulsion. The stability of the emulsion was slightly lower compared to that obtained using whey proteins at the same concentration at pH 7.0. The stability index was significantly improved with decreased droplet size when protein concentration was increased between 20 and 30 $mg\,mL^{-1}$.

6.4.4 Foaming Properties

The ability of a protein to form stable foams is important for the production of a large amount of food products. Food foams are complex systems which include a mixture of gases, liquids, and solids and can be defined as a system of air cells separated by a thin layer called the lamellar phase. Although proteins derived from egg white, soy, and milk together with gluten and collagen are the most commonly used for generating foams in the food industry, several plant-derived proteins have been shown to possess excellent foaming properties and to contribute to the uniform distribution of fine air cells in foods. For example, Boye et al. (2010) studied the foaming properties of proteins isolated by ultrafiltration or isoelectric precipitation from different pulses. The percentage of gas entrapped in the pulse concentrates, which indicates foaming capacity, ranged between 98% and 106% and was not significantly different for proteins extracted from pea, chickpea, or lentils. In addition, the foam expansion and foaming capacity properties of the samples were not affected by the extraction methodology: isoelectric precipitation or ultrafiltration.

Ogunwolu et al. (2009) determined the foam capacity and stability of a cashew nut protein isolate, a cashew nut protein concentrate, and a defatted cashew nut powder. At the natural pH of the sample, the foam capacity and stability of the cashew nut protein isolate (45% and 55%, respectively) were higher than those of both the protein concentrate (40% and 40%, respectively), and the defatted protein powder (14% and 8%, respectively). In this study, pH variations significantly affected the foaming properties of the protein samples. As mentioned previously, pH variations significantly affected the solubility of these proteins. The authors suggested that only the soluble protein fractions would be involved in foam formation.

6.5 Other Applications

Microencapsulation can be used to protect sensitive substances, develop controlled release properties, and mask unpleasant tastes and odours. It consists of the isolation of active substances to obtain products with spherical shape and micrometric size in which the active material is protected by a membrane.

The main techniques used for microencapsulation of active materials by vegetable proteins are spray drying and coacervation. Microencapsulation by coacervation is carried out by precipitation of materials that surround the active core under the effect of pH, temperature, or addition of a non-solvent (Nesterenko et al. 2013). Proteins extracted from plants are widely used for encapsulation of active substances, especially since the present 'green' trend in the food, cosmetics, and pharmaceutical industries started. The most common plants used for the generation of proteins which can be used in microencapsulation are pea, wheat, rice, sunflower, and especially soy. Soy proteins have been used to protect a probiotic, *Lactobacillus casei*, from gastrointestinal digestion. The probiotic stability was confirmed using Fourier-transform infrared spectroscopy and this, together with high thermal stability, made the encapsulated probiotic suitable for incorporation into functional foods or pharmaceuticals (Hadzieva et al. 2017). Other probiotics such as *Lactobacillus rhamnosus* have been efficiently encapsulated using soy proteins, and were stable against gastrointestinal conditions and fermentation during yogurt making (Li et al. 2016). Proteins extracted from lentils can also be used for microencapsulating interesting compounds such as canola oil. This was recently carried out by Chang et al. (2016), who obtained a very high entrapment efficiency (88%) of canola oil using lentil protein, maltodextrin, and sodium alginate. Other plant-derived proteins such as those extracted from oat do not have the properties needed for microencapsulation, but modifications such as enzymatic hydrolysis (Guan et al. 2007), acetylation or succinylation (Mohamed et al. 2009) could enhance their functionality, allowing them to be used as walls in microencapsulation.

Plant-derived proteins can be used for generating edible coatings. Edible coatings are employed in many food products to control moisture transfer, gas exchange, or oxidation processes in order to extend a product's shelf-life. For example, using edible coatings on minimally processed fruits and vegetables can have a similar effect to modified-atmosphere storage (Dhall 2013). One of the main advantages of edible coatings is that they can improve a product's organoleptic properties and can contain antimicrobial, bioactive, or nutritious compounds which would be consumed together with the food. Different compounds can be used for edible coatings, including lipids, polysaccharides, and proteins. Protein-based coatings have been shown to be efficient in, for example, preventing microbial deterioration of fruits. González-Estrada et al. (2017) recently evaluated the antifungal activity of postharvest treatments with soy protein isolate coatings forming solutions against blue mould decay on Persian lime (*Citrus latifolia*) artificially inoculated with *Penicillium italicum*. This protein plant-based coating was efficient in reducing water losses, maintaining quality characteristics of limes, and liberating active agents. Liu et al. (2017) investigated the preservation properties of soy protein isolate/nano-SiOx composite coating of apple fruits. The authors delayed the climacteric peak of apples to the fifth week and treated apples also had better physiological indices than the control group.

Plant proteins have even been studied as coatings in the manufacture of paperboard. Oviedo et al. (2014) tested rapeseed proteins in a pilot paperboard coating plant and obtained excellent results as the coated paperboard showed no negative effect on quality and industry painting experiments with the coated paperboard achieved the standard quality.

Another recently suggested application of plant-derived proteins is to use them as a sustainable solution to strengthen recycled paper and reduce disposal of contaminants during paper making. Tayeb et al. (2017) suggested that residual soy products represent an inexpensive, sustainable, and environmentally promising solution to enhance paper-making performance. Soy has also been used to form protein scaffolds for use as wound dressings (Har-el et al. 2014).

Acknowledgements

T. Lafarga is in receipt of a Juan de la Cierva contract awarded by the Spanish Ministry of Economy and Competitiveness (FJCI-2016-29541).

References

Akasha, I., Campbell, L., Lonchamp, J., and Euston, S.R. (2016). The major proteins of the seed of the fruit of the date palm (*Phoenix dactylifera* L.): characterisation and emulsifying properties. *Food Chemistry* 197: 799–806.

Aneiros, A. and Garateix, A. (2004). Bioactive peptides from marine sources: pharmacological properties and isolation procedures. *Journal of Chromatography B* 803 (1): 41–53.

Benkerroum, N. (2010). Antimicrobial peptides generated from milk proteins: a survey and prospects for application in the food industry. A review. *International Journal of Dairy Technology* 63 (3): 320–338.

Boland, M.J., Rae, A.N., Vereijken, J.M. et al. (2013). The future supply of animal-derived protein for human consumption. *Trends in Food Science and Technology* 29 (1): 62–73.

Boye, J., Aksay, S., Roufik, S. et al. (2010). Comparison of the functional properties of pea, chickpea and lentil protein concentrates processed using ultrafiltration and isoelectric precipitation techniques. *Food Research International* 43 (2): 537–546.

Brogden, K.A. (2005). Antimicrobial peptides: pore formers or metabolic inhibitors in bacteria? *Nature Reviews Microbiology* 3 (3): 238–250.

Calligaris, S., Manzocco, L., Anese, M., and Nicoli, M.C. (2016). Shelf-life assessment of food undergoing oxidation – a review. *Critical Reviews in Food Science and Nutrition* 56 (11): 1903–1912.

Carlos, Ruiz-Ruiz, J., Ramón-Sierra, J., Arias-Argaez, C. et al. (2017). Antibacterial activity of proteins extracted from the pulp of wild edible fruit of *Bromelia pinguin* L. *International Journal of Food Properties* 20 (1): 220–230.

Chandi, G.K. and Sogi, D. (2007). Functional properties of rice bran protein concentrates. *Journal of Food Engineering* 79 (2): 592–597.

Chang, C., Varankovich, N., and Nickerson, M.T. (2016). Microencapsulation of canola oil by lentil protein isolate-based wall materials. *Food Chemistry* 212: 264–273.

Cheng, Y., Xiong, Y.L., and Chen, J. (2010). Antioxidant and emulsifying properties of potato protein hydrolysate in soybean oil-in-water emulsions. *Food Chemistry* 120 (1): 101–108.

Chung, K.H., Shin, K.O., Hwang, H.J., and Choi, K.-S. (2013). Chemical composition of nuts and seeds sold in Korea. *Nutrition Research and Practice* 7 (2): 82–88.

Clarys, P., Deliens, T., Huybrechts, I. et al. (2014). Comparison of nutritional quality of the vegan, vegetarian, semi-vegetarian, pesco-vegetarian and omnivorous diet. *Nutrients* 6 (3): 1318–1332.

Cohen, J.E. (2003). Human population: the next half century. *Science* 302 (5648): 1172–1175.

Daniel, C.R., Cross, A.J., Koebnick, C., and Sinha, R. (2011). Trends in meat consumption in the USA. *Public Health Nutrition* 14 (4): 575–583.

Dave, L.A., Montoya, C.A., Rutherfurd, S.M., and Moughan, P.J. (2014). Gastrointestinal endogenous proteins as a source of bioactive peptides – an in silico study. *PLoS One* 9 (6): e98922.

Day, L. (2013). Proteins from land plants – potential resources for human nutrition and food security. *Trends in Food Science anf Technology* 32 (1): 25–42.

de Mejia, E.G. and Dia, V.P. (2010). The role of nutraceutical proteins and peptides in apoptosis, angiogenesis, and metastasis of cancer cells. *Cancer and Metastasis Reviews* 29 (3): 511–528.

Delgado, C.L. (2003). Rising consumption of meat and milk in developing countries has created a new food revolution. *Journal of Nutrition* 133 (11): 3907S–3910S.

Dhall, R. (2013). Advances in edible coatings for fresh fruits and vegetables: a review. *Critical Reviews in Food Science and Nutrition* 53 (5): 435–450.

Drucker, D.J. (2006). The biology of incretin hormones. *Cell Metabolism* 3 (3): 153–165.

FAO (1981). *Amino-Acid Content of Foods and Biological Data on Proteins*. Rome: Food and Agriculture Organisation of the United Nations.

Foegeding, E.A. and Davis, J.P. (2011). Food protein functionality: a comprehensive approach. *Food Hydrocolloids* 25 (8): 1853–1864.

Friedman, M. and Brandon, D.L. (2001). Nutritional and health benefits of soy proteins. *Journal of Agricultural and Food Chemistry* 49 (3): 1069–1086.

Fu, Y., Alashi, A.M., Young, J.F. et al. (2017). Enzyme inhibition kinetics and molecular interactions of patatin peptides with angiotensin I-converting enzyme and renin. *International Journal of Biological Macromolecules* 101: 207–213.

Fulop, T., Tessier, D., and Carpentier, A. (2006). The metabolic syndrome. *Pathologie Biologie* 54 (7): 375–386.

Ganz, T. (2003). Defensins: antimicrobial peptides of innate immunity. *Nature Reviews: Immunology* 3 (9): 710–720.

Gharsallaoui, A., Saurel, R., Chambin, O., and Voilley, A. (2012). Pea (*Pisum sativum*, L.) protein isolate stabilized emulsions: a novel system for microencapsulation of lipophilic ingredients by spray drying. *Food and Bioprocess Technology* 5 (6): 2211–2221.

González-Estrada, R.R., Chalier, P., Ragazzo-Sánchez, J.A. et al. (2017). Antimicrobial soy protein based coatings: application to Persian lime (*Citrus latifolia* Tanaka) for protection and preservation. *Postharvest Biology and Technology* 132: 138–144.

Goyal, A., Sharma, V., Upadhyay, N. et al. (2014). Flax and flaxseed oil: an ancient medicine and modern functional food. *Journal of Food Science and Technology* 51 (9): 1633–1653.

Guan, X., Yao, H., Chen, Z. et al. (2007). Some functional properties of oat bran protein concentrate modified by trypsin. *Food Chemistry* 101 (1): 163–170.

Guang, C., Phillips, R.D., Jiang, B., and Milani, F. (2012). Three key proteases – angiotensin-I-converting enzyme (ACE), ACE2 and renin – within and beyond the renin-angiotensin system. *Archives of Cardiovascular Diseases* 105 (6): 373–385.

Guo, Q. and Mu, T.H. (2011). Emulsifying properties of sweet potato protein: effect of protein concentration and oil volume fraction. *Food Hydrocolloids* 25 (1): 98–106.

Hadzieva, J., Mladenovska, K., Crcarevska, M.S. et al. (2017). *Lactobacillus casei* encapsulated in soy protein isolate and alginate microparticles prepared by spray drying. *Food Technology and Biotechnology* 55 (2): 173–186.

Hall, C., Tulbek, M.C., and Xu, Y. (2006). Flaxseed. *Advances in Food and Nutrition Research* 51: 1–97.

Hamden, K., Mnafgui, K., Amri, Z. et al. (2012). Inhibition of key digestive enzymes related to diabetes and hyperlipidemia and protection of liver-kidney functions by trigonelline in diabetic rats. *Scientia Pharmaceutica* 81 (1): 233–246.

Han, S.-W., Chee, K.-M., and Cho, S.-J. (2015). Nutritional quality of rice bran protein in comparison to animal and vegetable protein. *Food Chemistry* 172: 766–769.

Har-el, Y.-E., Gerstenhaber, J.A., Brodsky, R. et al. (2014). Electrospun soy protein scaffolds as wound dressings: enhanced reepithelialization in a porcine model of wound healing. *Wound Medicine* 5: 9–15.

Hatanaka, T., Inoue, Y., Arima, J. et al. (2012). Production of dipeptidyl peptidase IV inhibitory peptides from defatted rice bran. *Food Chemistry* 134 (2): 797–802.

Hernández-Ledesma, B. and Hsieh, C.-C. (2017). Chemopreventive role of food-derived proteins and peptides: a review. *Critical Reviews in Food Science and Nutrition* 57 (11): 2358–2376.

Hoffman, J.R. and Falvo, M.J. (2004). Protein – which is best. *Journal of Sports Science and Medicine* 3 (3): 118–130.

Jo, C., Khan, F.F., Khan, M.I., and Iqbal, J. (2017). Marine bioactive peptides: types, structures, and physiological functions. *Food Reviews International* 33 (1): 44–61.

Joshi, M., Adhikari, B., Aldred, P. et al. (2012). Interfacial and emulsifying properties of lentil protein isolate. *Food Chemistry* 134 (3): 1343–1353.

Joshi, M., Adhikari, B., Aldred, P. et al. (2011). Physicochemical and functional properties of lentil protein isolates prepared by different drying methods. *Food Chemistry* 129 (4): 1513–1522.

Juillerat-Jeanneret, L. (2013). Dipeptidyl peptidase IV and its inhibitors: therapeutics for type 2 diabetes and what else? *Journal of Medicinal Chemistry* 57 (6): 2197–2212.

Kaur, C. and Kapoor, H.C. (2001). Antioxidants in fruits and vegetables – the millennium's health. *International Journal of Food Science and Technology* 36 (7): 703–725.

Kim, J.Y., Woo, H.J., Ahn, C.-W. et al. (1999). Cytotoxic effects of peptides fractionated from bromelain hydrolyzates of soy-bean protein. *Food Science and Biotechnology* 8 (5): 333–337.

Kim, Y., Kee, J.I., Lee, S., and Yoo, S.-H. (2014). Quality improvement of rice noodle restructured with rice protein isolate and transglutaminase. *Food Chemistry* 145: 409–416.

Kitts, D.D. and Weiler, K. (2003). Bioactive proteins and peptides from food sources. Applications of bioprocesses used in isolation and recovery. *Current Pharmaceutical Design* 9 (16): 1309–1323.

Koc, A.B., Abdullah, M., and Fereidouni, M. (2011). Soybeans processing for biodiesel production. In: *Soybean – Applications and Technology*. London: InTech.

Korhonen, H. and Pihlanto, A. (2006). Bioactive peptides: production and functionality. *International Dairy Journal* 16 (9): 945–960.

Lacroix, I.M. and Li-Chan, E.C. (2012). Dipeptidyl peptidase-IV inhibitory activity of dairy protein hydrolysates. *International Dairy Journal* 25 (2): 97–102.

Lafarga, T., Aluko, R.E., Rai, D.K. et al. (2016). Identification of bioactive peptides from a papain hydrolysate of bovine serum albumin and assessment of an antihypertensive effect in spontaneously hypertensive rats. *Food Research International* 81: 91–99.

Lafarga, T. and Hayes, M. (2014). Bioactive peptides from meat muscle and by-products: generation, functionality and application as functional ingredients. *Meat Science* 98 (2): 227–239.

Lafarga, T. and Hayes, M. (2017). Bioactive protein hydrolysates in the functional food ingredient industry: overcoming current challenges. *Food Reviews International* 33 (3): 217–246.

Lamppa, J.W., Horn, G., and Edwards, D. (2014). Toward the redesign of nutrition delivery. *Journal of Controlled Release* 190: 201–209.

Lau, C.C., Abdullah, N., Shuib, A.S., and Aminudin, N. (2014). Novel angiotensin I-converting enzyme inhibitory peptides derived from edible mushroom *Agaricus bisporus* (J.E. Lange) Imbach identified by LC–MS/MS. *Food Chemistry* 148: 396–401.

Leitzmann, C. (2014). Vegetarian nutrition: past, present, future. *American Journal of Clinical Nutrition* 100 (Supplement 1): 496S–502S.

Li, C., Wang, C.L., Sun, Y. et al. (2016). Microencapsulation of *Lactobacillus rhamnosus* GG by transglutaminase cross-linked soy protein isolate to improve survival in simulated gastrointestinal conditions and yoghurt. *Journal of Food Science* 81 (7): M1726–M1734.

Liu, R., Liu, D., Liu, Y. et al. (2017). Using soy protein SiOx nanocomposite film coating to extend the shelf life of apple fruit. *International Journal of Food Science and Technology* 52 (9): 2018–2030.

Lynch, S.A., Mullen, A.M., O'Neill, E.E., and García, C.Á. (2017). Harnessing the potential of blood proteins as functional ingredients: a review of the state of the art in blood processing. *Comprehensive Reviews in Food Science and Food Safety* 16 (2): 330–344.

Marcone, S., Belton, O., and Fitzgerald, D.J. (2017). Milk-derived bioactive peptides and their health promoting effects: a potential role in atherosclerosis. *British Journal of Clinical Pharmacology* 83 (1): 152–162.

McCarthy, A.L., O'Callaghan, Y.C., and O'Brien, N.M. (2013). Protein hydrolysates from agricultural crops – bioactivity and potential for functional food development. *Agriculture* 3 (1): 112–130.

Meakin, P.J., Harper, A.J., Hamilton, D.L. et al. (2012). Reduction in BACE1 decreases body weight, protects against diet-induced obesity and enhances insulin sensitivity in mice. *Biochemical Journal* 441 (1): 285–296.

Megha, A. and Grant, D. (1986). Effect of heat on the functional properties of pea flour and pea protein concentrate. *Canadian Institute of Food Science and Technology Journal* 19 (4): 174–180.

Mine, Y., Ma, F., and Lauriau, S. (2004). Antimicrobial peptides released by enzymatic hydrolysis of hen egg white lysozyme. *Journal of Agricultural and Food Chemistry* 52 (5): 1088–1094.

Mohamed, A., Biresaw, G., Xu, J. et al. (2009). Oats protein isolate: thermal, rheological, surface and functional properties. *Food Research International* 42 (1): 107–114.

Nesterenko, A., Alric, I., Silvestre, F., and Durrieu, V. (2013). Vegetable proteins in microencapsulation: a review of recent interventions and their effectiveness. *Industrial Crops and Products* 42: 469–479.

Neto, V., Narain, N., Silva, J., and Bora, P. (2001). Functional properties of raw and heat processed cashew nut (*Anacardium occidentale*, L.) kernel protein isolates. *Molecular Nutrition and Food Research* 45 (4): 258–262.

Nongonierma, A.B. and Fitzgerald, R.J. (2015). Investigation of the potential of hemp, pea, rice and soy protein hydrolysates as a source of dipeptidyl peptidase IV (DPP-IV) inhibitory peptides. *Food Digestion: Research and Current Opinion* 6 (1–3): 19–29.

Norris, R. and FitzGerald, R.J. (2013). Antihypertensive peptides from food proteins. In: *Bioactive Food Peptides in Health and Disease* (ed. B. Hernández-Ledesma and C.C. Hsieh), 1550–1555. San Diego: Elsevier.

O'Sullivan, J., Murray, B., Flynn, C., and Norton, I. (2016). The effect of ultrasound treatment on the structural, physical and emulsifying properties of animal and vegetable proteins. *Food Hydrocolloids* 53: 141–154.

Ogunwolu, S.O., Henshaw, F.O., Mock, H.-P. et al. (2009). Functional properties of protein concentrates and isolates produced from cashew (*Anacardium occidentale* L.) nut. *Food Chemistry* 115 (3): 852–858.

Okezie, B.O. and Bello, A. (1988). Physicochemical and functional properties of winged bean flour and isolate compared with soy isolate. *Journal of Food Science* 53 (2): 450–454.

Ovando, C.A., de Carvalho, J.C., Pereira, G.V.D.M. et al. (2016). Functional properties and health benefits of bioactive peptides derived from Spirulina: a review. *Food Reviews International* 34 (1): 34–51.

Oviedo, J.P., Metz, R., Schulz, J. et al. (2014). Rapeseed proteins for paperboard coating. *Chemie Ingenieur Technik* 86 (8): 1249–1259.

Pihlanto, A. (2006). Antioxidative peptides derived from milk proteins. *International Dairy Journal* 16 (11): 1306–1314.

Rabetafika, H.N., van Remoortel, V., Danthine, S. et al. (2011). Flaxseed proteins: food uses and health benefits. *International Journal of Food Science and Technology* 46 (2): 221–228.

Rosendorff, C., Lackland, D.T., Allison, M. et al. (2015). Treatment of hypertension in patients with coronary artery disease. *Hypertension* 65 (6): 1372–1407.

Ruby, M.B., Heine, S.J., Kamble, S. et al. (2013). Compassion and contamination. Cultural differences in vegetarianism. *Appetite* 71: 340–348.

Sabaté, J. and Soret, S. (2014). Sustainability of plant-based diets: back to the future. *American Journal of Clinical Nutrition* 100 (Supplement 1): 476S–482S.

Segura Campos, M.R., Chel Guerrero, L.A., and Betancur Ancona, D.A. (2010). Angiotensin-I converting enzyme inhibitory and antioxidant activities of peptide fractions extracted by ultrafiltration of cowpea *Vigna unguiculata* hydrolysates. *Journal of the Science of Food and Agriculture* 90 (14): 2512–2518.

Shewry, P.R., Napier, J.A., and Tatham, A.S. (1995). Seed storage proteins: structures and biosynthesis. *Plant Cell Online* 7 (7): 945–956.

Sosulski, F. and McCurdy, A. (1987). Functionality of flours, protein fractions and isolates from field peas and faba bean. *Journal of Food Science* 52 (4): 1010–1014.

Steckelings, U.M., Rompe, F., Kaschina, E., and Unger, T. (2009). The evolving story of the RAAS in hypertension, diabetes and CV disease – moving from macrovascular to microvascular targets. *Fundamental and Clinical Pharmacology* 23 (6): 693–703.

Sun, J., He, H., and Xie, B.J. (2004). Novel antioxidant peptides from fermented mushroom *Ganoderma lucidum*. *Journal of Agricultural and Food Chemistry* 52 (21): 6646–6652.

Tam, J.P., Wang, S., Wong, K.H., and Tan, W.L. (2015). Antimicrobial peptides from plants. *Pharmaceuticals* 8 (4): 711–757.

Tayeb, A.H., Hubbe, M.A., Pal, L. et al. (2017). Soy proteins as a sustainable solution to strengthen recycled paper and reduce deposition of hydrophobic contaminants in papermaking: a bench and pilot-plant study. *ACS Sustainable Chemistry and Engineering* 5 (8): 7211–7219.

Velarde-Salcedo, A.J., Barrera-Pacheco, A., Lara-González, S. et al. (2013). In vitro inhibition of dipeptidyl peptidase IV by peptides derived from the hydrolysis of amaranth (*Amaranthus hypochondriacus* L.) proteins. *Food Chemistry* 136 (2): 758–764.

WHO (2013). *A Global Brief on Hypertension: Silent Killer, Hlobal Public Health Crisis.* Geneva: World Health Organization.

Wong, B.T., Zhai, J., Hoffmann, S.V. et al. (2012). Conformational changes to deamidated wheat gliadins and β-casein upon adsorption to oil–water emulsion interfaces. *Food Hydrocolloids* 27 (1): 91–101.

Wu, G. (2009). Amino acids: metabolism, functions, and nutrition. *Amino Acids* 37 (1): 1–17.

Yu, J., Ahmedna, M., and Goktepe, I. (2007). Peanut protein concentrate: production and functional properties as affected by processing. *Food Chemistry* 103 (1): 121–129.

Zasloff, M. (2002). Antimicrobial peptides of multicellular organisms. *Nature* 415 (6870): 389–395.

Zhang, H. and Tsao, R. (2016). Dietary polyphenols, oxidative stress and antioxidant and anti-inflammatory effects. *Current Opinion in Food Science* 8: 33–42.

7

Seaweed Proteins and Applications in Animal Feed

Marco García-Vaquero

University College Dublin (UCD), School of Veterinary Medicine, Dublin, Ireland

7.1 Introduction

The global human population is projected to reach 9.6 billion individuals by 2050 (Gerland et al. 2014). Given that the nutritional trends of the population are transitioning towards higher protein diets, protein requirements are expected to increase between 32% and 78% (Henchion et al. 2017). Currently, animal-derived products such as meat, dairy, fish, and shellfish represent 43% of the total protein supply for human consumption (Henchion et al. 2017). Within this scenario, animal production – aquaculture, ruminant, and monogastric livestock – is expected to increase by 70% and approximately 235% more animal feed will be needed (Herman and Schmidt 2016). Furthermore, the market for pet food is increasing due to the growth in pet ownership (particularly canine and feline animals) in developing countries.

The increased demands for quality protein, from human food and animal feed sectors, could be supplied by terrestrial plants; however, increased production of terrestrial crops raises environmental concerns related to water use, deforestation, and soil degradation. For example, soy production is currently driven by the need for animal feed – with 85% of its production destined for this market – and an increase in soy production is associated with deforestation and habitat loss, particularly in South America (Henchion et al. 2017).

Marine macroalgae or seaweeds could be used as an alternative biomass to supply the quality protein required by the food and feed markets. Seaweeds are a diverse group of marine organisms, with more than 10 000 different species described to date, capable of adapting to the harsh conditions of the marine environment by producing unique natural compounds and proteins different in composition from those found in terrestrial plants (Garcia-Vaquero et al. 2017a). Moreover, seaweeds offer numerous advantages over terrestrial crops such as higher photosynthetic efficiency, faster growth, and reproduction rates, higher productivity, less use of arable land for growth and an easier cultivation processes (Wang et al. 2017). The incorporation of macroalgae- and other seaweed-derived products in animal feed is receiving increased attention, particularly in aquaculture, with the development of integrated seaweed cultivation systems in the

Novel Proteins for Food, Pharmaceuticals, and Agriculture: Sources, Applications, and Advances, First Edition. Edited by Maria Hayes.

proximity of fish farms. The cultivation of seaweeds could serve as a source of protein or as a water purification mechanism to reduce the environmental impacts of intensive fish aquaculture (Ratcliff et al. 2016).

7.2 Macroalgae as a Source of Proteins, Peptides, and Amino Acids

Macroalgae have a highly variable protein composition, in terms of both total protein content and amino acid profiles, depending on the seaweed species and on environmental factors such as the place of collection, season, light intensity, temperature, and nutrient concentration in the sea water (Garcia-Vaquero and Hayes 2016).

The most commonly used method to quantify protein is based on the assessment of crude protein from the overall nitrogen content of the seaweeds using the traditional conversion factor of 6.25. However, macroalgae may contain non-protein nitrogen (i.e. free nitrates), often resulting in an overestimation of the algal protein content (Makkar et al. 2016). Novel conversion factors calculated from the total amino acid and total nitrogen contents of various seaweed species can give more reliable protein quantification values in algae (Biancarosa et al. 2017). The total nitrogen content of the macroalgae multiplied by the novel conversion factors 5.38, 4.92, and 5.13 offered a better estimation of the protein content in brown, red, and green seaweed species respectively (Makkar et al. 2016; Biancarosa et al. 2017).

Table 7.1 summarises the total protein concentrations (% dry weight) described in selected brown, red, and green seaweed species. Seasonal variations were appreciated in the protein contents of macroalgae, with higher protein concentrations in winter compared to summer (Galland-Irmouli et al. 1999). However, in general, the protein concentrations described in brown macroalgae were lower (1–16%) in comparison with green (11–26%) and red seaweed species (11–33%), with protein contents comparable to protein-rich foods such as soybean, cereals, eggs, and fish (Garcia-Vaquero and Hayes 2016).

In addition to the high protein content described in several seaweed species, most macroalgae are also considered a source of essential amino acids (Garcia-Vaquero et al. 2017b). The levels of arginine and lysine in red macroalgae and the contents of leucine, phenylalanine, threonine, tyrosine, and valine in all seaweed species were higher than protein-rich ingredients used in animal feed, such as soybean and fishmeals (Biancarosa et al. 2017). Previous studies showed higher contents of sulfur-rich amino acids (i.e. methionine and cysteine) in *Macrocystis pyrifera* and *Ulva* sp. than in soybean meal, concluding that both seaweed species could be used as a dietary supplement for wool-producing animals (Makkar et al. 2016).

Furthermore, high-protein foods are now being used to generate bioactive peptides or cryptides (Hayes et al. 2015). Bioactive peptides are sequences of 2–30 amino acids that could be released from the original protein – by enzymatic hydrolysis, fermentation, or other food processing technologies – and exhibit multiple biological properties different from those shown within their parent protein, such as antihypertensive, antioxidant, antitumour, and antimutagenic activity, amongst others (Garcia-Vaquero and Hayes 2016). Several macroalgal-derived bioactive peptides with promising

Table 7.1 Total protein concentrations (% dry weight) in selected brown, red, and green seaweed species.

Macroalgae (sp.)	Protein content (% dry weight)	References
Phaeophyta (brown macroalgae)		
Fucus sp.	3.9–4	Biancarosa et al. (2017)
Ascophyllum nodosum	3	Biancarosa et al. (2017)
	6.8	Schiener et al. (2017)
Laminaria sp.	1.5–5.1	Nielsen et al. (2016)
	6.6	Biancarosa et al. (2017)
	2.4–12.7	Manns et al. (2017)
Saccharina latissima	1.1–7.5	Nielsen et al. (2016)
	4.8–14.1	Manns et al. (2017)
	4–16.1	Mols-Mortensen et al. (2017)
	10.6	Stévant et al. (2017)
Eisenia arborea	9.2–12.5	Landa-Cansigno et al. (2017)
Alaria esculenta	10.6	Stévant et al. (2017)
Rhodophyta (red macroalgae)		
Porphyra sp.	13.5–20.6	Biancarosa et al. (2017)
Gracilaria sp.	9.2	Chan and Matanjun (2017)
	12.6	Chan and Matanjun (2017)
Chondrus crispus	11	Biancarosa et al. (2017)
Palmaria palmata	10.6	Biancarosa et al. (2017)
	10.9	Schiener et al. (2017)
Pyropia columbina	25.2	Lozano et al. (2016)
Chlorophyta (green macroalgae)		
Ulva sp.	26.7	Silva et al. (2015)
	13.6	Angell et al. (2017)
	11.2–15	Biancarosa et al. (2017)
	22.5	Bikker et al. (2016)
Cladophora rupestris	12	Biancarosa et al. (2017)

antihypertensive activities, such as IRLIIVLMPILMA and AKYSY, were obtained by enzymatic hydrolysis of protein extracted from *Palmaria palmata* and *Porphyra yezoensis* respectively (Saito and Hagino 2005; Fitzgerald et al. 2012). No studies are available to date on the use of seaweed-derived bioactive peptides in animal nutrition. However, several bioactive peptides generated from terrestrial plants and animal proteins have been included in the diet of calves, poultry, fish, and companion animals to improve nutritional status, gut function and resistance to infectious diseases (Hou et al. 2017).

The high protein levels described in macroalgae together with their amino acid profile and the possibility to generate bioactive peptides could suggest the incorporation of seaweeds as a novel ingredient in animal feed.

7.3 Seaweeds and Macroalgal-Derived Products in Animal Feed

One of the goals that the feed industry focuses on when exploring the use of novel ingredients in the diet of farmed animals is to enhance the efficiency of feed utilisation for the production of commercially valuable animal-derived products (i.e. fish, milk, meat, or eggs), while maintaining or improving the health of the animal (Hou et al. 2017). Furthermore, the humanisation of companion animals has led to the development of foods and treats for pets focusing on animal health, but also on the quality and palatability of the products (de Silva and Turchini 2008).

Recent studies evaluated the inclusion of different percentages of seaweeds in the diet of farmed animals and their effects on livestock' growth and feed utilisation (i.e. weight gain and feed conversion ratio) and on certain health parameters that could affect animal production (i.e. gut integrity and immune status) (Sweeney et al. 2016; Cabrita et al. 2017; Choi et al. 2017; Kulshreshtha et al. 2017; Singh et al. 2017; Sotoudeh and Mardani 2018). Moreover, the benefits of the inclusion of bioactive peptides (Hou et al. 2017), macroalgal-derived products (i.e. fermented seaweeds or algal waste) (Islam et al. 2016; Yang et al. 2016) and seaweed extracts (containing proteins and/or polysaccharides) (Sweeney et al. 2016; Boonsri et al. 2017) in the diet of farmed animals and pets have also been explored in recent studies.

The health benefits of the inclusion of macroalgae and other seaweed-derived products in the diet of farmed animals – aquaculture (shrimp and fish), ruminants (small and large animals), and monogastric livestock (poultry, swine, equine, and leporine animals) – and in the food and treats for companion animals (canine and feline animals) are detailed below.

7.3.1 Macroalgae in the Feed of Aquaculture Animals (Shrimp and Fish)

The global aquaculture sector has grown more than 10% per year over the last 40 years and currently accounts for more than 50% of all the shrimp and fish consumed in the market (Olmos Soto et al. 2015; Nadarajah and Flaaten 2017). Aquaculture mass production exceeded 100 million tonnes globally and fish farming represented 70% of this overall biomass, valued at US$ 165.8 billion in 2014 (Nadarajah and Flaaten 2017).

Expenditure on commercial feed for shrimp and fish accounts for 40–60% of the total costs in aquaculture (Olmos Soto et al. 2015). Animal protein sources, such as fishmeal, are extensively used in aquaculture feeds (Shpigel et al. 2017). However, their inclusion in animal feed is expected to decrease due to economic but also sanitary and environmental concerns (Olmos Soto et al. 2015). Seaweeds and other macroalgal-derived products are one of the most promising ingredients to replace fishmeal in aquaculture feed formulations, as seen in the extensive research activity performed in recent years (Table 7.2).

Table 7.2 Application of macroalgal- and seaweed-derived products in the feed of aquaculture (shrimp and fish species) and the main biological effects.

Livestock (sp.)	Macroalgae (sp.)	Level of inclusion in the feed	Main biological effects	References
Shrimp				
Juvenile black tiger shrimps (*Penaeus monodon*)	*Ulva* sp. and *Ulva* protein powder	30%	Protein extract from *Ulva* sp. showed higher apparent digestibility than seaweed powder.	Santizo and Serrano (2014)
White leg shrimps (*Litopenaeus vannamei*)	Nutrikelp (brown macroalgal meal: *Macrocystis* sp., *Lessoniaceae* sp. and *Lessonia* sp.) and Nutrigreen (green macroalgal meal: *Caulerpa* sp., *Enteromorpha* sp. and *Ulva* sp.).	4% and 8%	8% Nutrikelp showed the highest survival rate of juveniles compared to other supplemented and non-supplemented groups. Apparent digestibility coefficients of dry matter and crude protein were higher in all seaweed diets compared to control. No differences were found in final weight, feed conversion ratio, growth rate, protein efficiency and feed consumption amongst the different diets.	Cárdenas et al. (2015)
White shrimps (*Litopenaeus vannamei*)	*Ulva lactuca*	25%, 50%, 75%, and 100%	Commercial shrimp feed can be replaced by 50% with algae without significant losses of growth of shrimps.	Pallaoro et al. (2016)
Shrimps (*Penaeus vannamei*)	Protein extract from *Gracilaria fisheri*	0.025, 0.125, and 0.25 $mg\,g^{-1}$ of shrimp body weight	Shrimps receiving seaweed protein extract showed greater survival rates, more normalised hepatopancreas, and less toxin gene expression following an experimental infection with *Vibrio parahaemolyticus*.	Boonsri et al. (2017)
White leg shrimps (*Litopenaeus vannamei*)	*Undaria pinnatifida* and *Sargassum filipendula*	0.5%, 2%, and 4%	*Undaria pinnatifida* enhanced shrimps' immunity and specifically inhibit *Vibrio* sp., while *Sargassum filipendula* increased the resistance of shrimp to thermal fluctuations compared to non-supplemented animals.	Schleder et al. (2017)

(Continued)

Table 7.2 (Continued)

Livestock (sp.)	Macroalgae (sp.)	Level of inclusion in the feed	Main biological effects	References
Fish				
Rainbow trout (*Oncorhynchus mykiss*)	*Gracilaria vermiculophylla*	5% and 10%	Seaweed supplemented animals showed higher carotenoid concentration in the skin and lower in the flesh compared to control animals. Inclusion of more than 5% of seaweed showed impaired growth of the fish.	Araújo et al. (2016)
European seabass (*Dicentrarchus labrax*)	*Gracilaria* sp., *Ulva* sp. and *Fucus* sp.	2.5% and 7.5%	Seaweed supplementation improved the immune and antioxidant responses in juvenile fish without compromising the growth performance.	Peixoto et al. (2016)
Nile tilapia (*Oreochromis niloticu*)	*Ulva* sp. (mixture of *Ulva rigida* and *Ulva lactuca*)	5% and 10%	Seaweed meal increased total carotenoid contents in the skin and enhanced immune responses in supplemented fish. Seaweed supplementation showed no effects on the growth performance or the organoleptic properties of the flesh with respect to fish receiving a basal diet.	Valente et al. (2016)
Atlantic salmon (*Salmo salar*)	*Palmaria palmata*	5%, 10%, and 15%	Seaweed supplemented animals showed an improved hepatic function and a similar growth performance compared to animals receiving control diets.	Wan et al. (2016)
Red tilapia (*Oreochromis mossambicus* × *Oreochromis niloticus*)	Fermented *Enteromorpha prolifera*	1%, 2%, 3%, 4%, and 5%	Fermented seaweed had positive effects on the growth performance, the activity of the digestive enzymes and the immunity of the fish from the supplemented group with respect to control. The recommended optimum level of inclusion of this algal product in the diet of fish ranged between 3.7% and 4.1%.	Yang et al. (2016)

Barramundi (*Lates calcarifer*)	*Gracilaria pulvinata*	3%, 6%, and 9%	Seaweed supplementation up to 3% showed no negative effects on the growth performance and the body composition of the supplemented animals with respect to control. Increased levels of seaweed supplementation were negatively correlated with the serum triglycerides and cholesterol in the animals.	Morshedi et al. (2018)
Meagre (*Argyrosomus regius*)	*Gracilaria* sp. and *Alaria* sp.	5%	Seaweed supplementation did not affect the growth performance of fish and may confer advantages to animals in coping with biotic stressors.	Peixoto et al. (2017)
Gilthead seabream (*Sparus aurata*)	*Ulva lactuca*	3%, 8%, and 30%	100% of fishmeal was removed from a commercial diet by adding up to 14.6% of seaweed and 29.1% of poultry meat. The fish receiving the new feed formulation achieved similar growth performance with respect to those receiving a commercial diet. The price of the new feed was reduced by \$0.25 kg^{-1}.	Shpigel et al. (2017)
Rainbow trout (*Oncorhynchus mykiss*)	*Gracilaria pygmaea*	0.3%, 0.6%, 0.9%, and 1.2%	Animals supplemented with up to 0.9% of seaweed showed improvements in growth performance without affecting the antioxidant responses and the digestive function of the supplemented fish when compared to animals fed a basal diet.	Sotoudeh and Mardani (2017)
White spotted snapper (*Lutjanus stellatus*)	*Gracilaria lemaneiformis* and *Sargassum horneri*	5%, 10%, 15%, and 20%	Digestive enzymes such as pepsin, lipase, and amylase were reduced in the digestive tract of fish receiving 20% of seaweed in the diet with respect to other treatments and control groups. *Gracilaria lemaneiformis* and *Sargassum horneri* should be incorporated in the diets of juvenile fish to a maximum of 16.44% and 15% respectively.	Zhu et al. (2017)

The partial substitution of protein by seaweeds or the inclusion of different percentages of macroalgae in the diet of shrimps showed promising results, improving several productivity parameters (i.e. growth rates and survival of shrimp juveniles) and/or enhancing the health of the animals (i.e. resistance to thermal variations and enhanced immunity) (Cárdenas et al. 2015; Schleder et al. 2017). Pallaoro et al. (2016) replaced commercial shrimp feed by 50% with *Ulva lactuca* without negative effects on the growth performance of white shrimps (*Litopenaeus vannamei*) compared to animals receiving a control diet.

Protein extracts from *Gracilaria fisheri* were also included in small percentages, improving the survival rates of juvenile shrimp, and their response after an experimental infection with *Vibrio parahaemolyticus*, compared to animals receiving a basal diet (Boonsri et al. 2017), was positive. The addition of protein extracts from *Ulva* sp. (up to 30%) improved the digestibility of the feed in black tiger shrimp (*Penaeus monodon*) compared to animals receiving full algae (Santizo and Serrano 2014).

Different seaweed species were introduced in the feed of farmed fish at lower inclusion levels (0.3–30%) than in shrimp (Table 7.2). Levels of up to 5% of *Gracilaria vermiculophylla* did not affect growth and increased the carotenoid concentration in the skin of supplemented rainbow trout (*Oncorhynchus mykiss*) compared to animals receiving a basal diet (Araújo et al. 2016). The addition of *Gracilaria* sp. and *Alaria* sp. at 5% in the diet of meagre (*Argyrosomus regius*) showed positive effects, helping the supplemented animals to deal with biotic stressors without negative effects in productive performance (Peixoto et al. 2017). Furthermore, the incorporation of fermented seaweed (*Enteromorpha prolifera*) in the diet of red tilapia (*Oreochromis mossambicus* × *Oreochromis niloticus*) had positive effects on growth, activity of digestive enzymes and immunity of the supplemented fish compared to control animals, with an optimum level of inclusion of fermented seaweed in the feed ranging between 3.7% and 4.1% (Yang et al. 2016). Recently, Shpigel et al. (2017) substituted 100% of fishmeal in the feed of gilthead seabream (*Sparus aurata*) using seaweeds. The addition of up to 14.6% of seaweed (*Ulva lactuca*) and 29.1% of poultry meat reduced the price of the fish feed by $0.25 per kg and achieved similar growth rates to animals fed a commercial diet containing fishmeal.

7.3.2 Macroalgae in the Feed of Monogastric Animals (Poultry, Swine, Equine, and Leporine)

The addition of moderate to high amounts of brown macroalgae may be detrimental to monogastric species. For example, pigs supplemented with 10% *Ascophyllum nodosum* showed low weight compared to non-supplemented animals (Jones et al. 1979). As seen in Table 7.3, recent studies evaluating the use of seaweeds in monogastric species (poultry, swine, equine, and leporine animals) are based on the addition of low percentages of various seaweed species (0.05–4%) to evaluate their effects on animal health (i.e. gut health and immunological responses) or to improve the quality of the composition of livestock-derived products (i.e. increase the levels of iodine in meat), while maintaining or improving animal performance (i.e. weight gain and egg production).

In hens, dietary supplementation with seaweeds increased the abundance of beneficial bacteria while reducing the prevalence of *Clostridium perfringens* (Kulshreshtha et al. 2014), and reduced the negative effects on body weight and egg production of

Table 7.3 Application of macroalgal- and seaweed-derived products in the feed of monogastric animals (poultry, swine, equine, and leporine species) and main biological effects.

Livestock (sp.)	Macroalgae (sp.)	Level of inclusion in the feed	Main biological effects in the animals	References
Poultry				
Broiler chicken	*Ulva lactuca*	1% and 3%	Animals receiving 3% of seaweed in the diet showed higher dressing percentage and breast muscle yield in comparison to broilers receiving 1% of seaweed or a basal diet. Serum total lipids, cholesterol, and uric acid concentrations were significantly decreased in birds fed algal diets compared to animals from the control group.	Abudabos et al. (2013)
Lohmann brown classic laying hens	*Chondrus crispus* and *Sarcodiotheca gaudichaudii*	0%, 5%, 1%, and 2%	Hens receiving seaweed showed no significant effect on feed intake, body weight, egg production, faecal moisture content and blood serum profile compared to birds fed a basal diet. Algae supplemented animals showed an increase in the abundance of beneficial bacteria (*Bifidobacterium longum* and *Streptococcus salivarius*) and a decrease of the prevalence of *Clostridium perfringens* compared to animals receiving non-supplemented diets.	Kulshreshtha et al. (2014)
Indonesian pegging ducks	Fermented *Gracilaria* sp.	10%, 12.5%, and 15%	Meat antioxidant status improved at the highest levels of seaweed supplementation, while levels of 12.5% decreased the fat content of the supplemented animals compared to ducks fed a control diet. Ducks supplemented with seaweed showed no detrimental effect on meat productivity and resulted in a more profitable production than animals farmed with a basal diet.	Santoso et al. (2016)

(Continued)

Table 7.3 (Continued)

Livestock (sp.)	Macroalgae (sp.)	Level of inclusion in the feed	Main biological effects in the animals	References
Ross chicks	*Ascophyllum nodosum* extract	500 and 1000 mg kg^{-1} diet	Seaweed supplementation reduced *Campylobacter jejuni* counts in the caecum of the chicks compared to those receiving a basal diet. At the highest levels of inclusion of seaweed extract in the diet the animals showed an increased expression of genes associated with gut integrity.	Sweeney et al. (2016)
Lohmann Lite hens	*Chondrus crispus* and *Sarcodiotheca gaudichaudii*	2% and 4%	The incorporation of seaweed in the diet of hens reduced the negative effects on body weight and egg production of a challenge with *Salmonella enteritidis* compared to birds receiving a basal diet.	Kulshreshtha et al. (2017)
Swine				
Piglets (Seghers hybrid × Pietrain)	*Ascophyllum nodosum*	1% and 2%	Seaweed could be used in the diet of piglets to improve gut health, animal performance and the iodine content of pork derived products.	Dierick et al. (2009)
Sows and piglets (Large White x [Large White x Landrace])	Seaweed extract (no species mentioned)	10 g extract containing laminarin and fucoidan per day	The supplementation of the sows with seaweed extracts enhanced the immune response and improved the gut morphology of the suckling piglets, making them more immune competent to deal with post-weaning adversities than the piglets from the control group.	Heim et al. (2015)
Female pigs	Seaweed extract (no species mentioned)	180 mg laminarin and 340 mg fucoidan per kilogram feed	Seaweed extracts reduced the gene expression of pro-inflammatory cytokines in the colon of supplemented pigs after an experimental infection with *Salmonella Typhimurium* compared to animals receiving a basal diet.	Bouwhuis et al. (2017)

Table 7.3 (Continued)

Livestock (sp.)	Macroalgae (sp.)	Level of inclusion in the feed	Main biological effects in the animals	References
Weaned pigs (Landrace × Yorkshire × Duroc)	*Ecklonia cava*	0.05%, 0.1%, and 0.15%	Animals fed seaweed in the diet showed an increase in average daily gain compared to non-supplemented pigs. No effects were appreciated in other productive parameters and in the total anaerobic bacteria counts amongst the animals of the different treatment groups with respect to the control group.	Choi et al. (2017)
Equine				
Mature aged quarter horse stallions	*Ascophyllum nodosum* extract	1.50%	Dietary supplementation with seaweed extracts showed no effect on feed intake, body weight and sperm quality when comparing supplemented and non-supplemented horses during a 90 day trial.	Williams et al. (2017)
Leporine				
New Zealand white rabbits	*Lithothamnium* sp.	0.25%, 0.5%, 0.75%, and 1%	Seaweed inclusion did not affect feed intake, digestibility, weight gain and carcass yield. Levels of 1% of seaweed produced morphological changes in the ileal villi that could have negative effects for the supplemented animals compared to those receiving a control diet.	Euler et al. (2010)
Buck and doe rabbits	*Ulva lactuca*	1% and 2%	Seaweed supplementation improved the reproductive performance of rabbits by improving the semen fertility characteristics of bucks and the prolificacy characteristics of does compared to non-supplemented rabbits.	Okab et al. (2013)

Salmonella enteritidis (Kulshreshtha et al. 2017). Pigs supplemented with *A. nodosum* (1–2%) showed improvements in gut health, animal performance and increased levels of iodine in meat compared to those receiving a non-supplemented diet (Dierick et al. 2009). Improvements in performance of weaned pigs were also appreciated by feeding *Ecklonia cava* (0.05–0.15%) with respect to animals fed a basal diet (Choi et al. 2017). In rabbits, the inclusion of *U. lactuca* (1–2%) improved the reproductive performance of the supplemented animals compared to those receiving a control diet (Okab et al. 2013).

The inclusion of fermented macroalgal products showed promising results in animal nutrition. Fermented *Gracilaria* sp. supplemented at levels of 10–15% in the diet of ducks improved the antioxidant status of their meat and decreased the accumulation of fat while maintaining growth, resulting in a more profitable animal production system compared to ducks receiving a non-supplemented diet (Santoso et al. 2016).

Recently, the use of seaweed extracts containing a mixture of polysaccharides and proteins showed beneficial results in monogastric animals. *A. nodosum* extracts reduced the counts of *Campylobacter jejuni* in the caecum of supplemented chicks compared to animals receiving a basal diet (Sweeney et al. 2016). The supplementation of sows with seaweed extracts containing polysaccharides (laminarin and fucoidan) enhanced the immune response and improved the gut morphology of suckling piglets compared to animals receiving a basal diet (Heim et al. 2015). Furthermore, similar seaweed extracts reduced the gene expression of proinflammatory cytokines in the colon of supplemented pigs after experimental infection with *Salmonella typhimurium* (Bouwhuis et al. 2017). However, the use of *A. nodosum* extracts at levels of 1.5% for 90 days showed no effect on feed intake, body weight, and sperm quality of supplemented quarter horse stallions compared to animals receiving a basal diet (Williams et al. 2017).

7.3.3 Macroalgae in the Feed of Ruminants (Small and Large)

One of the main differences between ruminants and monogastric animals is the ability of small and large ruminants to use non-protein nitrogen in microbial processes that take place in the rumen. Briefly, the ruminal bacteria transform the inorganic nitrogen into ammonia to synthesise microbial proteins that will later be absorbed in the small intestine (Garcia-Vaquero and Hayes 2016). Seaweeds are a rich source of both protein and non-protein nitrogen, so researchers included several seaweed species at higher levels (up to 40%) in the feed of ruminants compared to monogastric animals (Table 7.4).

Recent scientific studies added macroalgae to the feed of small ruminants at percentages ranging from 10% to 40% and compared the ruminal degradability of seaweeds to that of traditional forages such as alfalfa hay (Cabrita et al. 2017) and barley grass (Rjiba-Ktita et al. 2017). The addition of *A. nodosum* at low percentages (1–5%) to the diet of ewes (Caroprese et al. 2016) and rams (Zhou et al. 2017) showed improvements in the fatty acid content of milk and reductions in the population of ruminal *Escherichia coli* respectively.

Seaweeds were studied in large ruminants at inclusion levels ranging from 2% to 20% (Table 7.4). The addition of 2% of *Undaria pinnatifida* to the feed of Hanwoo steers resulted in meat products with lower cholesterol levels than those found in the meat of steers fed a non-supplemented diet (Hwang et al. 2014). Dairy cattle receiving 20% of *Sargassum wightii* showed an increase in milk production and total fat content in milk compared to animals receiving a control diet (Singh et al. 2017). Recently, the addition

Table 7.4 Application of macroalgal- and seaweed-derived products in the feed of ruminants (small and large animals) and main biological effects.

Livestock (sp.)	Macroalgae (sp.)	Level of inclusion in the feed	Main biological effects in the animals	References
Small ruminants				
Comisana ewes	*Ascophyllum nodosum*	1%	*Ascophyllum nodosum* combined with flaxseed was suggested to maintain the production and the fatty acid profile of the milk during the summer season.	Caroprese et al. (2016)
Merino sheeps	*Ulva rigida* and *Gracilaria vermiculophylla*	25%	Seaweeds showed lower dry matter digestibility than alfalfa hay. No differences were appreciated in the immunological status between supplemented and non-supplemented animals.	Cabrita et al. (2017)
Barbarine rams	*Chaetomorpha linum* and *Ulva lactuca*	10%, 20%, 30%, and 40%	Both seaweeds showed similar ruminal degradability than a medium quality forage (i.e. barley grass). The inclusion of seaweed at levels greater than 20% resulted in a decrease of the ruminal degradation of the mixture in vitro.	Rjiba-Ktita et al. (2017)
Rams	*Ascophyllum nodosum*	1%, 3%, and 5%	Total ruminal *Escherichia coli* population was linearly reduced in seaweed supplemented rams compared to animals from the control group.	Zhou et al. (2017)
Large ruminants				
Hanwoo steers	*Undaria pinnatifida*	2%	Seaweed supplemented steers showed an improvement in growth, immunity and fatty acid profile in the meat, with significantly reduced cholesterol compared to non-supplemented animals.	Hwang et al. (2014)
Jersey dairy cows	*Ascophyllum nodosum*	57, 113, or 170 g day^{-1}	Feeding seaweed exponentially increased milk iodine concentration, but had no effect on animal performance compared to cows from the control group.	Antaya et al. (2015)
Holstein dairy cows	Brown seaweed by-products (no species mentioned)	2% and 4%	Seaweed supplementation did not compromise ruminal fermentation and animal performance, measured as milk production and composition, with respect to cows receiving a basal diet.	Hong et al. (2015)

(Continued)

Table 7.4 (Continued)

Livestock (sp.)	Macroalgae (sp.)	Level of inclusion in the feed	Main biological effects in the animals	References
Hanwoo cows and calves	Fermented *Undaria pinnatifida*	10%	Suckling calves from the seaweed supplemented group showed an increase in weaning weight, average daily gain and serum immunoglobulin G that could be related to the increase in immunoglobulin G, moisture, crude fat and crude protein in the colostrum of the seaweed supplemented cows compared to the control group.	Islam et al. (2016)
Sahiwal dairy cows	*Sargassum wightii*	20%	Seaweed supplemented dairy cows showed an increase in fat content and milk yield compared to animals receiving non-supplemented diets.	Singh et al. (2017)

of 10% of fermented *U. pinnatifida* showed beneficial effects in Hanwoo cows and calves due to an increased level of immunoglobulin G in the colostrum of the animals compared to livestock receiving a control diet (Islam et al. 2016).

Previous studies in large ruminants have researched the beneficial effects of polysaccharide-rich extracts from *A. nodosum* (Tasco®). The addition of Tasco to the diet of feed-lot steers improved the characteristics of the beef carcase, increased the colour stability of the muscle and extended the shelf-life of meat products compared to livestock receiving a basal diet (Evans and Critchley 2014).

7.3.4 Macroalgae in Pet Food (Canine and Feline Animals)

Pet ownership brings many physiological and psychological benefits (McCusker et al. 2014). Approximately 68% of USA households owned a pet in 2016, which accounts for 89.7 million dogs and 94.2 million cats in this country alone (APA 2017). Moreover, there is increased humanisation of pets, leading to concerns about pets' general health and well-being far beyond that directed towards livestock (de Silva and Turchini 2008). This trend is reflected in the market by the purchase of high-quality food for companion animals. In 2016 'premium foods' (described as those containing more nutrients, fewer preservatives, and more efficiently metabolised) together with 'foods with added vitamins or minerals' (containing calcium or vitamin E) and 'natural foods' (those with no synthetic materials) represented 75% of the total canine and feline food preferences amongst pet owners in the USA (APA 2017).

Macroalgae and seaweed extracts could have potential applications in the pet food market as a novel source of proteins, amino acids, and other micronutrients (i.e. minerals and vitamins). For example, the level of taurine is limited in feline food. Taurine is a sulfur-containing amino acid essential for strict carnivores such as cats, and its deficit was linked to the development of retinal degeneration and cardiomyopathy (McCusker et al. 2014). Minimum dietary levels of 0.2–0.27% dry matter of taurine in cat food are recommended by the European Pet Food Industry Federation (FEDIAF 2017). The taurine concentrations in seaweeds varied between 0.01 and 6.28 mg per g of dry matter and were higher than the levels described in terrestrial plants (McCusker et al. 2014).

Moreover, the addition of seaweeds in pet food was also explored for its technofunctional properties. Seaweeds can act as emulsifying agents and texture modifiers and can modify the whipping properties of foods, contributing to the taste and texture of the final product (Garcia-Vaquero et al. 2017a). Recently, full seaweed powder was used as a binding agent in pet treats due to its gelling properties (Weinberg and Saxe 2017).

7.4 Challenges Concerning the Use of Macroalgae in Animal Feed

Researchers foresee a future where the seaweed industry will grow due to the multiple uses of the biomass, including food and feed applications, but also as raw materials for the production of extracts (rich in proteins and polysaccharides) that could be used as valuable ingredients in food and animal feed (Garcia-Vaquero et al. 2017b), cosmeceuticals (Martins et al. 2014), nutraceuticals (Hayes 2015), pharmaceuticals (Rajauria et al. 2017), and other agronomic applications (Craigie 2011).

The increasing market for seaweed represents a challenge for the feed industry (i.e. legislation on the incorporation of seaweeds and other algal-derived ingredients, feed formulation and percentages of inclusion in animal feed) and also for the seaweed producers (i.e. sustainability of the supply of raw materials and cultivation strategies).

7.4.1 Legislation on the Use of Macroalgal- and Seaweed-Derived Products in Animal Feed

Seaweeds and other macroalgal-derived products offer huge potential for incorporation in animal feed and future industrial exploitation as a source of protein and other micronutrients. However, one of the main challenges that the industry faces when introducing new products in Europe is the legal requirements concerning the placement of feed in the market and also regulations on feed hygiene and chemical safety of the feed products, summarised in Figure 7.1.

Regulations EC 178/2002 and EC 767/2009 set the framework for European Union legislation on feed, describing and defining key terms and legal responsibilities at all stages of feed production. Additionally, Regulation EC 769/2009 defines novel feed as a feed material not in place in the EU market before 1 September 2010. Feed business operators have to notify their intention to market a novel feed material to the representatives of the European feed business. The novel feed will be registered and details will be published for information purposes in the Feed Materials Register. In the case of macroalgal proteins, no authorisation is needed for their use as feed, but they are considered novel food ingredients if intended for human use (Spiegel et al. 2013).

Directive 2002/32/EC sets the maximum levels for certain contaminants in animal feed. Calcareous marine algae used in animal feed should have a maximum level of 15 and 1000 ppm of lead and fluorine respectively. Seaweeds and algal-derived products

OVERVIEW OF EUROPEAN LEGISLATION ON ANIMAL FEED

GENERAL REGULATIONS

Regulation EC 178/2002. General food law.

Art. 2. Definition of animal feed.
Art. 17. Responsibilities of feed business operators (FBO)
Art. 18. Feed traceability

Regulation EC 767/2009. Placing on the market and use of feed.

Annex III. List of materials prohibited for feed.
Recommendation 2011/25/EU. Definitions.
Regulation EU 68/2013. Feed Catalogue.

CHEMICAL FEED SAFETY

Directive 2002/32/EC. Undesirable substances in animal feed.

Annex I. Maximum arsenic in seaweed meal and feed materials derived from seaweed. Upon request, analysis of inorganic arsenic must be performed and be lower than 2 ppm, specially in *Hizikia fusiforme* (*Sargassum fusiforme*).

Regulation EC 396/2005. Maximum residue levels (MLRs) pesticides in feed of plant/animal origin.

Article 2.1. Fresh, processed and/or composite feed in or on which pesticide residues may be present (Annex I). This list was amended by **Regulation (EU) 212/2013** including seaweed (code number 0290000).

FEED HYGIENE

Regulation EC 183/2005. Feed hygiene requirements.

Article 6. Hazard analysis and critical control points (HACCP) system.
Article 9. Official controls, notification and registration.
Annex I. Hygiene and records in primary feed production.
Annex II. Requirements for other production.

Figure 7.1 Regulations on animal feed products in the European Union.

are specifically mentioned in Annex I of this Directive, as they must contain less than 2 ppm of inorganic arsenic if those analyses are requested, paying special attention to feed products derived from *Sargassum fusiforme*. Pesticide residues in seaweeds are mentioned in Regulation EC 212/2013 that modifies Annex I of Regulation EC 396/2005, but the maximum residue limits (MRLs) will not apply if they are used for feed use until separate MRLs have been established. Furthermore, Regulation EC 183/2005 establishes definitions, responsibilities, and guidelines concerning good feed hygiene practices from primary production to placement of the feed in the market.

7.4.2 Sustainability of Seaweed Supply for Use in Animal Feed

Seaweeds could be an alternative source of macro- and micronutrients in animal feed in substitution for terrestrial plants. However, European, Canadian, and Latin American seaweed industries still largely rely on harvesting natural resources (Rebours et al. 2014). Overharvest of these marine stocks could lead to negative environmental impacts and problems in the sustainability of the supply.

Science-based management plans can maintain a sustainable, but relatively limited and stable collection of biomass over several decades (Buschmann et al. 2017) due to the higher growth rates of seaweeds compared to those of terrestrial plants (Kraan 2013). Recently developed macroalgal cultivation systems offer an alternative production strategy to the seaweed industry and could potentially solve several issues related to the supply of biomass for a growing and demanding market (Buschmann et al. 2017). Seaweed cultivation represents a challenge due to the need to develop the technology necessary to achieve large-scale cultivation, to identify the appropriate seaweed species and to tailor the selection of these algal strains based on specific properties of the biomass as a source of food, chemicals or fuel (Li et al. 2016). Several cultivation systems have been successfully developed for *Laminaria* sp. (offshore cultivation), *Ulva* sp. (tidal flat farm), and *Sargassum* sp. (floating cultivation) (Kraan 2013). In 2014, the leading seaweed farming countries, China and Indonesia, produced more than 20 million tonnes of cultivated macroalgae, while most European countries produced limited quantities (Buschmann et al. 2017).

The co-cultivation of macroalgae with other livestock, such as fish and molluscs, could be a viable commercial alternative (Burg et al. 2013). Integrated multi-trophic aquaculture (IMTA) aims to create a balanced ecosystem, taking into account unused inputs from livestock (feed and waste), that cannot be buffered by the natural host ecosystem, for use as a nutrient by the seaweed crop. This system produces a valuable macroalgal biomass that could be commercialised and reduce the ecological impacts of the co-existing aquaculture cultivation (Ratcliff et al. 2016).

Acknowledgements

This chapter did not receive any specific grant from funding agencies in the public, commercial or non-for-profit sectors.

References

Abudabos, A.M., Okab, A.B., Aljumaah, R.S. et al. (2013). Nutritional value of green seaweed (*Ulva lactuca*) for broiler chickens. *Italian Journal of Animal Science* 12 (2): e28.

Angell, A.R., Paul, N.A., and de Nys, R. (2017). A comparison of protocols for isolating and concentrating protein from the green seaweed *Ulva ohnoi*. *Journal of Applied Phycology* 29 (2): 1011–1026.

Antaya, N.T., Soder, K.J., Kraft, J. et al. (2015). Incremental amounts of *Ascophyllum nodosum* meal do not improve animal performance but do increase milk iodine output in early lactation dairy cows fed high-forage diets1. *Journal of Dairy Science* 98 (3): 1991–2004.

APA (2017). *2017–2018 American Pet Products Association (APA) National Pet Owners Survey*. Greenwich: American Pet Products Association.

Araújo, M., Rema, P., Sousa-Pinto, I. et al. (2016). Dietary inclusion of imta-cultivated *Gracilaria vermiculophylla* in rainbow trout (*Oncorhynchus mykiss*) diets: effects on growth, intestinal morphology, tissue pigmentation, and immunological response. *Journal of Applied Phycology* 28 (1): 679–689.

Biancarosa, I., Espe, M., Bruckner, C. et al. (2017). Amino acid composition, protein content, and nitrogen-to-protein conversion factors of 21 seaweed species from Norwegian waters. *Journal of Applied Phycology* 29 (2): 1001–1009.

Bikker, P., van Krimpen, M.M., van Wikselaar, P. et al. (2016). Biorefinery of the green seaweed *Ulva lactuca* to produce animal feed, chemicals and biofuels. *Journal of Applied Phycology* 28 (6): 3511–3525.

Boonsri, N., Rudtanatip, T., Withyachumnarnkul, B., and Wongprasert, K. (2017). Protein extract from red seaweed Gracilaria fisheri prevents acute hepatopancreatic necrosis disease (AHPND) infection in shrimp. *Journal of Applied Phycology* 29 (3): 1597–1608.

Bouwhuis, M.A., Mcdonnell, M.J., Sweeney, T. et al. (2017). Seaweed extracts and galacto-oligosaccharides improve intestinal health in pigs following *Salmonella typhimurium* challenge. *Animal* 11 (9): 1488–1496.

Burg, V.D.S., Stuiver, M., Veenstra, F. et al. (2013). *A Triple P review of the feasibility of sustainable offshore seaweed production in the North Sea*. Wageningen: Wageningen UR.

Buschmann, A.H., Camus, C., Infante, J. et al. (2017). Seaweed production: overview of the global state of exploitation, farming and emerging research activity. *European Journal of Phycology* 52 (4): 391–406.

Cabrita, A.R.J., Correia, A., Rodrigues, A.R. et al. (2017). Assessing in vivo digestibility and effects on immune system of sheep fed alfalfa hay supplemented with a fixed amount of *Ulva rigida* and *Gracilaria vermiculophylla*. *Journal of Applied Phycology* 29 (2): 1057–1067.

Cárdenas, J.V., Gálvez, A.O., Brito, L.O. et al. (2015). Assessment of different levels of green and brown seaweed meal in experimental diets for whiteleg shrimp (*Litopenaeus vannamei, Boone*) in recirculating aquaculture system. *Aquaculture International* 23 (6): 1491–1504.

Caroprese, M., Ciliberti, M.G., Marino, R. et al. (2016). Polyunsaturated fatty acid supplementation: effects of seaweed *Ascophyllum nodosum* and flaxseed on milk production and fatty acid profile of lactating ewes during summer. *Journal of Dairy Research* 83 (3): 289–297.

Chan, P.T. and Matanjun, P. (2017). Chemical composition and physicochemical properties of tropical red seaweed, *Gracilaria changii*. *Food Chemistry* 221: 302–310.
Choi, Y., Hosseindoust, A., Goel, A. et al. (2017). Effects of Ecklonia cava as fucoidan-rich algae on growth performance, nutrient digestibility, intestinal morphology and caecal microflora in weanling pigs. *Asian-Australasian Journal of Animal Sciences* 30 (1): 64.
Craigie, J.S. (2011). Seaweed extract stimuli in plant science and agriculture. *Journal of Applied Phycology* 23 (3): 371–393.
De Silva, S.S. and Turchini, G.M. (2008). Towards understanding the impacts of the pet food industry on world fish and seafood supplies. *Journal of Agricultural and Environmental Ethics* (21, 5): 459–467.
Dierick, N., Ovyn, A., and de Smet, S. (2009). Effect of feeding intact brown seaweed *Ascophyllum nodosum* on some digestive parameters and on iodine content in edible tissues in pigs. *Journal of the Science of Food and Agriculture* 89 (4): 584–594.
Directive 2002/32/EC. (2002) On undesirable substances in animal feed. Official Journal of the European Communities.
Euler, A.C.C., Ferreira, W.M., Teixeira, E.D.A. et al. (2010). Desempenho, digestibilidade e morfometria da vilosidade ileal de coelhos alimentados com diferentes níveis de inclusão de "*Lithothamnium*". *Revista Brasileira de Saúde e Produção Animal* 11 (1): 91–103.
Evans, F. and Critchley, A. (2014). Seaweeds for animal production use. *Journal of Applied Phycology* 26 (2): 891–899.
FEDIAF (2017). Nutritional guidelines for complete and complementary pet food for cats and dogs. In: . Bruxelles: FEDIAF.
Fitzgerald, C.N., Mora-Soler, L., Gallagher, E. et al. (2012). Isolation and characterization of bioactive pro-peptides with in vitro renin inhibitory activities from the macroalga *Palmaria palmata*. *Journal of Agricultural and Food Chemistry* 60 (30): 7421–7427.
Galland-Irmouli, A.-V., Fleurence, J., Lamghari, R. et al. (1999). Nutritional value of proteins from edible seaweed *Palmaria palmata* (Dulse). *Journal of Nutritional Biochemistry* 10 (6): 353–359.
Garcia-Vaquero, M. and Hayes, M. (2016). Red and green macroalgae for fish and animal feed and human functional food development. *Food Reviews International* 32 (1): 15–45.
Garcia-Vaquero, M., Lopez-Alonso, M., and Hayes, M. (2017a). Assessment of the functional properties of protein extracted from the brown seaweed Himanthalia elongata (Linnaeus) S. F. Gray. *Food Research International* 99 (Part 3): 971–978.
Garcia-Vaquero, M., Rajauria, G., O'Doherty, J., and Sweeney, T. (2017b). Polysaccharides from macroalgae: recent advances, innovative technologies and challenges in extraction and purification. *Food Research International* 99: 1011–1020.
Gerland, P., Raftery, A.E., Ševčíková, H. et al. (2014). World population stabilization unlikely this century. *Science* 346 (6206): 234–237.
Hayes, M. (2015). Seaweeds: a nutraceutical and health food. In: *Seaweed Sustainability. Food and Non-Food Applications* (ed. B.K. Tiwari and D. Troy). London: Academic Press.
Hayes, M., García-García, M., Fitzgerald, C., and Lafarga, T. (2015). Seaweed and milk derived bioactive peptides and small molecules in functional foods and cosmeceuticals. In: *Biotechnology of Bioactive Compounds*, 669–691. Chichester: Wiley Blackwell.
Heim, G., O'Doherty, J.V., O'Shea, C.J. et al. (2015). Maternal supplementation of seaweed-derived polysaccharides improves intestinal health and immune status of suckling piglets. *Journal of Nutritional Science* 4: e27.

Henchion, M., Hayes, M., Mullen, A.M. et al. (2017). Future protein supply and demand: strategies and factors influencing a sustainable equilibrium. *Foods* 6 (7): 53.

Herman, E.M. and Schmidt, M.A. (2016). The potential for engineering enhanced functional-feed soybeans for sustainable aquaculture feed. *Frontiers in Plant Science* 7: 440.

Hong, Z., Kim, E., Jin, Y. et al. (2015). Effects of supplementing brown seaweed by-products in the diet of Holstein cows during transition on ruminal fermentation, growth performance and endocrine responses. *Asian-Australasian Journal of Animal Sciences* 28 (9): 1296.

Hou, Y., Wu, Z., Dai, Z. et al. (2017). Protein hydrolysates in animal nutrition: industrial production, bioactive peptides, and functional significance. *Journal of Animal Science and Biotechnology* 8 (1): 24.

Hwang, J., Islam, M., Ahmed, S. et al. (2014). Seamustard (*Undaria pinnatifida*) improves growth, immunity, fatty acid profile and reduces cholesterol in Hanwoo steers. *Asian-Australasian Journal of Animal Sciences* 27 (8): 1114.

Islam, M.M., Ahmed, S.T., Mun, H.-S. et al. (2016). Effect of fermented seaweed by-product supplementation on reproduction of Hanwoo cows and growth and immunity of their calves. *Animal Production Science* 56 (11): 1828–1833.

Jones, R., Blunden, G., and Probert, A. (1979). Effects of dietary *Ascophyllum nodosum* on blood parameters of rats and pigs. *Botanica Marina* 22: 393–394.

Kraan, S. (2013). Mass-cultivation of carbohydrate rich macroalgae, a possible solution for sustainable biofuel production. *Mitigation and Adaptation Strategies for Global Change* 18 (1): 27–46.

Kulshreshtha, G., Rathgeber, B., Macisaac, J. et al. (2017). Feed supplementation with red seaweeds, *Chondrus crispus* and *Sarcodiotheca gaudichaudii*, reduce *Salmonella enteritidis* in laying hens. *Frontiers in Microbiology* 8: 567.

Kulshreshtha, G., Rathgeber, B., Stratton, G. et al. (2014). Feed supplementation with red seaweeds, *Chondrus crispus* and *Sarcodiotheca gaudichaudii*, affects performance, egg quality, and gut microbiota of layer hens. *Poultry Science* 93 (12): 2991–3001.

Landa-Cansigno, C., Hernández-Carmona, G., Arvizu-Higuera, D.L. et al. (2017). Bimonthly variation in the chemical composition and biological activity of the brown seaweed *Eisenia arborea* (Laminariales: Ochrophyta) from Bahía Magdalena, Baja California Sur, Mexico. *Journal of Applied Phycology* 29 (5): 2605–2615.

Li, X., Zhang, Z., Qu, S. et al. (2016). Breeding of an intraspecific kelp hybrid Dongfang no. 6 (*Saccharina japonica*, Phaeophyceae, Laminariales) for suitable processing products and evaluation of its culture performance. *Journal of Applied Phycology* 28 (1): 439–447.

Lozano, I., Wacyk, J.M., Carrasco, J., and Cortez-San Martín, M.A. (2016). Red macroalgae *Pyropia columbina* and *Gracilaria chilensis*: sustainable feed additive in the *Salmo salar* diet and the evaluation of potential antiviral activity against infectious salmon anemia virus. *Journal of Applied Phycology* 28 (2): 1343–1351.

Makkar, H.P., Tran, G., Heuzé, V. et al. (2016). Seaweeds for livestock diets: a review. *Animal Feed Science and Technology* 212: 1–17.

Manns, D., Nielsen, M.M., Bruhn, A. et al. (2017). Compositional variations of brown seaweeds *Laminaria digitata* and *Saccharina latissima* in Danish waters. *Journal of Applied Phycology* 29 (3): 1493–1506.

Martins, A., Vieira, H., Gaspar, H., and Santos, S. (2014). Marketed marine natural products in the pharmaceutical and cosmeceutical industries: tips for success. *Marine Drugs* 12 (2): 1066–1101.

McCusker, S., Buff, P.R., Yu, Z., and Fascetti, A.J. (2014). Amino acid content of selected plant, algae and insect species: a search for alternative protein sources for use in pet foods. *Journal of Nutritional Science* 3, e39.
Mols-Mortensen, A., Ortind, E.Á.G., Jacobsen, C., and Holdt, S.L. (2017). Variation in growth, yield and protein concentration in *Saccharina latissima (Laminariales, Phaeophyceae)* cultivated with different wave and current exposures in the Faroe Islands. *Journal of Applied Phycology* 29 (5): 2277–2286.
Morshedi, V., Nafisi Bahabadi, M., Sotoudeh, E. et al. (2018). Nutritional evaluation of *Gracilaria pulvinata* as partial substitute with fish meal in practical diets of barramundi (*Lates calcarifer*). *Journal of Applied Phycology* 30 (1): 619–628.
Nadarajah, S. and Flaaten, O. (2017). Global aquaculture growth and institutional quality. *Marine Policy* 84 (Supplement C): 142–151.
Nielsen, M.M., Manns, D., d'Este, M. et al. (2016). Variation in biochemical composition of *Saccharina latissima* and *Laminaria digitata* along an estuarine salinity gradient in inner Danish waters. *Algal Research* 13: 235–245.
Okab, A.B., Samara, E.M., Abdoun, K.A. et al. (2013). Effects of dietary seaweed (*Ulva lactuca*) supplementation on the reproductive performance of buck and doe rabbits. *Journal of Applied Animal Research* 41 (3): 347–355.
Olmos Soto, J., Paniagua-Michel, J.D.J., Lopez, L., and Ochoa, L. (2015). Functional feeds in aquaculture. In: *Springer Handbook of Marine Biotechnology* (ed. S.K. Kim), 1303–1319. Berlin: Springer.
Pallaoro, M.F., do Nascimento Vieira, F., and Hayashi, L. (2016). Ulva lactuca (Chlorophyta Ulvales) as co-feed for Pacific white shrimp. *Journal of Applied Phycology* 28 (6): 3659–3665.
Peixoto, M.J., Salas-Leitón, E., Brito, F. et al. (2017). Effects of dietary *Gracilaria* sp. and *Alaria* sp. supplementation on growth performance, metabolic rates and health in meagre (*Argyrosomus regius*) subjected to pathogen infection. *Journal of Applied Phycology* 29 (1): 433–447.
Peixoto, M.J., Salas-Leitón, E., Pereira, L.F. et al. (2016). Role of dietary seaweed supplementation on growth performance, digestive capacity and immune and stress responsiveness in European seabass (*Dicentrarchus labrax*). *Aquaculture Reports* 3: 189–197.
Rajauria, G., Foley, B., and Abu-Ghannam, N. (2017). Characterization of dietary fucoxanthin from *Himanthalia elongata* brown seaweed. *Food Research International* 99 (Part 3): 995–1001.
Ratcliff, J.J., Wan, A.H.L., Edwards, M.D. et al. (2016). Metal content of kelp (*Laminaria digitata*) co-cultivated with Atlantic salmon in an integrated multi-trophic aquaculture system. *Aquaculture* 450 (Supplement C): 234–243.
Rebours, C., Marinho-Soriano, E., Zertuche-González, J.A. et al. (2014). Seaweeds: an opportunity for wealth and sustainable livelihood for coastal communities. *Journal of Applied Phycology* 26 (5): 1939–1951.
Regulation (EC) 178/2002. (2002) Laying down the general principles and requirements of food law. Official Journal of the European Communities.
Regulation (EC) 183/2005. (2005) Laying down requirements for feed hygiene. Official Journal of the European Communities.
Regulation (EC) 212/2013. (2013) Replacing Annex I to Regulation (EC) No 396/2005 of the European Parliament and of the Council as regards additions and modifications with

respect to the products covered by that Annex. Official Journal of the European Communities.

Regulation (EC) 396/2005. (2005) On maximum residue levels of pesticides in or on food and feed of plant and animal origin and amending Council Directive 91/414/EEC. Official Journal of the European Communities.

Regulation (EC) 767/2009.(2009) On the placing on the market and use of feed, amending European Parliament and Council Regulation (EC) No 1831/2003 and repealing Council Directive 79/373/EEC, Commission Directive 80/511/EEC, Council Directives 82/471/EEC, 83/228/EEC, 93/74/EEC, 93/113/EC and 96/25/EC and Commission Decision 2004/217/EC. Official Journal of the European Communities.

Regulation (EC) 769/2009. (2009) Establishing the standard import values for determining the entry price of certain fruit and vegetables. Official Journal of the European Communities.

Rjiba-Ktita, S., Chermiti, A., Bodas, R. et al. (2017). Aquatic plants and macroalgae as potential feed ingredients in ruminant diets. *Journal of Applied Phycology* 29 (1): 449–458.

Saito, M. and Hagino, H. (2005). Antihypertensive effect of oligopeptides derived from nori (*Porphyra yezoensis*) and Ala-Lys-Tyr-Ser-Tyr in rats. *Journal of the Japanese Society of Nutrition and Food Science* doi: doi 10.4327/jsnfs.58.177.

Santizo, R., Serrano, J.R., and Corre, V. (2014). Proximate composition and dry matter digestibility of *Ulva lactuca* in the black tiger shrimp *Penaeus monodon*. *ABAH Bioflux* 6 (1): 75–83.

Santoso, S., Suprijatna, E., Setiadi, A., and Susanti, S. (2016). Effect of duck diet supplemented with fermented seaweed wastes on carcass characteristics and production efficiency of indigenous Indonesian ducks. *Indian Journal of Animal Research* 50 (5).

Schiener, P., Zhao, S., Theodoridou, K. et al. (2017). The nutritional aspects of biorefined *Saccharina latissima*, *Ascophyllum nodosum* and *Palmaria palmata*. *Biomass Conversion and Biorefinery* 7 (2): 221–235.

Schleder, D.D., da Rosa, J.R., Guimarães, A.M. et al. (2017). Brown seaweeds as feed additive for white-leg shrimp: effects on thermal stress resistance, midgut microbiology, and immunology. *Journal of Applied Phycology* 29 (5): 2471–2477.

Shpigel, M., Guttman, L., Shauli, L. et al. (2017). *Ulva lactuca* from an integrated multi-trophic aquaculture (IMTA) biofilter system as a protein supplement in gilthead seabream (*Sparus aurata*) diet. *Aquaculture* 481: 112–118.

Silva, D., Valente, L., Sousa-Pinto, I. et al. (2015). Evaluation of IMTA-produced seaweeds (*Gracilaria*, *Porphyra*, and *Ulva*) as dietary ingredients in Nile tilapia, *Oreochromis niloticus L.*, juveniles. Effects on growth performance and gut histology. *Journal of Applied Phycology* 27 (4): 1671–1680.

Singh, B.K., Chopra, R.C., Rai, S.N. et al. (2017). Nutritional evaluation of seaweed on nutrient digestibility, nitrogen balance, milk production and composition in Sahiwal cows. *Proceedings of the National Academy of Sciences, India Section B: Biological Sciences* 87 (2): 437–443.

Sotoudeh, E. and Mardani, F. (2018). Antioxidant-related parameters, digestive enzyme activity and intestinal morphology in rainbow trout (*Oncorhynchus mykiss*) fry fed graded levels of red seaweed, *Gracilaria pygmaea*. *Aquaculture Nutrition* 24 (2): 777–785.

Spiegel, V.D.M., Noordam, M., and Fels-Klerx, H. (2013). Safety of novel protein sources (insects, microalgae, seaweed, duckweed, and rapeseed) and legislative aspects for their application in food and feed production. *Comprehensive Reviews in Food Science and Food Safety* 12 (6): 662–678.

Stévant, P., Marfaing, H., Rustad, T. et al. (2017). Nutritional value of the kelps *Alaria esculenta* and *Saccharina latissima* and effects of short-term storage on biomass quality. *Journal of Applied Phycology* 29 (5): 2417–2426.

Sweeney, T., Meredith, H., Ryan, M.T. et al. (2016). Effects of *Ascophyllum nodosum* supplementation on *Campylobacter jejuni* colonisation, performance and gut health following an experimental challenge in 10 day old chicks. *Innovative Food Science and Emerging Technologies* 37: 247–252.

Valente, L.M.P., Araújo, M., Batista, S. et al. (2016). Carotenoid deposition, flesh quality and immunological response of Nile tilapia fed increasing levels of IMTA-cultivated *Ulva* spp. *Journal of Applied Phycology* 28 (1): 691–701.

Wan, A.H.L., Soler-Vila, A., O'Keeffe, D. et al. (2016). The inclusion of *Palmaria palmata* macroalgae in Atlantic salmon (*Salmo salar*) diets: effects on growth, haematology, immunity and liver function. *Journal of Applied Phycology* 28 (5): 3091–3100.

Wang, S., Xia, Z., Hu, Y. et al. (2017). Co-pyrolysis mechanism of seaweed polysaccharides and cellulose based on macroscopic experiments and molecular simulations. *Bioresource Technology* 228: 305–314.

Weinberg, B. and Saxe, L. 2017. Formed jerky treats formulation and method. Google Patents.

Williams, S.J., Jones, T.N., Lambert, B., and Harp, R. (2017). Effects of Tasco supplementation on stallion semen characteristics: a pilot study. *Journal of Equine Veterinary Science* 52: 91.

Yang, H., Li, Z.-B., Chen, Q. et al. (2016). Effect of fermented *Enteromopha prolifera* on the growth performance, digestive enzyme activities and serum non-specific immunity of red tilapia (*Oreochromis mossambicus* × *Oreochromis niloticus*). *Aquaculture Research* 47 (12): 4024–4031.

Zhou, M., Huenerberg, M., Chen, Y. et al. (2017). 612 effects of the seaweed *Ascophyllum nodosum* on the rumen microbiome and fecal pathogenic *Escherichia coli* serotypes in sheep. *Journal of Animal Science* 95 (Supplement 4): 299–300.

Zhu, D., Wen, X., Li, S. et al. (2017). Evaluation of the red alga *Gracilaria lemaneiformis* and brown alga *Sargassum horneri* as ingredients in diets for white spotted snapper *Lutjanus stellatus Akazaki* juveniles. *Journal of Applied Phycology* 29 (6): 3211–3219.

8

Marine By-Products as a Source of Proteins for Potential Food, Pharma, and Agricultural Feed Use

Maria Hayes

Teagasc Food Research Centre, Food BioSciences Department, Dublin, Ireland

8.1 Introduction

Production of seafood exceeded 160 million tons in 2014 (FAO 2016) and due to the increasing global population, the demand for seafood products is likely to increase further. However, presently, only 50–60% of a fish caught is utilised for food consumption while the remainder of the fish remains unprocessed or is used for the development of fertilisers, animal and fish feeds. By-products from marine processing include blood, fish trimmings, viscera, skin, shell, fins, and others (Gómez-Guillén et al. 2011). By-products should really be termed co-products or rest raw materials as they are often rich in protein, essential amino acids (EAAs), polyunsaturated fatty acids (PUFAs), pigments, and other compounds that may be used for the development of functional food ingredients if handled in a food-grade and hazard analysis and critical control points (HACCP) manner during processing. The type of by-product generated during marine processing varies depending on the type and quantity of fish species being processed. However, in general, by-products from all processing sectors, including whitefish, pelagic fish, shellfish, and demersal species, are rich in proteins although different processing methods may be required for different by-products (Ottesen et al. 2016).

Proteins that may be generated from fish co-products that are commercially available currently include gelatine and collagen. Gelatine has several applications in food products and can be found in confectionery, photography, cosmetics, and pharmaceutical items. Although gelatine for commercial use is largely sourced from bovine and porcine by-products, it has been generated from several fish species and by-products, including salmon, tuna trimmings, blue whiting, boarfish, and several other pelagic species. Moreover, several studies have examined the potential health beneficial effects of salmon gelatine hydrolysates and peptides have been identified with antidiabetic, angiotensin-1-converting enzyme (ACE-1), renin, antimicrobial, and other health benefits.

Another potential source of valuable proteins is fish blood. Fish blood is not currently utilised commercially but the quantities generated may warrant investment in the recovery and utilisation of this by-product for protein generation. For example, in

Novel Proteins for Food, Pharmaceuticals, and Agriculture: Sources, Applications, and Advances, First Edition. Edited by Maria Hayes.

Alaska in 2013, 200 000 tons of salmon were processed. Blood represents about 5% of the weight of a salmon and therefore approximately 20 million pounds of blood could be recovered. This blood is a source of plasma which has been shown previously to contain effective inhibitors of protease enzymes, which can be useful for extending the shelf-life of meat products (Fowler and Park 2015). Indeed, technology for utilisation of warm-blooded animal blood exists already and several bioactive peptides and proteins have been isolated from beef and porcine blood (Lafarga et al. 2015).

This chapter will discuss issues related to the use of fish-derived by-products, including fish blood in food, animal feed, and pharmaceuticals.

In addition, shell waste generated from processing of shellfish such as crab, prawns, and krill is a rich source of the polysaccharide chitin which can be processed further into the more soluble forms chitosan and chitosan oligosaccharides (COSs) (Hayes et al. 2008). However, shell material and shellfish that are too small for commercial use are also a rich source of proteins and pigments and utilisation of novel processing methods could result in the total utilisation of the catch/by-catch in the future. This chapter will also discuss proteins and peptides that are derived from these sources.

8.2 Biological Activities of Marine-Derived Proteins

Biological activities relate to the production of bioactive peptides and amino acids from fish proteins. Bioactive peptides are sequences of amino acids between 2 and 30 in length that impart a health benefit to the consumer that goes above and beyond basic human nutrition. Bioactive peptides can be produced from any food protein source using commercially available enzymes through controlled hydrolysis, lactic acid bacteria (LAB) through fermentation or hydrolysis with acids/bases. The protein from which the peptide is derived is inactive within the parent protein and must be released through hydrolysis/fermentation to have a biological effect. Biological activities that can result for peptides include antimicrobial activities, heart health benefits including inhibition of enzymes important in the human renin-angiotensin-aldosterone system (RAAS) that control blood pressure and salt water balance, including ACE-1 and renin, inhibition of enzymes important in the development of type 2 diabetes including dipeptidyl peptidase IV (DPP-IV), and mental health beneficial peptides including those that inhibit the enzymes prolyl oligopeptidase and others (Rustad and Hayes 2012).

8.2.1 Angiotensin-Converting Enzyme Inhibition (ACE-1)

Angiotensin-converting enzyme 1 is a key enzyme in the RAAS which is responsible for the conversion of angiotensin (Ang) 1 (a vasodilator) into Ang 2 (a vasoconstrictor) which leads to elevated blood pressure in humans and other animals, including dogs and cats. ACE-1 inhibitory peptides derived from food proteins have shown potential in the prevention and management of high blood pressure and several fish-derived protein hydrolysates have been identified as containing ACE-1 inhibitory peptides. In addition, this action prevents the breakdown of the vasodilator bradykinin. In addition to the classic inhibition of ACE-1, emerging evidence suggests that food peptides can exert antihypertensive activity through upregulation of ACE-2 which is an ACE-1 homologue that counterbalances the detrimental effect of elevated ACE-1, which can result in improvement of endothelial function, as well as reduced vascular oxidation and inflammation (Wu et al. 2017).

There is a total of 891 ACE-1 inhibitory peptides reported in BIOPEP (www.uwm.edu.pl/biochemia/index.php/en/biopep). ACE-1 inhibitory peptides have been identified previously from fish by-products (Rustad 2003), milk (Hayes et al. 2007), and other cereal sources including barley (Gangopadhyay et al. 2016). ACE-1 inhibitory peptides can be competitive, non-competitive or mixed inhibitors of ACE-1.

Previously, three hypotensive rapeseed protein-derived peptides with the amino acid sequences TF, LY, and RALP were characterised and their interactions with renin and ACE-1 enzyme characterised using molecular docking, fluorescent, and CD spectra techniques in conjunction with enzyme kinetic studies (He et al. 2014). Renin has two catalytic aspartic acid residues, Asp32 and Asp 215, that are located in the active site cleft and can fit up to seven amino acid residues. ACE-1 has a conserved HEXXH zinc-binding motif for controlling blood pressure. Inhibitors can work by filling the active site of these enzymes and binding to crucial amino acid residues, preventing substrate binding. In addition, peptides can change the structure of the ACE-1 or renin proteins around the active site from molecular collisions with inhibitors. As bonding is important for these inhibition strategies, the charge of the amino acids within a peptide and their location are important factors when studying the mechanism of action of ACE-1 inhibitory peptides. Figure 8.1 shows the ideal amino acid composition of ACE-1 and renin inhibitory peptides.

8.2.2 Structure of Peptides Important in the Inhibition of Enzymes Related to the Development of Type 2 Diabetes

The DPP-IV enzyme can inactivate incretins, glucose-dependent insulinotropic polypeptide (GIP), and glucagon-like peptide-1 (GLP-1). These are gut-derived hormones that play a significant role in glucose regulation by stimulating pancreatic glucose-dependent insulin, suppressing glucagon release, promoting beta-cell proliferation and survival, retarding gastric emptying, and modulating appetite (Lacroix and Li-Chan 2013). The administration of orally available DPP-IV inhibitors such as

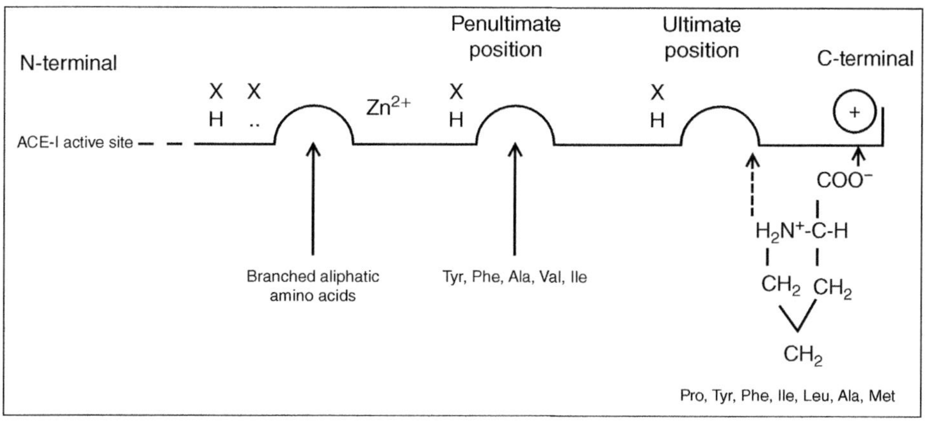

Figure 8.1 Ideal amino acid composition of an ACE-1 and renin inhibitory peptide.

sitagliptin, vildagliptin, and saxagliptin can extend the half-life of these hormones and may be a useful strategy in the treatment and control of type 2 diabetes. 408 DPP-IV inhibitory peptides are found in BIOPEP and these were derived from proteins including Atlantic salmon skin gelatine hydrolysate (Li-Chan 2012), bovine, caprine (Lacroix and Li-Chan 2013), and camel milk as well as from egg yolk protein (Zambrowicz et al. 2015), meat (Gallego et al. 2014), and rice bran.

Dipeptidyl dipeptidase-IV inhibitors, which are also known as gliptins, are designed to mimic the N-terminal dipeptide that is cleaved by DPP-IV. Commercially available DPP-IV inhibitors, including vildagliptin (Galvus®, Novartis) and saxagliptin (Onglyza™, Bristol Myers Squibb) which are approved for use in the USA and Europe, operate in this way. A typical competitive inhibitor will compete with the natural substrate for DPP-IV for binding to the active site. Proline-containing peptides such as Ile-Pro are excellent competitive inhibitors of DPP-IV and have IC_{50} values at very low concentrations (Hatanaka et al. 2012). Long peptides with up to 30 amino acids can act as non-competitive inhibitors of DPP-IV by blocking formation of the DPP-IV active dimer.

Bioactive peptides have also been reported that can regulate factors important in the control of cholesterol, including platelet activating factor acetylhydrolase (PAF-AH) and beta-site APP-cleaving (BACE)-1 enzymes. The vascular inflammatory role of PAF-AH is thought to be due to the formation of lysophosphatidyl choline and oxidised non-esterified fatty acids. This enzyme is considered a promising therapeutic target for the prevention of atherosclerosis and there is a need to expand the available chemical templates of PAF-AH inhibitors (Fitzgerald et al. 2013).

8.3 Fish Protein Hydrolysates

There is no single definition of what constitutes a marine by-product but often they are categorised according to their end use as products that can be used for human consumption, animal feed, fertiliser, and actual waste which is now often incinerated (Rustad 2003). Fish by-products include fish offal, backbone, head, liver, gonads, skins, fins, and all materials that are produced as by-products of the filleting process.

The use of these by-products as food, feed, and fertiliser is not a new area of research. Indeed, in Scandinavia, in 1903, cod heads were collected for production of feed and fertiliser and today in Iceland and Norway, companies such as Matís are active in research concerning utilisation of this product. Also, several traditional Scandinavian and Asian food products result from fermentation of fish for preservation purposes (Table 8.1).

Furthermore, the use of by-products in the manufacture of fish sauces has a long tradition in both Asia and Scandinavia and fish sauce was eaten in ancient Greece and Rome (Rustad 2003). Fish sauce is manufactured using salt and fish raw materials stored at ambient temperatures for a period of up to six months. Endogenous enzymes found in the fish material and microbial fermentation help to degrade the fish proteins, resulting in a protein-rich liquid. Moreover, fish silage is also produced from fish by-products and is used for animal feed. It is manufactured using acids such as formic acid at concentrations of between 2% and 3%. The resulting fish mince is stored until all the by-product material is degraded by endogenous enzymes such as fish pepsin and, if stored correctly, the fish silage will have a pH of 3–4. This material is then used as animal feed and can also be processed further to separate oils and proteins (Rustad 2003).

Table 8.1 Traditional fermented fish products found in Scandinavia and Asia.

Fermented food product	Local name	Fish species used	Fermentation time	Country of origin
Fermented *Cyprinidae* sp.	Prahoc	*Cyprinidae* sp.	Variable	Cambodia
Fermented *Ctenops* sp.	Ketjap-ikan	*Ctenops* sp.	Variable	Indonesia
Fermented shrimp	Trassi	Shrimp	Variable	Indonesia
Katsuwonus pelamis	Qunaga	*Katsuwonus pelamis*	Variable	Japan
Sardinella perforate (Sardines)	Patis	Sardines	6–12 months	Philippines
Atlantic herring (*Clupea harengus*)	Gaffelbitar	Atlantic herring	12–18 months	Scandinavia
Atlantic herring (*Clupea harengus*)	Tidbits	Atlantic herring	12–18 months	Scandinavia
Atlantic herring (*Clupea harengus*)	Surströmming (sour herring)	Atlantic herring	12–18 months	Sweden
Trout (*Salmo trutta*)	Rakfisk	Trout	3–12 months	Norway and Sweden
Shark	Hákarl	Shark	Variable	Iceland
Several fish species	Gravadlax	Several fish species	Variable	Germany

Fish protein hydrolysates have been made from viscera, fish frames, flesh from backbones, and heads (Gildberg 1992; Rustad 2003) and from fish that are not suitable for processing such as boarfish and in some instances blue whiting (Hayes et al. 2016a,b). Hydrolysates may be defined as proteins that are broken down by chemical or enzymatic degradation to peptides with varying sizes. It is a suitable technology for adding value to fish by-products as the hydrolysates often have improved technofunctional and/or bioactive properties due to the presence of peptides less than 3 kDa in size. In addition, fish protein hydrolysates can result from autolytic process of the digestive enzymes present in the fish by-product itself and products such as fish paste and soup are often produced using this method (Theary et al. 2013).

If using commercially available enzymes for production of a fish protein hydrolysate, it is often necessary to deactivate native fish enzymes using heat treatment prior to carrying out the hydrolysis with the commercial enzyme of choice. Several enzymes are reported in the literature which, have been used to produce fish protein hydrolysates. These include alcalase®, protamex, esperase, neutrase, flavourzyme, papain, and others (Kristinsson and Rasco 2000; Hayes et al. 2016a,b; Vazquez et al. 2017). This method has several advantages over autolysis: it can be controlled and the activity of peptides produced in the hydrolysate can be predicted based on the protein sequence of the by-product and the activity of the selected enzyme using *in silico* analysis.

Fish protein hydrolysates have potential for use as food ingredients due to their technofunctional activities, including emulsifying and foaming, water holding activities, oil holding activities, improved solubility, and water activities as well as health beneficial activities such as inhibition of enzymes important in the development of cardiovascular disease, such as ACE-1 and renin, and type 2 diabetes, such as DPP-IV. Indeed, in Japan, the home of functional foods, several fish protein hydrolysate products exist that have health claims related to improved heart health and other health attributes (Table 8.2).

There are some sensory disadvantages associated with fish protein hydrolysates, however, including bitter taste due to the production of peptides less than 5 kDa. Enzymes (usually proteases) act by cleaving proteins into smaller peptides and free amino acids that are more water soluble and which can alter the sensory and surface active properties compared to the intact fish protein. A major drawback of the use of fish protein hydrolysates is the production of bitter and unpalatable flavours due to exposure of hydrophobic amino acids during the hydrolysis process (Aspevik 2016). Sensory issues can be overcome by the use of membrane technologies, generation of hydrolysates at slightly alkaline conditions and other methods such as use of 2-butanol partitioning to separate the hydrophobic amino acids out of the hydrolysate, although all methods add cost to the production process. Bitterness can be related to the specificity of the enzyme used, the degree of hydrolysis (DH), and the molecular weight distribution of the resultant fish protein hydrolysate (Aspevik 2016).

8.4 Fish Blood Proteins

Fish blood contains valuable proteins, including blood plasma. Fish blood plasma isolated from salmon was found to contain a number of protease inhibitors (PIs), including alpha-2-macroglobulin, a protein that inhibits several classes of proteases (Barrett 1981; Fowler and Park 2015). Plasma from animal sources has also been investigated for PI activities and pig plasma was found to improve the gel strength of surimi manufactured from big-eye snapper (Benjakul et al. 2001).

8.4.1 Preparation of Fish Plasma from Salmon

Collection of salmon blood and preparation of plasma has been described by Fowler and Park (Fowler and Park 2015). Briefly, blood was collected by bleeding fish into bottles containing 3.8% sodium citrate solution as an anticoagulant and gently mixing at a ratio of 9:1 (v:v) blood to sodium citrate solution. Blood was kept on ice and then centrifuged for 15 minutes at 1500× g at 4 °C using a Beckman J6-MI centrifuge. The supernatant was regarded as salmon plasma and concentrated. The salmon plasma was dried using a freeze-drier and subsequently stored.

8.4.2 Concentration of Fish Plasma from Salmon

Concentration was carried out at 4 °C using a tangential flow filtration system from Millipore. Plasma was recirculated through the membrane at a feed pressure of 30 psi and a retentate pressure of 10 psi until the permeate flow was very low. The system was then cleaned using sodium hydroxide at a concentration of 0.1 N. The concentration of protein in the salmon plasma was increased from 23.88 ± 1.07 mg/mL to 36.66 ± 0.36 mg/mL using this method. The protein content of plasma from salmon is

Table 8.2 Fish protein hydrolysate products on the market with claimed health benefits.

Product name	Type of product	Composition	Health benefit	Application	Country
Seacure®	Natural product	Hydrolysate of fish protein (Pacific whiting)	Intestinal health	Sold as a nutraceutical	Proper Nutrition, USA
Nutripeptide®	Natural product	Hydrolysate of fish protein	Postprandial blood glucose control	Sold as a nutraceutical	SILAB, France
Hydro MN Peptide®	Natural product	Hydrolysate of fish protein (marine cartilage [collagen] peptide product)	Postprandial blood glucose control	Sold as a nutraceutical	Celergen, Swiss
Gabolysat PC 60®	Natural product	Hydrolysate of fish protein	Anxiolytic action	Sold as a nutraceutical	France
Stabilium®	Natural product	Hydrolysate of fish protein (*Garum armoricum*® fish hydrolysate)	Anxiolytic action	Sold as a nutraceutical	France
Procalm®	Natural product	Hydrolysate of fish protein	Anxiolytic action	Sold as a nutraceutical; pet health (canine anxiety disorders)	France
Fish gelatine	Natural product	Hydrolysis of fish gelatine	Natural supplement; bone health	Sold as a nutraceutical	France/China/Japan/USA
Soluble fish protein hydrolysate	Natural product	Hydrolysate of blue whiting	ACE-1 inhibitory, antiproliferative, satiety	Powder, nutraceutical, ingredient	Ireland
Valtyron®	Natural product	Hydrolysate of sardine muscle	ACE-1 inhibitory, antihypertensive, novel food status	Powder, supplement, nutraceutical	Japan

thought to be less during the spawning phase and therefore this would be an important consideration when harvesting plasma. The salmon plasma isolated by Fowler and Parker was found to contain both cysteine and serine PIs and was found to be an effective PI of proteolysis in both Pacific whiting surimi and salmon mince. Salmon plasma could be used in salmon fillets to inhibit protease-mediated texture softening (Fowler and Parker 2015).

8.4.3 Protease Inhibitors from Fish Blood

Protease inhibitors are well known to have several applications in medicine and biotechnology. Ammonium sulfate fractionation can be used to salt out PIs and fish blood has been found to contain PIs. PIs can target or inhibit proteases – protein-digesting enzymes. These proteases play a crucial role in many biological events including digestion, blood coagulation, and apoptosis. Regardless of their crucial roles, they need to be checked regularly by PIs as their excess may possibly damage the host organism. They are classified on the basis of the amino acid composition of PIs where protease-PI enzymatic reactions occur, for example serine, cysteine, and aspartic acid.

Nowadays, various PIs are being researched to fight various parasitic or viral diseases including malaria, schistosomiasis, colds, and flu viruses. Some PIs can principally be used for treatment of hypertension and congestive heart failure by blocking conversion of angiotensin 1 to angiotensin 2, for example angiotensin-converting enzyme inhibitors (ACEIs). Also PIs target amyloid beta-peptide (Abeta) levels in the brain which is primarily responsible for the development of Alzheimer's disease (AD) (Shamsi et al. 2016). Also, PIs inhibit enzymatic activity of the HIV-1 protease receptor (PR) by preventing cleavage events in Gag and Gag-Pol that result in production of non-virulent virus particles. They can also be used in food manufacture to extend the shelf-life of fish products such as surimi (Fowler and Park 2015).

Rainbow trout plasma was found to increase gel strength in Alaskan pollock surimi (Li et al. 2008a) and a cysteine PI was also isolated from chum salmon plasma (Li et al. 2008b).

8.4.4 Clotting Agents from Fish Blood

Recent work in this area was carried out as part of a PhD at the engineering department of the University of Pennsylvania. The aim of this PhD was to isolate clotting factors from salmon blood. Haemorrhage due to trauma is the leading cause of preventable death among American soldiers, according to the National Institute of Trauma. Uncontrollable bleeding is also seen regularly in civilian incidences of trauma and is a common major surgical complication.

The human blood clotting process involves a complex cascade of tightly regulated enzymatic reactions. Two of the most important proteins in this cascade are fibrinogen and thrombin. Thrombin is an enzyme that activates fibrinogen monomers to form a polymeric fibrin network, which is the basis of a blood clot. During trauma, a state of consumptive coagulopathy, the body depletes these two proteins, causing severe bleeding. A novel way to counteract haemorrhage is to supply additional thrombin and fibrinogen to the focal injury site. However, fibrinogen has proved technically challenging to produce using recombinant technologies and mammalian-based proteins carry the risk of pathogen transmission and immune response.

8.4.5 Salmon Blood Protein Used in Clotting Bandage Manufacture

Salmon-derived proteins, on the other hand, overcome both of these obstacles. DiamondStat is a novel haemostatic bandage that delivers fibrinogen and thrombin purified from salmon blood. It is a 4 × 4 in. adhesive bandage that delivers 10 mg cm^{-2} of fibrinogen and 90 units cm^{-2} of thrombin to effectively stop haemorrhage. The production of DiamondStat bandages begins with harvesting blood from salmon. Through a series of centrifugations and precipitations, prothrombin (a zymogen precursor to thrombin) and fibrinogen are extracted from the blood. Additional precipitations and filtrations further purify the fibrinogen solution to pharmaceutical grade. The prothrombin is converted to thrombin in an immobilised snake venom catalyst column. Thrombin is then purified through an affinity column and ultrafiltration. Both thrombin and fibrinogen solutions are run through endotoxin removal columns and then sprayed onto pieces of gauze. The proteins are lyophilised onto the gauze and the final bandage, which consists of fibrinogen and thrombin gauze pieces and an adhesive backing, is assembled. Bandages are sterilised via gamma irradiation and are then ready for use. At a capacity of 300 000 bandages per year and $800 per bandage, DiamondStat production is very profitable. It will yield an internal rate of return of 289.76% and a net present value of over $489 million.

8.4.6 Potential Applications

Along with the potential applications described above, several products derived from animal blood are commercially available (Table 8.3) and used in the manufacture of food products.

Currently, no commercial fish processors recover or harvest blood using GMP although companies including Matís (Iceland) and several research projects are looking at this topic. If technologies for recovery of fish blood are improved and implemented, it is likely that processing companies can find markets for recovered fish blood proteins

Table 8.3 Blood products derived from blood harvested from warm-blooded animals.

Blood fraction	Use	Food products	References
Whole blood	Protein	Spanish blood sausage	Santos et al. 2003
Animal plasma	Fat replacer	Fermented Bologna sausage	Cofrades et al. 2000
Bovine globin and plasma	Fat replacer	Ham pâté	Viana et al. 2005
Bovine haemoglobin	Iron fortification	Cookies	Walters et al. 1993
Porcine plasma	Protease inhibitor	Surimi	Visessanguan et al. 2000
Chicken plasma	Protease inhibitor	Surimi	Rawdkuen et al. 2004
Porcine transglutaminase, fibrinogen, thrombin	Binder	Restructured meat products	Tseng et al. 2006

that compete with blood proteins derived from warm-blooded animals such as cows and pigs. Furthermore, fish blood-derived proteins would be considered halal, kosher and suitable for use by vegetarians, thereby offering expanded markets for marine processors and ingredient developers.

8.5 Fish Testes

Testes from fish such as cod contain several interesting compounds that are discarded as waste. In countries such as Iceland, companies such as Matis are exploring the use of cod testes as a source of positive nuclear proteins, nucleotides, and phospholipids. Proteins from cod testes may stimulate the immune system and could potentially be used as ingredients in the manufacture of fish food (Khora 2013; Ottesen et al. 2016). Nuclear proteins have a rich proportion of positive amino acids, including taurine and arginine. Research is currently being carried out that looks at the use of high and low pH values for recovery of proteins (pH shift approach) which was developed by Kristinsson and colleagues (Kristinsson et al. 2005). The proteins recovered using this method contain less fat, have antioxidant properties, and have less odour than proteins generated using hydrolysis.

8.6 Fish Collagen and Gelatine

Fish collagen can be isolated from fish by-products including skins, bones, and fins. By-products from fish processing represent a serious environmental and economic problem due to inadequate disposal options and/or costs associated with disposal in landfills. Processing leftovers including trimmings, fins, frames, heads, shells, skin, viscera, and stickwater/effluent are currently used in Ireland for the production of fishmeal, fish oil, fertiliser, and animal feeds (Hayes and Mckeon 2014). Another important source of by-products is the solid waste from surimi processing, which can amount for 50–70% of the original raw material (Gómez-Guillén et al. 2011).

Gelatine is a soluble protein obtained by partial hydrolysis of collagen, the main fibrous protein constituent in the bones, cartilages, and skins of animals (Gómez-Guillén et al. 2011). Collagen is the most abundant protein in mammals and is the major protein constituent of skin, cartilage tissues, blood vessels, and teeth. It is found with other proteins such as elastin and proteoglycans around the cells in tissues where it forms the extracellular matrix. The collagen molecule is a triple helix, with three alpha-chains that adopt a three-dimensional structure suitable for intramolecular hydrogen bonding (Hayes and Mckeon 2014). It contains all of the 20 natural amino acids, with a particularly high percentage of glycine, hydroxyproline, and proline. Collagen mostly consists of tripeptides, with frequent repetitions of the sequence Gly-Pro-X or Gly-X-Hyp, and the distribution of polar and non-polar amino acid residues at the X position determines the order of aggregation of the molecule. Denaturation causes total or partial separation of the collagen chains due to destruction of the hydrogen bonds, causing loss of the triple helix conformation, and following denaturation, the polymers adopt a coiled form (Karim and Bhat 2009). Gelatine stability is thought to be influenced by the proportion of total amino acids and these can vary depending on the source of collagen.

The manufacture of gelatine includes treatment of raw material (fish skins, bones) with dilute acid or alkali, which causes a partial cleavage of the cross-links in collagen structure, resulting in formation of 'warm water-soluble collagen', that is, gelatine (Schrieber and Gareis 2007). It is known that various marine processing by-products, such as fish skin, bones, scales, surimi production discharge waste, and squid skin, represent a good source of gelatine (Karim and Bhat 2009; Norziah et al. 2009; Nagarajan et al. 2012).

The manufacture of fish gelatine hydrolysates using commercially available enzymes offers ingredient companies the opportunity to develop bioavailable, odourless food powder ingredients for use as functional foods with health benefits, including prevention of sarcopenia, heart health benefits and others (Ranawana and Kaur 2013). Challenges related to the use of fish gelatine include removal of odour, colour, gelling, and film forming properties as well as consistency in gelatine amino acid composition. Compared to mammalian sources, gelatines derived from fish show notable differences in physical and chemical properties. Gelatine is rated based on parameters including solubility, transparency, colour, odour, and taste, and functional properties including rheology, moisture, ash, protein, pH, setting point and time, melting point and time, gel strength, and viscosity.

During production of gelatine, native collagen must be pretreated before it can be converted into a form suitable for extraction (Gómez-Guillén et al. 2011). This is achieved by heating in water at temperatures higher than 45 °C. A chemical pretreatment is intended to break non-covalent bonds in order to disorganise the protein structure, and produce adequate swelling and collagen solubilisation (Karim and Bhat 2009; Gómez-Guillén et al. 2011). Since gelatine is obtained by denaturation of collagen, its properties are greatly influenced not only by the species or tissue from which it is extracted but also by the extraction process, which may depend on pH, temperature, and time during both pretreatment and extraction phases (Gómez-Guillén et al. 2002). Pretreatment with acids is normally used.

8.6.1 Fish Collagen Hydrolysates

For food and functional food use, collagen can be broken down into gelatine which can be broken down further into hydrolysed collagen. Hydrolysed collagen is a polypeptide composite made by further hydrolysis of denatured collagen or gelatine and the molecular weights are within the range approximately 500–25 000 Da (Mohammad et al. 2014). The molecular mass and the size of the peptide molecules are deliberately reduced by hydrolysis using specific proteolytic enzymes. This results in a product that can be easily dissolved in cold water and which loses its gelling properties but retains its surface active properties. The mechanisms involved in processing hydrolysed collagen are demineralisation, extraction of collagen to gelatine, enzymatic hydrolysis to obtain hydrolysed collagen, ion exchange, filtration, evaporation, sterilisation, and finally drying (Mohammad et al. 2014). A number of studies have used enzymatic hydrolysis of collagen or gelatine for the production of bioactive peptides.

Hydrolysed gelatine and collagen have an attractive molecular profile. This contributes to their wide range of application. The molecular profile is dependent on the raw materials and especially the manufacturing process used. By employing precisely controlled enzymatic hydrolysis, a product can be obtained with a mean molecular weight within a specified range.

Fish skin can be an important by-product in the fisheries industry. For example, fish processors produce skinned and deboned fish, which results in substantial amounts of fish skins and bones as by-products. One of the problems associated with these by-products is their heterogeneity and the hydroproline content of the by-product collagen (Blanco et al. 2017). Beneficial effects of collagen hydrolysates which are produced by enzymatic degradation of gelatine extracted from collagen (which is extracted from fish by-products using a number of methods including hot-water extraction) include antioxidant, antiageing, antiatherosclerotic, antiosteoportic, antiosteoarthritic, anti-inflammatory, antitumour, wound healing, antiobesity, and hypoglycaemic activities (Song and Li 2017). Few human dietary intervention studies have been carried out concerning this but one human study performed using the collagen hydrolysate Pure Gold Collagen supplement, which recruited 300 people aged 18–74, found that it can reduce the signs of ageing (Borumand and Sibilla 2014). Fish protein hydrolysates and collagen protein hydrolysates are shown in Table 8.4.

8.7 Stickwater Proteins Recovered using Membrane Filtration

Processing of fish by-products can generate an aqueous stream known as stickwater that contains a large proportion of solids and which is often discharged into the marine environment, posing both a potential environmental problem and loss of revenue (Pedersen et al. 2010). Stickwater samples recovered from cod and pollock processing were found to contain approximately 6% protein on a wet weight basis, and after freeze drying, samples contained 70.5–86.2% protein and 10.6–13.9% ash, with variable amounts of lipid respectively (Bechtel 2005).

Membrane filtration can be used to recover the solids from stickwater while producing water that has a minimal environmental impact. Membrane technology has been used quite successfully in the dairy processing sector for recovery of solids from process waters. However, stickwater is gelatinous in nature and has high viscosity which leads to a tendency towards fouling of membranes. Use of ceramic and polymeric membranes can assist in overcoming these challenges. Pilot studies have found that membranes with a pore size of 0.01 μm would retain 70% of the solids contained in stickwaters (Pedersen et al. 2010). Tests carried out in Alaska found that spiral wound polymeric membranes with a molecular weight cut-off (MWCO) of 300 000 Da showed promise in the recovery of solids with protein contents of up to 80% from stickwater and were less expensive than ceramic membrane filters (Pedersen et al. 2010). Drying of these stickwater proteins could be achieved using vacuum methods, freeze or spray-drying. The recovered proteins could have potential for use as binder agents or as attractant or appetite enhancers for aquaculture purposes (Pedersen et al. 2010).

8.8 Functional Applications of By-Product Protein Hydrolysates

Functional properties are physicochemical characteristics that influence the performance of the protein in different food systems during different stages from processing

Table 8.4 Fish protein hydrolysates and collagen/gelatines obtained from fish by-product proteins.

Fish raw material	Enzyme used	Characteristics	References
Clupea harengus protein hydrolysate (raw herring)	Alcalase	Protein 87.9%, fat 4%, moisture 4.7%, ash 12.5%	Hoyle and Merritt 1994
Mallatus villosus protein hydrolysate	Alcalase	Protein 72.4%, fat 0.18%, moisture 6.34%, ash 20.8%	Shahidi et al. 1995
Merluccius productus solid waste protein hydrolysates	Alcalase	Protein 82.25%, fat 3.94%, ash 13.62%	Benjakul and Morrissey 1997
Chlamys islandica processing waste		Protein 81.3%, fat 0.56%, moisture 5.5% and ash 12.6%	Mukhin et al. 2001
Salmo salar head protein hydrolysate		Protein 82.3%, fat 0.8%, moisture 5.3% and ash 10.4%	Gbogouri et al. 2004
Oncorhynchus nerka head protein hydrolysate	Alcalase Flavourzyme Palatase	Protein 63.3%, fat 23.7%, moisture 5.9% and ash 7.1%	Sathivel et al. 2006
Tuna by-product protein hydrolysate		Protein 66.40%, fat 2.37%, moisture 7.25% and ash 25.94%	Nilsang et al. 2005
Acipenser persicus visceral protein hydrolysate		Protein 65.82%, fat 0.18%, moisture 4.45% and ash 7.67%	Ovissipour et al. 2009
Ctenopharyngodon idella skin protein hydrolysate		Protein 92.9%, fat 0.2%, moisture 2.87% and ash 5.86%	Wasswa et al. 2007
Spanish mackerel (*S. niphonius*) acid collagen hydrolysate	Acid and pepsin	Protein 78.37%, fat 10.33%, moisture 0.72% and ash 10.58%	Chi et al. 2014
Megrim (*Lepidorhombus boscii*) skin waste	Acetic acid	Gelatine	Gómez-Guillén et al. 2002
Dover sole skin (*Solea vulgaris*)	Acetic acid	Gelatine	Gómez-Guillén et al. 2002
Tilipia fish skins	Sodium hydroxide, butyl alcohol	Collagen	Tang et al. 2015

to consumption (Šližytė et al. 2005). Hydrolysis can improve the performance of fish proteins in food systems as it alters the isoelectric point and changes the structure of the protein molecule. Factors including the composition of the substrate, selection of the enzyme, degree of hydrolysis, and further downstream processing steps such as drying and filtration methods can affect the functional characteristics. Protein functional properties are characterised according to hydrodynamic and surface active properties including foaming, emulsifying, solubility, gelation, water, and oil holding capacities, and solubility and degree of hydrolysis are directly related with a greater DH, resulting in improved solubility (de Castro and Sato 2013).

Conditions used for digestion of peptides

In vitro conditions	Mouth	Stomach	Small intestine
Enzymes added	α-amlyase	Pepsin and gastric lipase	Trypsin, chymotrypsin pancreatic lipase
Temperature	37°C	37°C	37°C
pH	7	1.3–2.5 (~3)	6.5–6.8 (~7)
Simulated uided	Simulated salivary fluid **uid**(SSF)*	Simulatedgastricfluid (SGF)*	Simulated intestinal fluid (SIF)*
Action	Chewing and deglutition	Storage, grinding, mixing gastric emptying (duodenum)	Intestinal transit, nutritional absorption
Time length	2 mins	2 hrs	2 hrs

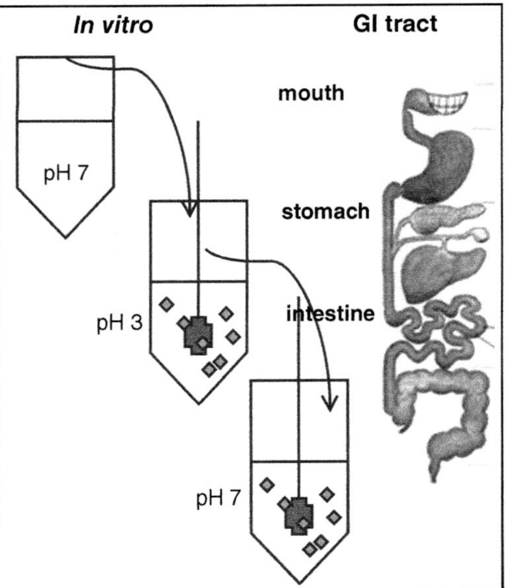

Figure 8.2 The INFOGEST gastrointestinal digestion model.

8.8.1 Solubility

The increased solubility of fish protein hydrolysates compared to that of original proteins is due to the loss of the secondary and tertiary structure and the release of small peptides. Proteins and protein hydrolysates have the lowest solubility at the isoelectric point (pI) (Kristinsson and Rasco 2000). Solubility can be attributed to both the net charge of peptides that increases as pH moves away from pI and surface hydrophobicity which promotes aggregation and hydrophobic interaction.

8.8.2 Water Holding Capacity

Fish protein hydrolysates in general have excellent water holding capacity (WHC) and can be used to increase the cooking yield of products and can be used in intermediate moisture foods as binding agents and to improve texture (Chiang et al. 1999).

8.8.3 Oil Absorbing Capacity

Hydroxyproline content affects oil absorbing capacity and hydrolysates with a higher proportion of charged amino acids, such as aspartic acid, glutamic acid, lysine, and arginine, are able to absorb more fat which is an important attribute in the manufacture of confectionery products (Šližyté et al. 2009).

8.8.4 Emulsifying Properties

Based on the work of Chobert et al. (1998), fish protein hydrolysates with small peptides with low molecular weights may not have good emulsifying properties as the peptides may be too amphiphilic. Other researchers have suggested, however, that the physical and chemical properties of peptides play an important role in functional performance (Kristinsson and Rasco 2000). The amino acid composition and sequence are more important for emulsifying properties than the length of generated peptides (Rahali 2000). Environmental factors including pH, temperature, and degree of hydrolysis therefore all have an effect on emulsifying properties. In addition, in general, fish protein hydrolysates have a greater proportion of EAAs compared to non-essential amino acids (nEAAs) compared to unhydrolysed proteins.

8.8.5 Bioavailability of Fish-Derived Hydrolysates and Peptides

The bioavailability of peptides is correlated with bioaccessibility for gastrointestinal tract absorption, gut wall transport coefficient, and resistance to systemic metabolism (Ting et al. 2015). Bioaccessibility can be modelled *in vitro* using the INFOGEST and TNO Gastro-Intestinal Model (TIM-1) methods where the effect of pH, ionic interaction, enzyme activity, temperature, and solubility of the peptide can be determined using static, semi-static, or dynamic models (Minekus 2015). These methods are limited, however, due to the variability of the microbiome in the gastrointestinal tract of humans and animals. However, they can provide a useful guide for researchers and industry towards profiling of a peptide containing hydrolysate prior to assessing bioavailability in animal models such as rats or pigs. The INFOGEST model has been used to determine the potential bioaccessibility of proteins and lipids generated using the pH shift process from the microalga *Nannochloropsis oculata*. An overview of the INFOGEST model is shown in Figure 8.2.

There are a number of advantages and disadvantages to using simulated gastrointestinal digestion models and cell models with Caco-2 cells . An excellent review carried out by Carbonell-Capella et al. discusses these. Briefly, advantages include that the assays are simple and relatively easy to carry out, are inexpensive and a number of samples can be screened simultaneously. Furthermore, reference samples can be used to validate the method. Disadvantages include that they are not truly representative of the animal or human gut, the oral and large intestine phases are often not included and it is difficult to extrapolate and relate to the *in vivo* data (Carbonell-Capella et al. 2014). Other models use transcytosis assays using cells including the hCMEC/D3 cell line to assess whether peptides can cross the blood–brain barrier (BBB) (Hayes et al. 2016a,b).

8.9 Challenges and Conclusions

The major challenges concerning the use of fish protein hydrolysates and bioactive peptides for health include (i) characterisation costs associated with the laboratory, industry, animal, and human models, (ii) costs associated with production, including MWCO filtration and up-scaling, (iii) sensory issues concerning bitterness and taste (Lafarga and Hayes 2017) as well as challenges concerning marketing and compliance with regulations, as outlined earlier. However, identification and characterisation methods including HPLC and mass spectrometry are becoming more mainstream but the costs associated with animal and human dietary intervention trials remain high. The use of by-products from fish food processing seems to be the most feasible option economically and environmentally.

References

Aspevik, T. 2016. Fish protein hydrolysates based on Atlantic salmon by-products: enzyme cost-efficiency and characterisation of sensory, surface active and nutritional properties. PhD thesis, University of Bergen.

Barrett, A.J. (1981). Alpha 2-macroglobulin. *Methods in Enzymology* 80 (1): 737–754.

Bechtel, P.J. (2005). Properties of stickwater from fish processing byproducts. *Journal of Aquatic Food Product Technology* 14 (2): 25–38.

Benjakul, S. and Morrissey, M.T. (1997). Protein hydrolysates from Pacific whiting solid wastes. *Journal of Agricultural and Food Chemistry* 45 (9): 3423–3430.

Benjakul, S., Srivilai, C., and Vissessanguam, W. (2001). Porcine plasma protein as a proteinase inhibitor in bigeye snapper muscle and surimi. *Journal of the Science of Food and Agriculture* 81: 1039–1046.

Blanco, M., Vázquez, J.A., Pérez-Martín, R.I., and Sotelo, C.G. (2017). Hydrolysates of fish skin collagen: an opportunity for valorizing fish industry byproducts. *Marine Drugs* 15 (5): 131–137.

Borumand, M. and Sibilla, S. (2014). Daily consumption of the collagen supplement pure gold collagen® reduces visible signs of aging. *Clinical Interventions in Aging* 9: 1747–1758.

Carbonell-Capella, J.M., Buniowska, M., Barba, F.J. et al. (2014). Analytical methods for determining bioavailability and bio-accessibility of bioactive compounds from fruit and vegetables: a review. *Comprehensive Reviews in Food Science and Food Safety* 13 (2): 155–171.

Chi, C.-F., Cao, Z.-H., Wang, B. et al. (2014). Antioxidant and functional properties of collagen hydrolysates from Spanish mackerel skin as influenced by average molecular weight. *Molecules* 19: 11211–11230.

Chiang, W.-D., Shih, C.-J., and Chu, Y.-H. (1999). Functional properties of soy protein hydrolysate produced from a continuous membrane reactor system. *Food Chemistry* 65 (2): 189–194.

Chobert, J.-M., Nacka, F., Burova, T. et al. (1998). Inductions of new physicochemical and functional properties by the glycosylation of whey proteins. *Journal of Protein Chemistry* 17 (5): 495–503.

Cofrades, S., Guerra, M., Carballo, J. et al. (2000). Plasma protein and soy fiber content effect on bologna sausage properties as influenced by fat level. *Journal of Food Science* 65: 281–287.

De Castro, R.J.S. and Sato, H.H. (2013). Antioxidant activities and functional properties of soy protein isolate hydrolysates obtained using microbial proteases. *International Journal of Food Science and Technology* 49 ((2): 317–328.

FAO (2016). The state of world fisheries and aquaculture. *Contributing to Food Security and Nutrition for all* 1–204.

Fitzgerald, C., Gallagher, E., O'Connor, P. et al. (2013). Development of a seaweed derived platelet activating factor acetylhydrolase (PAF-AH) inhibitory hydrolysate, synthesis of inhibitory peptides and assessment of their toxicity using the Zebrafish larvae assay. *Peptides* 50: 119–124.

Fowler, M.R. and Park, J.W. (2015). Salmon blood plasma: effective inhibitor of protease-laden Pacific whiting surimi and salmon mince. *Food Chemistry* 176: 448–454.

Gangopadhyay, N., Wynne, K., O'Connor, P. et al. (2016). *In silico* and *in vitro* analyses of the angiotensin-I-converting enzyme inhibitory activity of hydrolysates generated from crude barley (Hordeum vulgare) protein concentrates. *Food Chemistry* 203: 367–374.

Gbogouri, G.A., Linder, M., Fanni, J., and Parmentier, M. (2004). Influence of hydrolysis degree on the functional properties of salmon by product hydrolysates. *Journal of Food Science* 69: 615–622.

Gallego, M., Aristoy, M.C., and Toldrá, F. (2014). Dipeptidyl peptidase IV inhibitory peptides generated in Spanish dry-cured ham. *Meat Science* 96 (2): 757–761.

Gildberg, A. (1992). Recovery of proteinases and protein hydrolysates form fish viscera. *Bioresource Technology* 39 (3): 271–276.

Gómez-Guillén, M.C., Turnay, J., Fernández-Díaz, M.D. et al. (2002). Structural and physical properties of gelatin extracted from marine species: a comparative study. *Food Hydrocolloids* 16: 25–34.

Gómez-Guillén, M.C., Giménez, B., López-Caballero, M.E., and Montero, M.P. (2011). Functional and bioactive properties of collagen and gelatin from alternative sources: a review. *Food Hydrocolloids* 25 (8): 1813–1827.

Hatanaka, T., Inoue, Y., Arima, J. et al. (2012). Production of dipeptidyl peptidase IV inhibitory peptides from defatted rice bran. *Food Chemistry* 132 (2): 792–802.

Hayes, M. and Mckeon, K. (2014). Advances in the processing of marine discard and by-products. In: *Seafood Processing By-Products: Trends and Applications* (ed. S. Kim), 125–139. New York: Springer.

Hayes, M., Stanton, C., Slattery, H. et al. Casein fermentate of *Lactobacillus animalis* DPC6134 contains a range of novel propeptide angiotensin-converting enzyme inhibitors. *Applied and Environmental Microbiology* 73 (14): 4658–4667.

Hayes, M., Carney, B., Slater, J., and Brúck, W. (2008). Mining marine shellfish waste for bioactive molecules: chitin and chitosan, part a: extraction methods. *Biotechnology Journal* 3 (7): 871–877.

Hayes, M., Mora, L., Hussey, K., and Aluko, R.E. (2016a). Boarfish protein recovery using the pH-shift process and generation of protein hydrolysates with ACE-I and antihypertensive bioactivities. *Innovative Food Science and Emerging Technologies* 37: 253–260.

Hayes, M., Moen, L.F., Auty, M.A.E., and Lea, T.E. (2016b). Transport of prolyl endopeptidase inhibitory peptides across the blood-brain barrier demonstrated using the hCMEC/D3 cell line transcytosis assay. *Journal of Agricultural and Food Chemistry* 64: 146–150.

He, R., Aluko, R.E., and Ju, X.-R. (2014). Evaluating molecular mechanism of hypotensive peptides interactions with renin and angiotensin converting enzyme. *PLoS One* 9 (3): e91051.

Hoyle, N.T. and Merritt, J.H. (1994). Quality of fish protein hydrolysates from herring (*Clupea harengus*). *Journal of Food Science* 59 (1): 76–79.

Karim, A.A. and Bhat, R. (2009). Fish gelatin: properties, challenges, and prospects as an alternative to mammalian gelatins. *Food Hydrocolloids* 23 (3): 563–576.

Khora, S.S. (2013). Marine fish-derived bioactive peptides and proteins for human therapeutics. *International Journal of Pharmacy and Pharmaceutical Sciences* 5: 31–37.

Kristinsson, H.G. and Rasco, B.A. (2000). Fish protein hydrolysates: production, biochemical and functional properties. *Critical Reviews in Food Science and Nutrition* 40: 43–81.

Kristinsson, H.G., Theodore, A.E., Demir, N., and Ingadottir, B. (2005). A comparative study between acid and alkali aided processing and surimi processing for the recovery of proteins from channel catfish muscle. *Journal of Food Science* 70 ((4): C298–C306.

Lacroix, I.M.E. and Li-chan, E.C.Y. (2013). Inhibition of dipeptidyl peptidase (IV) (DPP-IV) and α-glucosidase activities by pepsin treated whey proteins. *Journal of Agricultural and Food Chemistry* 61 (31): 7500–7506.

Lafarga, T. and Hayes, M. (2017). Bioactive protein hydrolysates in the functional food ingredient industry: overcoming current challenges. *Food Reviews International* 33: 217–246.

Lafarga, T., Rai, D.K., O'Connor, P., and Hayes, M. (2015). A bovine fibrinogen enriched fraction as a source of peptides with in vitro renin and angiotensin-I-converting enzyme inhibitory activities. *Journal of Agricultural and Food Chemistry* 63: 8786–8684.

Li, D.K., Lin, H., and Kim, S.M. (2008a). Effect of rainbow trout (*Oncorhynchus mykiss*) plasma protein on the gelation of Alaska pollock (*Theragra chalcogramna*) Surimi. *Journal of Food Science* 73 ((4): C227–C234.

Li, D.K., Lin, H., and Kim, S.M. (2008b). Purification and characterisation of a cysteine protease inhibitor from chum salmon (*Oncorhynchus keta*) plasma. *Journal of Agricultural and Food Chemistry* 50 (1): 106–111.

Li-Chan, E.C., Huang, S.L., Jao, C.L. et al. (2012). Peptides derived from Atlantic salmon skin gelatin as dipeptidyl-peptidase IV inhibitors. *Journal of Agricultural and Food Chemistry* 60 (4): 973–978.

Minekus, M. (2015). The TNO Gastro-intestinal model (TIM). In: *The Impact of Food Bioactives on Health* (ed. K. Verhoeckx et al.). New York: Springer.

Mohammad, A.W., Suhimi, N.M., Aziz, A.G.K.A., and Johim, J. (2014). Process for production of hydrolysed collagen from agricultural resources: potential for further development. *Journal of Applied Sciences* 14: 1319–1323.

Mukhin, V.A., Novikov, V.Y., and Ryzhikova, L.S. (2001). A protein hydrolysate enzymatically produced from the industrial waste of processing Icelandic scallop *Chlamys islandica*. *Applied Biochemistry and Microbiology* 37 (3): 292–296.

Nagarajan, M., Benjakul, S., Prodpran, T. et al. (2012). Characteristics and functional properties of gelatin from splendid squid (*Loligo formosana*) skin as affected by extraction temperatures. *Food Hydrocolloids* 29: 389–397.

Nilsang, S., Lertsiri, S., Suphantharika, M., and Assavanig, A. (2005). Optimization of enzymatic hydrolysis of fish soluble concentrate by commercial proteases. *Journal of Food Engineering* 70 (4): 571–578.

Norziah, M.H., Al-Hassan, A., Khairulnizam, A.B. et al. (2009). Characterisation of fish gelatin from surimi processing wastes: thermal analysis and effect of transglutaminase on gel properties. *Food Hydrocolloids* 23: 1610–1616.

Ottesen, O., Árnason, J., Smárason, B. O., Zhuravleva, N. and Bjornsdóttir, R. 2016. Values from waste. Matís Icelandic Food and Biotech R&D, Report no. 15-16.

Ovissipour, M., Abediam, A.M., Motamedzadegan, A. et al. (2009). The effect of enzymatic hydrolysis time and temperature on the properties of protein hydrolysates from the Persian sturgeon (*Acipenser persicus*) viscera. *Food Chemistry* 115: 238–242.

Pedersen, L.D., Smiley, S., Bechtel, P.J., and Spengier, C. (2010). Stickwater processing by membrane filtration. In: *A Sustainable Future: Fish Processing by-Products* (ed. P.J. Bechtel and S. Smiley), 121–131. Alaska Sea Grant, University of Alaska Fairbanks.

Rahali, V., Chobert, J.M., Haertlé, T., and Guéguen, J. (2000). Emulsification of chemical and enzymatic hydrolysates of beta-lactoglobulin: characterization of the peptides adsorbed at the interface. *Nahrung* 44 (2): 89–95.

Ranawana, V. and Kaur, B. (2013). Role of proteins in insulin secretion and glycemic control. *Advances in Food and Nutrition. Research* 70: 1–47.

Rawdkuen, S., Benjakul, S., Visessanguan, W., and Lanier, T. (2004). Chicken plasma protein: proteinase inhibitory activity and its effect on surimi gel properties. *Food Research International* 37 (6): 156–165.

Rustad, T. (2003). Utilisation of marine by-product. *Electronic Journal of Environmental, Agricultural and Food Chemistry* 2 (4): 458–463.

Rustad, T. and Hayes, M. (2012). Marine bioactive peptides and protein hydrolysates: Generation, isolation procedures, and biological and chemical characterisations. In: *Marine Bioactive compounds: Sources, Characterization and Applications* (ed. M. Hayes), 55–99. New York: Springer.

Santos, E.M., González-Fernández, C., Jaime, I., and Rovira, J. (2003). Physicochemical and sensory characterisation of Morcilla de Burgos, a traditional Spanish blood sausage. *Meat Science* 65: 893–898.

Sathivel, I.S., Bechtel, P.J., and Prinyawiwatkul, W. (2006). Physicochemical and Rhelogical properties of Salmon protein powders. *International Journal of Food Engineering* 2 (2): 1556–3758.

Schrieber, R. and Gareis, H. (2007). *Gelatine Handbook: Theory and Industrial Practice.* New York: Wiley Blackwell.
Shahidi, F., Han, X.-Q., and Synowiecki, J. (1995). Production and characterisation of protein hydrolysates from capelin (*Mallotus villosus*). *Food Chemistry* 53 (3): 285–293.
Shamsi, T.N., Parveen, R., and Fatima, S. (2016). Characterisation, biomedical and agricultural applications of protease inhibitors: a review. *International Journal of Biological Macromolecules* 91: 1120–1133.
Šližytė, R., Daukser, E., Falch, E. et al. (2005). Characteristics of protein fractions generated from hydrolysed cod (*Gadus morhua*) by products. *Process Biochemistry* 40: 2021–2033.
Šližytė, R., Mozuraitytė, R., Martínez-Alvarez, O. et al. (2009). Functional, bioactive and antioxidative properties of hydrolysates obtained from cod (*Gadus morhua*) backbones. *Biochemistry* 44: 668–677.
Song, H. and Li, B. (2017). Beneficial effects of collagen hydrolysate: a review on recent developments. *Biomedical Journal of Scientific and Technical Research* doi: 10.26717/BJSTR.2017.01.000217.
Tang, L., Chen, S., Su, W. et al. (2015). Physicochemical properties and film forming ability of fish skin collagen extracted from different freshwater species. *Process Biochemistry* 50 (1): 148–155.
Ting, Y., Zhao, Q., Xia, C., and Huang, Q. (2015). Using *in vitro* and *in vivo* models to evaluate bioavailability of nutraceuticals. *Journal of Agricultural and Food Chemistry* 63 (5): 1332–1338.
Theary, C., Panagides, D., Laillou, A. et al. (2013). Fish sauce, soy sauce and vegetable oil fortification in Cambodia: where do we stand to date? *Food Nutrition Bulletin* 34: S62–S71.
Tseng, T.-F., Tsai, C.-M., Yang, J.-H., and Chen, M.-T. (2006). Porcine blood plasma transgluatominase combined with thrombin and fibrinogen as a binder in restructured meat. *Asian-Australasian Journal of Animal Science* 19 (7): 1054–1058.
Vazquez, J.A., Blanco, M., Massa, A.E. et al. (2017). Production of fish protein hydrolysates from *Scyliorhinus canicula* discards with antihypertensive and antioxidant activities by enzymatic hydrolysis and mathematical optimisation using response surface methodology. *Marine Drugs* 15 (10): 306.
Viana, F.R., Silva, V.D.M., Delvivo, F.M. et al. (2005). Quality of ham pâté containing bovine globin and plasma as fat replacers. *Meat Science* 70 (1): 153–160.
Visessanguan, N., Benjakul, S., and An, H. (2000). Porcine plasma proteins as a surimi protease inhibitor: effects on actomyosin gelation. *Food and Chemical Toxicology* 65 (4): 607–611.
Wasswa, J., Tang, J., Gu, X.-H., and Yuan, X.-Q. (2007). Influence of the extent of enzymatic hydrolysis on the functional properties of protein hydrolysate from grass carp. (*Ctenopharynogodon idella*) skin. *Food Chemistry* 104 (4): 1698–1704.
Walter, T., Hertrampf, E., Pizarro, F. et al. (1993). Effect of bovine-hemoglobin fortified cookies on iron status in school children: a nationwide program in Chile. *American Journal of Clinical Nutrition* 57 (2): 190–194.
Wu, J., Liao, W.a.U., and U, C. (2017). Revisiting the mechanisms of ACE inhibitory peptides from food proteins. *Trends in Food Science and Technology* 69: 214–219.
Zambrowicz, A., Pokora, M., Setner, B. et al. (2015). Multifunctional peptides derived from an egg yolk protein hydrolysate: isolation and characterisation. *Amino Acids* 47 (2): 369–380.

9

Bioavailability, Bioaccessibility, and Nutritional Measurement of Proteins

Maria Hayes

Teagasc Food Research Centre, Food BioSciences Department, Dublin, Ireland

9.1 Introduction

Once a protein enters the mouth or buccal cavity, it is in the gastrointestinal tract (GIT) environment and subject to digestion. Depending on its structure and the bacterial composition of the gut, the protein is broken down into smaller oligopeptides, peptides, and amino acids (AAs) by gut and bacterial enzymes for potential absorption and use at sites of action within the body. The gastrointestinal environment is complex and consists of endogenous enzymes and proteins and peptides known as gut endogenous proteins (GEPs) as well as bile and stomach acids. The determination of the oral bioavailability and bioaccessibility of novel proteins is therefore important in order to determine if a protein is available for use as a nutrient or, in some instances, if it is broken down to bioactive, toxic, or allergenic peptides and can impart a positive or negative health effect to the consumer that goes above and beyond basic human nutrition.

Several *in vitro* methods to mimic human gastrointestinal digestion have been developed in the last 10 years which employ commercially available enzymes and different physical structures that mimic peristalsis and display bioaccessibility of the protein or food ingredient with cell culture. Cell culture usually employs human epithelial cell lines such as Cancer coli-2 (Caco-2) that, if cultured correctly, are a useful model of the intestinal epithelial barrier (Lea 2015).

Furthermore, *in vivo* models in living organisms have been used to determine protein bioavailability and are considered the 'gold standard' in the determination of protein quality. Methods include the Protein Digestibility-Corrected Amino Acid Score (PDCAAS) and the Digestible Indispensable Amino Acid Score (DIAAS) which are both recognised by the Food and Agricultural Organisation (FAO) as protein measurement methodologies (Rutherfurd et al. 2015). This chapter will discuss the different *in vitro* and *in vivo* methods that are available and developments in recent years in this research area.

Novel Proteins for Food, Pharmaceuticals, and Agriculture: Sources, Applications, and Advances,
First Edition. Edited by Maria Hayes.

9.2 Measurement of Protein Content in Foods

Different analytical methods have been developed to measure the content of proteins in foods. Results from different methods may vary, however, due to a variety of factors, including using established analytical procedures in laboratories, lack of analytical infrastructure or high economic costs associated with certain methods. A recent review by Angell et al. (2016) demonstrated that 52% of all studies on the protein content of seaweeds used nitrogen determination with subsequent conversion, using a nitrogen-to-protein conversion factor of 6.25. However, many studies have documented that this method leads to an overestimation of the protein content in most foods (Jones 1941; Imafidon and Sosulski 1990; Mariotti et al. 2008).

Analytical methods commonly used to determine protein content include amino acid analysis, the Kjeldahl method (Kjeldahl 1983) and the Bradford method as well as a modified version of the Lowry method. The Kjeldahl method is performed according to method 981.10 of the AOAC International (Latimer 2016). The modified Lowry protein measurement is carried out according to the method described by Hartree (1972b). The Bradford assay is conducted according to the method described by Bradford (1976) and amino acid analysis is normally performed using a high-performance liquid chromatography (HPLC) method. A recent review by Maehre et al. (2018) details the positive and negative attributes of each assay method.

9.3 Bioaccessibility, Bioavailability, and Bioactivity of Proteins

Despite the development of different digestion models that mimic the complex physicochemical and physiological conditions of the GI tract, comparison of results obtained from different studies is difficult as no definitive model for bioaccessibility and bioavailability currently exists. In order to determine which model is best to use for proteins, it is first necessary to understand and define the terms bioaccessibility and bioavailability.

Bioaccessibility can be defined as the quantity or fraction of a protein which is released from the food matrix in the GI tract and which becomes available for absorption (Heaney 2001; Carbonell-Capella et al. 2014); that is, the availability of a compound for absorption through the gut wall. Development of an *in vitro* model to predict the bioaccessibility of a protein should therefore include digestive transformations of food into material ready for assimilation, absorption/assimilation steps that predict absorption into the intestinal epithelial cells and the presystemic metabolism – hepatic and intestinal (Carbonell-Capella et al. 2014). In the laboratory, bioaccessibility assessment of a protein involves digestion of the food using selected procedures normally based on the INFOGEST method (Minekus et al. 2014; Egger et al. 2016) which simulate gastric and small intestinal digestion and subsequently uptake by epithelial cells such as Caco-2 cells (Courraud et al. 2013).

Bioavailability refers to the digestion, assimilation, and bioactivity of a protein or compound/nutrient. It includes utilisation of the protein and can be defined as the fraction of ingested protein that reaches the systemic circulation and is used. It includes digestion in the GI tract, absorption, metabolism, distribution in tissue, and,

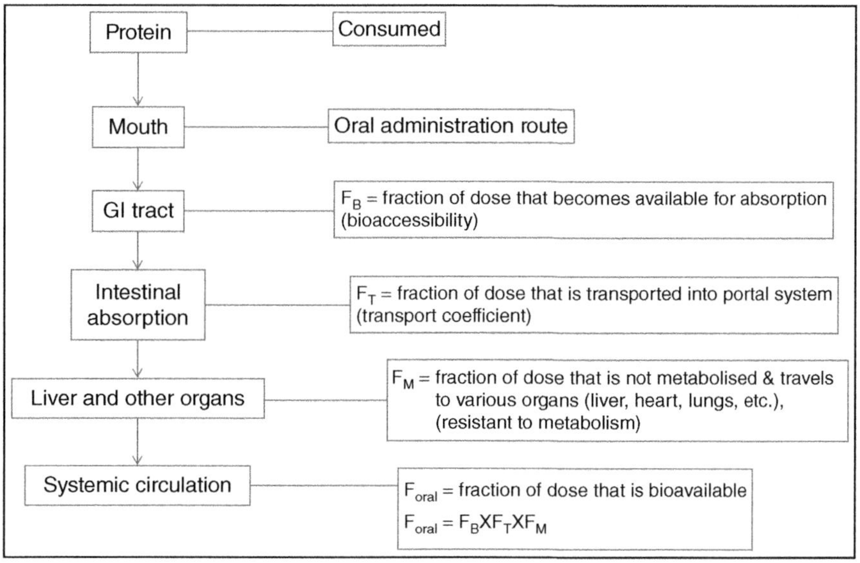

Figure 9.1 Bioavailability of ingested proteins. GI, gastrointestinal.

sometimes, bioactivities (Wood 2005; Carbonell-Capella et al. 2014). Bioavailability of a protein refers to the fraction of protein that can reach the systemic circulation (Holst and Williamson 2008). Bioavailability is usually determined using animal models from the blood plasma concentration of the protein obtained after administration of an acute or chronic dose of a protein (Rein et al. 2013).

Oral bioavailability and bioactivities are intricately linked. Indeed, sufficient dosage at the target site of action is critical for a protein, peptide, or amino acid to generate the desirable biological activities required to be termed 'bioactive'. In order to obtain this desirable biological activity, the bioavailability of the bioactive compound is critical to ensure that the correct dose is administered. Oral bioavailability is dependent on the ability of the protein, peptide, or amino acid to survive structural alteration from preabsorption degradation and postabsorption metabolism (Ting et al. 2015). Figure 9.1 represents the process of orally ingested proteins becoming bioavailable and factors that can influence bioavailability and subsequently bioactivities.

Bioactivities of proteins can include their nutritional properties, beneficial and non-nutritive compounds, storage proteins, and potential health benefits that may be derived through digestion of the proteins into bioactive peptides (BPs) or their enzymatic or microbial degradation into the same through controlled enzyme hydrolysis and fermentation (Pihlanto et al. 2017).

9.4 Protein Hydrolysates

Protein hydrolysates are protein fragments produced by hydrolysis with enzymes or acids or through fermentation that contain amino acids and peptides of different sizes. Some hydrolysates can be synthesised or produced through bioengineering, depending on their structure. Hydrolysates have improved nutritional value compared to native proteins and can also possess bioactivities due to the generation of BPs. BPs may be defined as sequences of 2–30 amino acids in size that may impart a health benefit to the consumer that goes above and beyond basic human nutrition (Fitzgerald 2014).

Proteins like collagen may be hydrolysed by enzymes to improve their bioavailability. Indeed, several types of food processing by-products are hydrolysed to change non-edible proteins into edible peptides (Colla et al. 2014; Aspevik et al. 2017). Proteins are also hydrolysed by acids or bases to produce amino acids. Protein hydrolysates have various structures and can be produced in large amounts. They serve as important nutrients and food ingredients, as well as playing other roles such as improving transport of iron in the diet (iron-chelating protein hydrolysates) (Li et al. 2017). The positive effects of protein hydrolysates on the absorption of minerals, such as iron or other metals, have been demonstrated *in vivo* and *in vitro*.

The chelation ability of protein hydrolysates is thought to be a key factor in the promotion of iron absorption. Through metal chelation, peptides or amino acids increase the solubility and bioavailability of metals. Therefore, iron-chelatable protein hydrolysates are potential promoters of iron absorption. We will focus on peptides and amino acids which have the ability to chelate iron.

Certain special amino acids have strong iron chelation abilities, and peptides containing these amino acids have higher iron chelation abilities than other peptides. The activity of these iron chelates is related to these special amino acids, which can also determine whether they prefer Fe^{3+} or Fe^{2+}. Peptides containing glutamic acid (Glu) and aspartic acid (Asp) have higher affinities for Fe^{3+}, and peptides containing arginine (Arg) and asparagine (Asn) prefer to form chelates with Fe^{2+} (Li et al. 2017).

The degree of hydrolysis (DH) of a protein is defined as the proportion of cleaved peptide bonds in a protein hydrolysate and methods used to measure this include the trinitrobenzenesulfonic acid (TNBS), o-phthaldialdehyde (OPA), trichloroacetic acid soluble nitrogen (SN-TCA), and titration methods. These methods have been reviewed by Rutherfurd (2010).

9.5 *In Vitro* Models

In order to determine the bioaccessibility of a protein without using animal models, a number of *in vitro* models have been developed which range from simple to complex but all based on the activities of the GI tract (Ting et al. 2015). Parameters that must be controlled within these bioaccessibility models include pH, ionic interaction, enzyme activities, the ratio of food bolus to digestive fluids and temperature as well as luminal solubility and, more recently, the effect of GEPs must also be taken into account (Dave et al. 2016). Events in the GI tract encompass a complex biochemical process where enzyme activities along with lipids, starches, and GEPs can cause structural changes to proteins and affect their absorption kinetics (Ting et al. 2015). Hydrophobic protein absorption, for example, can be improved in the presence of lipids (Trevaskis et al. 2008).

9.6 INFOGEST Method

In vitro gastrointestinal digestion is widely employed to assess the potential of proteins as human trials are often costly, resource intensive, and ethically disputable. *In vitro* digestion methods can estimate/determine endpoints such as the bioaccessibility of macronutrients including lipids, proteins, and carbohydrates as well as other nutritional and antinutritional components. A large number of enzymes from different sources, including porcine, rabbit, and human origin, may be used which differ in activities and characterisation. As mentioned already, differences in pH, mineral type, ionic strength, and digestion time can alter enzyme activities and results.

The INFOGEST method is a static *in vitro* digestion that was developed by Minekus et al. (2014) as part of a COST action called INFOGEST. The INFOGEST method proposed a frameset of parameters including the oral, gastric, and small intestinal digestion methods that should be used to simulate *in vivo* conditions *in vitro*. The protocol focuses on physiological enzyme activities and was validated in several interlaboratory studies using skim milk powder (SMP) and demonstrated improved harmonisation compared with previous experiments with individual *in vitro* digestion protocols (Egger et al. 2016). The enzyme activities and salt concentrations are based on human data. The method was compared to an *in vivo* pig experiment with the same SMP and hydrolysis of proteins was determined using gel electrophoresis and mass spectrometry. Proteins detected after gastric *in vitro* digestion corresponded to gastric and duodenal *in vivo* samples from the pig and the intestinal *in vitro* digestion results compared favourably with the distal jejunal samples, thereby validating the *in vitro* digestion method for the study of protein hydrolysis (Egger et al. 2017). Figure 9.2 shows a brief outline of the steps involved in the *in vitro* digestion of proteins using this standardised method.

Briefly, the protein is combined with alpha-amylase, simulated salivary fluid (SSF) and minerals and water and incubated for two minutes at 37 °C. Subsequently, the mixture is exposed to simulated gastrointestinal fluid (SGF), water, minerals, and pepsin to mimic the gastric phase of digestion, carried out at 37 °C. The sample is then mixed with simulated intestinal fluid (SIF), bile and pancreatin, sodium hydroxide and water and minerals and incubated at 37 °C for two hours. This mimics what happens in the gut. The method can be adapted to include lipases but not gut microflora, which is a limitation currently. Following the simulated digestion, the mixture may be applied to Caco-2 or other epithelial cells to determine absorption, bioavailability, and transport across the gut–blood barrier.

Limitations of this method are that it is a static model and does not simulate the physicochemical condition of peristalsis motion, nor does it incorporate transitional change in the physiological environment or elimination from the absorption site. This can result in overestimation of the oral efficiency of protein or other compounds.

This method is relatively easy to perform, is useful when determining initial dosing efficiency, is cost-effective and fast and allows for large sample throughput.

9.6.1 Multifactorial *In Vitro* Bioaccessibility Models: The TNO Gastrointestinal Model (TIM)

The TNO Gastrointestinal Model (TIM) is a multicompartmental model, designed to realistically simulate conditions in the lumen of the GIT, and has been used to study the behaviour of foods in the GI tract successfully (Minekus 2015). The TIM uses computer

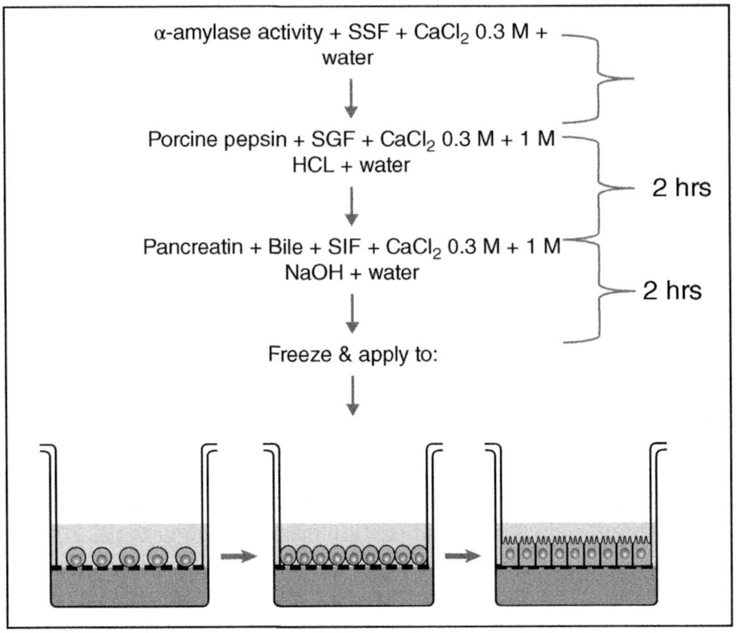

Figure 9.2 Steps involved in the *in vitro* digestion of proteins using the standardised INFOGEST method. Samples are combined with the enzymes/salts shown. SGF, simulated gastrointestinal fluid; SIF, simulated intestinal fluid; SSF, simulated salivary fluid.

simulation of the digestive conditions in the lumen of the gut during transit and digestion of a meal *in vivo*; parameters that are controlled include gastric and small intestinal transit, flow rates and composition of digestive fluids, pH values, and removal of water and metabolites. Digestion models in young adults and the elderly as well as for dogs, pigs, and pre-rumen calves have been developed. TIM results typically allow the user to decipher the availability of a compound for absorption through the gut wall – the bioaccessibility (Minekus 2015).

9.6.2 Transport Coefficient

To pass across the intestinal lining, the transcellular and paracellular routes are options. Via the transcellular route, proteins are absorbed from the lumen into enterocytes by passive diffusion which results from a concentration gradient or by active transport through transcellular protein channels. Enterocytes limit what is released into the portal circulation. In addition, amino acids and small peptides less than 200 Da can pass through tight junctions. The rate of protein transport into the portal circulation depends on the permeability to the cellular membrane and the resistibility to intestinal metabolism. A model that uses a dialysis step following *in vitro* digestion is the easiest to use to measure membrane transport rates of dissolved proteins (Bouayed et al. 2011).

9.7 Cell Culture Models

Human epithelial cell lines are widely used to model the intestinal epithelial barrier and the most widely used is the Caco-2 cell line that was first established from a human colorectal adenocarcinoma by Jorgen Fogh (Fogh et al. 1977; Lea 2015). Caco-2 cells express several morphological and functional properties and they can acquire apical brush borders with microvilli. Tight junctions may form between adjacent cells and they can express enzyme activities typical of enterocytes. They take between 14 and 21 days to reach confluence and have high electrical resistance (Lea 2015). They can be cultured on permeable filters to mimic the conditions of the intestine and can be used for measurement of drug absorption after oral intake in humans.

The primary advantages of using the Caco-2 cell model to determine absorption of proteins or transport of protein across the intestinal barrier include its reproducibility and simplicity compared to *in vivo* models. Furthermore, if cultured and used correctly, this model can provide valuable information on the absorption of food components and drugs. However, there are several limitations to its use, including that as a model of the intestinal epithelium, it is not 100% accurate as the intestinal epithelium contains more than one cell type and not just enterocytes. Furthermore, a mucus layer is not present when culturing just Caco-2 cells and in addition, the bacteria that could be found in the intestine are not usually included in the Caco-2 cell model.

9.7.1 Transcytosis Assays Using Human Cerebral Microvascular Endothelial Cell Line (hCMEC/D3)

The ability of a protein or peptide to pass through the blood–brain barrier (BBB) may be assessed using transcytosis *in vitro* models and human endothelial cell lines such as human cerebral microvascular endothelial cell line (hCMEC/D3). The BBB is a barrier between the blood and brain in cerebral capillaries and impedes therapeutic and pathological agents from eliciting an effect at an attainable dose (Hayes et al. 2016). Breakdown products from proteins such as peptides may have beneficial effects in the brain. The ability of peptides to reach the brain/central nervous system and cross the BBB has been assessed using the peptide PPL and a transcytosis model using hCMEC/D3 cells (Hayes et al. 2016). The hCMEC/D3 cell line derived from a primary cell culture through co-expression of hTERT and the SV40 T large T antigen via an efficient lentiviral vector system retains most of its morphological and functional characteristics of brain endothelial cells even without co-culture of glial cells and therefore can be used as a model for the human BBB (Zlokovic 1995; Fernandéz et al. 2013).

9.7.2 Bioactivities of Protein Breakdown Products – Bioactive Peptides

Bioactive peptides are encrypted in the structure of the parent proteins and are released mainly by enzymatic processes while others can be prepared through chemical synthesis which is often the preferred option as it is cheaper. BPs can play a significant role in human health and affect different systems of the body, including the digestive, endocrine, cardiovascular, immune, and nervous systems. They are considered the new generation of biologically active regulators; they can prevent oxidation and microbial degradation in foods and also improve the treatment of various diseases (Sánchez and Vázquez 2017).

9.7.3 Effects on Diseases Linked to Development of Metabolic Syndrome

Metabolic syndrome (MS) consists of disease states including diabetes, centripetal obesity, hypertension, dyslipidaemia (elevated triglycerides, dense low-density lipoproteins, and low high-density lipoproteins), and alterations in the thrombotic potential that are related to hyperinsulinaemia and insulin resistance (Hollander and Mechanick 2008).

Peptides structurally related to the insulin receptor (IR)-binding protein mcIRBP-19 have been identified in various plants that have IR kinase-activating abilities similar to mcIRBP-19, which has a proven blood glucose-lowering activity that exhibits IR-binding potential. Alpha-glucosidase and dipeptidyl peptidase IV (DPP-IV) are enzymes that play a significant role in the development of type 2 diabetes (T2D). Some studies have shown that dairy and marine proteins could be a natural source of these enzymes. The beneficial effects of these inhibitory peptides can be explained through several mechanisms, such as the satiety response, regulation of incretin hormones, insulinaemia levels, and reducing the activity of carbohydrate-degrading digestive enzymes (Patil et al. 2015). Peptides that inhibit DPP-IV can be screened for in the laboratory using simple, commercially available bioassay kits available from suppliers such as the Caymann Chemical Company, USA. The assay employs a positive control, sitagliptin, and is a fluorometric method. A number of studies have identified inhibitors of DPP-IV using this method from by-product proteins, including meat by-product proteins (Lafarga and Hayes 2014), milk and other sources (Lacroix et al. 2017).

9.7.4 Anti-Inflammatory Peptides

Proteins and hydrolysates of egg, milk, soy, and plant sources have shown anti-inflammatory properties. Ovotransferrin is an egg white protein well known for its antibacterial activity (Wu et al. 2010). This protein is known to have immunomodulating effects on chicken macrophages and heterophil-granulocytes and can inhibit proliferation of mouse spleen lymphocytes (Xie et al. 2002; Sánchez and Vázquez 2017). Immunomodulatory peptides derived from hydrolysates of rice and soybean proteins act to stimulate reactive oxygen species (ROS), which triggers non-specific immune defence systems. Anti-inflammatory activities are usually determined using *in vivo* assays and determination of the effect of anti-inflammatory peptides on markers of inflammation such as C-reactive protein (CRP) which has been shown to present in future cases of cardiovascular disease.

9.7.5 Antioxidant Activities

Oxidation is one of the leading causes of disease and pathogenesis in humans. Glutathione, a tripeptide (gamma-Glu-Cys-Gly), protects cells from free radicals. The dipeptide carnosine (beta-alanyl-histidine), and its related dipeptides anserine (beta-alanyl-1-methylhistidine) and balenine (beta-alanyl-3-methylhistidine), have high antioxidant and pH-buffering properties in muscle cells and are derived from food sources such as chicken and red meat (Sánchez and Vázquez 2017). Peptides derived from dairy proteins have also shown antioxidant activity as well as numerous other food protein sources including egg, cheese, milk, meat, and fish by-products. The addition of a leucine (Leu) or proline (Pro) residue to the N-terminus of a His-His dipeptide will enhance antioxidant activity and facilitate further synergy with non-peptide antioxidants (Kitts and Weiler 2003). Other peptides include angiotensin-1-converting enzyme (ACE-1) inhibitory peptides, renin inhibitory peptides, and peptides important in mental health, including prolyl oligopeptidase inhibitory peptides, which are discussed elsewhere in this book.

9.7.6 Protein Digestibility Corrected Amino Acid Score (PDCAAS) Method for Protein Evaluation

Proteins are not all equal and vary depending on their origin, individual amino acid composition, and level of amino acid bioactivity. High-quality proteins are those that are readily digestible and contain the dietary essential amino acids in quantities that correspond to human requirements. Evaluation of food proteins is fundamental in determining the appropriateness in the human diet. Factors including the protein efficiency ratio, biological value, protein digestibility, bioavailability and bioaccessibility, and antinutritional factors are needed to market a protein-containing product.

In 1989, the joint FAO/WHO expert consultation on protein quality evaluation recommended use of the PDCAAS for evaluating protein quality in humans (FAO 2011). This method uses the limiting amino acid score which is multiplied by protein digestibility to assess how well dietary protein can match the demand for amino acids and allows for the prediction of dietary protein utilisation. The PDCAAS method is based on an estimate of crude protein (CP) digestibility determined over the total digestive tract in a rat or pig model, and values stated using this method generally overestimate the amount of amino acids absorbed. Some food products may claim a high protein content but since the small intestine does not absorb all amino acids in the same way, they are not providing the same contribution to a human's nutritional requirements. Furthermore, the PDCAAS method does not adjust for the susceptibility of foods to damage from processing and antinutritional factors, which can render some amino acids unavailable for absorption in the GI tract.

9.7.7 Digestible Amino Acid Score (DIAAS) Method for Protein Evaluation

The FAO decided that in protein quality evaluation, dietary amino acids should be treated as individual nutrients and wherever possible, data for digestible or bioavailable amino acids should be given in food tables on an individual amino acid basis (FAO 2011).

The DIAAS method determines amino acid digestibility at the end of the small intestine, providing a more accurate measure of the amounts of amino acids absorbed by the body and the protein's contribution to human amino acid and nitrogen requirements. Using the DIAAS method, researchers are now able to differentiate protein sources by their ability to supply amino acids for use by the body. For example, the DIAAS method was able to demonstrate the higher bioavailability of dairy proteins compared to plant-based protein sources. Recently, researchers compared the PDCAAS and DIAAS methods using oat protein concentrate as a substrate (Abelilla et al. 2018). This protein source is often used in human food but the quality was not previously characterised. The group determined the standardised ileal digestibility (SID) of CP and amino acids in oat protein concentrate and also determined the differences in protein quality estimates between the PDCAAS and DIAAS assessments using growing pigs for both measurements. They found that the DIAAS value for oat protein concentrate was close to the calculated value for PDCAAS, but below the recommended intake for protein. This study concluded that in order to reach the daily human amino acid requirement, oat protein should be consumed with other proteins of higher quality and specifically with greater lysine concentrations (Abelilla et al. 2018).

9.8 Conclusion

There are several methods available for protein determination in the laboratory, all with advantages and disadvantages. However, this makes direct comparison between laboratory studies difficult and the choice of analytical method should thus be justified depending on the purpose of the study. Furthermore, as highlighted in this chapter, there are several methods which can be used to determine the bioaccessibility and bioavailability of proteins and their breakdown products and, indeed, there are several potential bioactivities that extend beyond basic human nutrition associated with peptides generated from proteins in the GI tract or through hydrolysis and fermentation. Moreover, the *in vitro* and *in vivo* assessment methods used to determine protein quality vary and results obtained from studies should be interpreted with care. The FAO has issued recommendations regarding protein quality assessment and now recommends the DIAAS method. However, this is not available in every laboratory or industrial setting and can be a costly method, involving the use of animals. *In vitro* transcytosis models can provide estimations of the bioavailability and potential quality of proteins, are animal free and relatively inexpensive to establish and can easily be combined with *in vitro* protein digestion methods such as INFOGEST and TIM.

References

Abelilla, J., Liu, Y., and Stein, H.H. (2018). Digestible indispensable amino acid score (DIAAS) and protein digestibility corrected amino acid score (PDAAS) in oat protein concentrate measured in 20 to 30 kilogram pigs. *Journal of the Science of Food and Agriculture* 98 (1): 410–414.

Angell, A.R., Mata, L., de Nys, R., and Paul, N.A. (2016). The protein content of seaweeds: a universal nitrogen-to-protein conversion factor of five. *Journal of Applied Phycology* 28 (1): 511–524.

Aspevik, T., Otherhals, Å., Rønning, S.B. et al. (2017). Valorisation of proteins and co-and by-products from the fish and meat industry. *Topics in Current Chemistry* 375: 53–60.

Bouayed, J., Hoffmann, L., and Bohn, T. (2011). Total phenolics, flavonoids, anthocyanins and antioxidant activity following simulated gastrointestinal digestion and dialysis of apple varieties: bio-accessibility and potential uptake. *Food Chemistry* 128 (1): 14–21.

Bradford, M.M. (1976). Rapid and sensitive method for quantitation of microgram quantities of protein utilizing principle of protein-dye binding. *Analytical Biochemistry* 72: 248–254.

Carbonell-Capella, J.M., Buniowska, M., Barba, F.J. et al. (2014). Analytical methods for determining bioavailability and bio-accessibility of bioactive compounds from fruits and vegetables: a review. *Comprehensive Reviews in Food Science and Food Safety* 13: 155–170.

Colla, G., Rouphael, Y., Canaguier, R. et al. (2014). Bio-stimulant action of a plant-derived protein hydrolysate produced through enzymatic hydrolysis. *Frontiers in Plant Science* 5: 448–452.

Courraud, J., Berger, J., Cristol, J.P., and Avallone, S. (2013). Stability and bio-accessibility of different forms of carotenoids and vitamin A during in vitro digestion. *Food Chemistry* 136 (2): 871–877.

Dave, L.A., Hayes, M., Mora, L. et al. (2016). Gastrointestinal endogenous protein-derived bioactive peptides: an in vitro study of their gut modulatory potential. *International Journal of Molecular Sciences* 17 (4): 482–490.

Egger, L., Ménard, O., Delgado-Andrade, C. et al. (2016). The harmonized INFOGEST *in vitro* digestion method: from knowledge to action. *Food Research International* 88 (B): 217–225.

Egger, L., Schlegel, P., Baumann, C. et al. (2017). Physiological comparability of the harmonized INFOGEST *in vitro* digestion method to *in vivo* pig digestion. *Food Research International* 102: 567–574.

FAO. 2011. Dietary protein quality evaluation in human nutrition. Report of an FAO Expert Consultation, FAO Food and Nutrition Paper, 92.

Fernandéz, D.A., Louzao, M.C., Vilarino, N. et al. (2013). The kinetics, mechanistic and cytomorphological effects of palytoxin in human intestinal cells (Caco-2) explain its lower than parenteral oral toxicity. *FEBS Journal* 280: 3906–3918.

Fitzgerald, C. P. (2014). Development of bioactive bread enriched with a seaweed peptide fraction with potential heart-health effects. PhD thesis, University College London.

Fogh, J., Wright, J.C., and Loveless, J.D. (1977). Absence of HeLa cell contamination in 169 cell lines derived from human tumours. *Journal of the National Cancer Institute* 21: 393–408.

Hartree, E.F. (1972). Determination of protein modification of Lowry method that gives a linear photometric response. *Analytical Biochemistry* 48: 422–427.

Hayes, M., Moen, L.F., Auty, M.A., and Lea, T.E. (2016). Transport of a prolyl endopeptidase inhibitory peptide across the blood brain barrier demonstrated using the hCMEC/D3 cell line transcytosis assay. *Journal of Agricultural and Food Chemistry* 13 (64, 146): –150.

Heaney, R.P. (2001). Factors influencing the measurement of bioavailability, taking calcium as a model. *Journal of Nutrition* 131 (4): 1344–1348.

Hollander, J.M. and Mechanick, J.I. (2008). Complementary and alternative medicine and the management of the metabolic syndrome. *Journal of the American Dietetic Association* 108: 495–509.

Holst, B. and Williamson, G. (2008). Nutrients and phytochemicals: from bioavailability to bio-efficacy beyond antioxidants. *Current Opinion in Biotechnology* 19: 73–82.

Imafidon, G.I. and Sosulski, F.W. (1990). Non-protein nitrogen contents of animal and plant foods. *Journal of Agricultural and Food Chemistry* 38: 114–118.

Jones, D.B. (1941). *Factors for Converting Percentages of Nitrogen in Foods and Feeds into Percentages of Protein*. Washington, DC: US Department of Agriculture.

Kitts, D.D. and Weiler, K. (2003). Bioactive proteins and peptides from food sources. Applications of bioprocesses used in isolation and recovery. *Current Pharmaceutical Design* 9 (16): 1309–1323.

Kjeldahl, J. (1983). Neue Methode zur Bestimmung des Stickstoffs in organischen Körpern. *Fresenius Journal of Analytical Chemistry* 22: 366–382.

Lacroix, I.M.E., Chen, X.-M., Kitts, D.D., and Li-Chan, E.C.Y. (2017). Investigation into the bioavailability of milk protein-derived peptides with dipeptidyl peptidase IV inhibitory activity using Caco-2 cell monolayers. *Food and Function* 8: 701–709.

Lafarga, T. and Hayes, M. (2014). Bioactive peptides from meat muscle and by-products: generation, functionality and application as functional ingredients. *Meat Science* 98: 227–239.

Latimer, G.W. (2016). *Official Methods of Analysis of AOAC International*. Gaithersburg: AOAC International.

Lea, T.E. (2015). Caco-2 cell line. In: *The Impact of Food Bio-Actives on Gut Health* (ed. K. Verhoeckx et al.), 103–111. New York: Springer.

Li, Y., Jiang, H., and Huang, G. (2017). Protein hydrolysates as promoters of non-Haem iron absorption. *Nutrients* 9 (6): 609–615.

Maehre, H.K., Dalheim, L., Edvinsen, G.K. et al. (2018). Protein determination – method matters. *Foods* 7 (1): 5.

Mariotti, F., Tome, D., and Mirand, P.P. (2008). Converting nitrogen into protein – beyond 6.25 and Jones' factors. *Critical Reviews in Food Science* 48: 177–184.

Minekus, M. (2015). The TNO gastro-intestinal model (TIM). In: *The Impact of Food Bioactives on Health* (ed. K. Verhoeckx et al.). New York: Springer.

Minekus, M., Alminger, M., Alvito, P. et al. (2014). A standardised static *in vitro* digestion method suitable for food – an international consensus. *Food and Function* 5 (6): 1113–1124.

Patil, P., Mandal, S., Tomar, S.K., and Anand, S. (2015). Food protein-derived bioactive peptides in management of type 2 diabetes. *European Journal of Nutrition* 54: 863–880.

Pihlanto, A., Mattila, P., Mäkinen, S., and Pajani, A.-M. (2017). Bioactivities of alternative protein sources and their potential health benefits. *Food and Function* 10: 3443–3458.

Rein, M.J., Renouf, M., Cruz-Hernandez, C. et al. (2013). Bioavailability of bioactive food compounds: a challenging journey to bioefficacy. *British Journal of Clinical Pharmacology* 75 (3): 588–602.

Rutherfurd, S.M. (2010). Methodology for determining degree of hydrolysis of proteins in hydrolysates: a review. *Journal of AOAC International* 93 (5): 1515–1522.

Rutherfurd, S.M., Cui, J., Goroncy, A.K., and Moughan, P.J. (2015). Dietary protein structure affects endogenous Ileal amino acids but not true Ileal amino acid digestibility in growing male rats. *Journal of Nutrition* 145 (2): 193–198.

Sánchez, A. and Vázquez, A. (2017). Bioactive peptides: a review. *Food Quality and Safety* 1 (1): 29–46.

Ting, Y., Zhao, Q., Xia, C., and Huang, Q. (2015). Using *in vitro* and *in vivo* models to evaluate the oral bioavailability of Nutraceuticals. *Journal of Agricultural and Food Chemistry* 63: 1332–1338.

Trevaskis, N.L., Charman, W.N., and Porter, C.J.H. (2008). Lipid-based delivery systems and intestinal lymphatic drug transport: a mechanistic update. *Advanced Drug Delivery Reviews* 60 (6): 702–716.
Wood, R.J. (2005). Bioavailability: definition, general aspects and fortificants. In: *Encyclopedia of HumanNutrition*, 2e (ed. B. Caballero, A. Prentice and L. Allen), 1253–1260. Oxford: Elsevier.
Wu, J., Majumder, K., and Gibbons, K. (2010). Bioactive proteins and peptides from egg proteins. In: *Bioactive Proteins and Peptides as Functional Foods and Nutraceuticals* (ed. Y. Mine, E. Li-Chan and B. Jiang), 247–263. Oxford: Wiley Blackwell.
Xie, H., Huff, G.R., Huff, W.E. et al. (2002). Identification of ovotransferrin as an acute phase protein in chickens. *Poultry Science* 81: 112–120.
Zlokovic, B.V. (1995). Cerebrovascular permeability of peptides: manipulations of transport systems at the blood brain barrier. *Pharmaceutical Research* 12: 1395–1406.

10

Protein from Vegetable Sources: A Focus on Pea Protein

Catherine Lefranc-Millot and Virginie Teichman-Dubois

Roquette, Lestrem, France

10.1 Introduction

The growing worldwide population is currently approximately 7.4 billion people (http://data.worldbank.org/indicator/SP.POP.TOTL), meaning that there are already twice the number of humans that our planet can sustainably provide with food. The ecological footprints and biocapacities vary depending on region but each year sees the earlier arrival of Earth Overshoot Day (in 2017, it was August 2), which is the day when we will have used more from nature (ecological resources and services, overfishing, overharvesting forests, carbon dioxide emission) than our planet can renew in the whole year (www.overshootday.org).

The first point to discuss is regarding the environment, climate change, and sustainable sourcing of proteins. Today, around one-third vegetal versus two-thirds animal proteins are consumed in developing countries while international recommendations are to consume half of each source. From an ecological point of view, 5 kg of vegetal protein are necessary to produce 1 kg of animal protein, meaning that plant-based protein efficiency is higher. The production of animal-sourced foods is at the heart of world agriculture today (FAO 2013). A quarter of the earth's terrestrial surface is used for ruminant grazing and a third of global arable land is used to grow feed for livestock, accounting for 40% of total cereal production. Animal agriculture uses far more land resources than any other human activity. Yet livestock are increasingly recognised as a main contributor to climate change by emitting greenhouse gases, either directly from enteric fermentation or indirectly from deforestation and other activities related to feed production, including production, transport, storage, cooking, and wastage of food (Scarborough *et al.* 2014). There is consequently a need to find new alternatives to meat as a source of protein. Substituting plant-based proteins for animal ones has economic and sustainability advantages, including aiding the reduction of global warming, eutrophication, acidification, and land use (Davis *et al.* 2010).

From another point of view, obesity and its co-morbidities are a growing public health issue worldwide (http://data.worldbank.org/indicator/SP.POP.TOTL). According to the WHO (2014), the prevalence of obesity more than doubled between 1980 and 2014 and

Novel Proteins for Food, Pharmaceuticals, and Agriculture: Sources, Applications, and Advances,
First Edition. Edited by Maria Hayes.

almost 2.8 million people die each year as a result of being overweight or obese. In 2013, 42 million children under the age of five were overweight or obese. In 2014, more than 1.9 billion adults were overweight, of which 600 million were obese. In such an alarming situation, different avenues for fighting this trend might be considered, such as decreasing caloric intake, managing the feeling of hunger, and finally achieving satiety. One way to reach this last goal is by increasing the quantity of protein consumed.

An increasing proportion of the population is becoming vegan, vegetarian, or flexitarian. According to different studies (Fox and Ward 2008; de Backer and Hudders 2014), the main motivators for this include health concerns or the ethical treatment of animals, but they can vary. Animal rights and ecological concerns, together with taste preferences, promote vegetarianism/veganism, while an increase in health motives increases the odds of being semi-vegetarian/flexitarian (mildly or significantly reducing meat intake).

The increased numbers of the human population, the need for reduction of or alternatives to animal protein consumption, and the need to reduce CO_2 emissions are driving consumers towards higher consumption of plant-based proteins. Mainly present in vegetables or pulses (lentils, fava beans, peas, dry beans, soy), plant-based proteins are beneficial for health (cardiovascular disease prevention, glycaemic control, provision of minerals, vitamins, fibre) and cheaper for consumers (Nadathur *et al.* 2017).

10.2 The Advantages of Leguminous Plants

10.2.1 Sustainable Vegetable Protein Crops

Leguminous plants, among them peas, have low water requirements and their root nodules fix nitrogen and enrich the soil; they are therefore an excellent selection for crop rotations. Their use allows a reduction in the use of chemical fertilisers, one of the main sources of greenhouse gas in the farming sector and a major contributor in many areas of the world. Peas increase the yield of crops planted after them and, moreover, less acreage is required to produce 1 kg of protein from peas than many other plants. Beyond that, pea protein can satisfy the nitrogen needs of humans (Mahe *et al.* 1997) and demonstrates good digestibility and nitrogen retention properties (Gausseres *et al.* 1997).

10.2.2 Cultivation of Peas and Composition of Seeds

Peas (*Pisum sativum L.*) are also known as garden peas or common peas. They belong to the Fabaceae family (also called Papilionaceae), commonly known as the legume family, including peanuts, soybeans, lentils, lupins, chickpeas, and beans. This pea, traditionally sold after dehulling as 'split pea', has long been consumed in soups and purées. It is one of the plants that initiated agriculture in the Mediterraneo-European region together with wheat, barley, lentils, and vetches. Peas have been a staple of our diet for centuries. Many varieties (cultivars) of peas exist, with extensive genetic variability, for example, in plant height, leaf type, number of branches, and pods (Tzitzikas 2005). Some of the most common varieties include Alaska, Thomas Laxton/Laxton's Progress/Progress, Mr. Big, and Little Marvel.

Dry peas are primarily grown in three regions: Canada, Russia, and China, respectively producing 30%, 13%, and 13% of the global dry pea market in 2014 (Janzen *et al.* 2014;

FAOSTAT 2017). Peas contain good amounts of minerals, vitamins, fibre, carbohydrates, and protein. The major component of peas is starch, which accounts for up to 50% of the seed dry matter. Proteins and total dietary fibre account for approximately 24% and 20% of the seed dry matter, respectively. Lipids are present in lower amounts (2.5% of the seed dry matter). High variation in starch and protein contents is often observed, depending on the variety, whereas variations in the contents of other components are usually lower. Studies have determined that the average pea protein content might vary between 24.2% and 27.5% (Hood-Niefer *et al.* 2012) while older investigations (Coyne *et al.* 2005) had revealed values ranging from 30.93% to 12.38% (mean 20.93% – values calculated with a 5.44 conversion factor instead of the common 6.25 factor).

10.2.3 Processing of Pea Crops and Proteins Obtained from Processing

Pea crops are usually processed through cleaning and splitting mechanical steps. The processing of dry peas is designed to ensure a consistent, high-quality product. In recent years, there has been increasing interest in legumes as useful ingredients in a range of foods. The agri-food industry develops vegetal proteins as ingredients, obtained from wheat, soybean, pea or lupin. These 'vegetal raw materials' can play a role as partial or global substitutes for animal protein sources, and provide additional functional or nutritional properties.

Techniques for fractionation of pulses include clean label technologies, including dry milling (a mechanical process used to reduce the particle size of split or whole peas into coarse or fine flours, using roller or pin mills), and air classification (with reverse pressure air stream), generally used to fractionate pulses into a light or fine fraction (protein concentrate) and a heavy or coarse fraction (starch concentrate). The low purity of the protein fraction obtained using this process (38–65%) (Boye *et al.* 2010) necessitates further processing, such as isoelectric precipitation, for the preparation of extracts with higher protein purity (>70%). Techniques such as membrane separation can yield protein with improved functionality and can also be used effectively to remove some antinutritional components (Mondor *et al.* 2009), including protease and amylase inhibitors, lectins, and polyphenols (Singh 1988).

As an example, NUTRALYS® pea protein (Roquette, France) is obtained via a dry clean process (no solvents) to produce pea flour. The flour is then hydrated and the pea starch and internal fibre are extracted separately. The protein fraction is then coagulated for further purification and, finally, carefully dried in a multistage spray dryer. The resulting highly purified pea protein powder contains on average 85% protein, 7% fat, 3% carbohydrate, and 5% other components on a dry matter basis (Yang *et al.* 2012).

Peas are a very good source of nutrients. Pea fibre or pea starch, for example, are now widely used in the processed meat industry for fortification instead of more traditional flours (Asgar *et al.* 2010). The final enhanced protein concentration depends on the types of product needed, with differing nutritional or functional (ingredients) applications. The plant-based proteins obtained have enhanced digestibility compared to the raw material by modifying, for example, accessibility of the protein, or even by eliminating antinutritional factors. Hydrolysates are proteins which have been further hydrolysed by enzymes, most frequently, or chemicals or lactic acid bacteria (LAB). In terms of protein quality (amino acid composition and ratios, specifically), pea proteins are complementary to those of cereals.

Lastly, attention today is also focused on the efficiency – in other words, the speed – with which amino acids are used by the body. Related to this characteristic, protein composed of amino acids that are used slowly by the body is called 'slow protein' while protein composed of amino acids that are used quickly by the body is called 'fast protein' (Boirie 2004).

10.3 Quality of Pea Protein

10.3.1 Global Composition of the Pea Seed

Plant seed proteins are classified into three major categories: biologically active, structural, and storage proteins which are the most abundant proteins present and are used as nitrogen sources by the new embryos after seed germination (Tzitzikas *et al.* 2006; de Backer and Hudders 2014). Seed storage proteins are classified into different groups: albumins (water soluble), globulins (salt soluble), prolamins (aqueous alcohol soluble), and glutelins (alkali or acid soluble) (Shewry *et al.* 1995). The globulins are the most widely distributed group of storage proteins. They can be divided into two groups based on their sedimentation coefficients (S20.w): the 7S vicilin-type globulins (e.g. vicilin and convicilin) and the 11S legumin-type globulins (e.g. legumin) (Shewry *et al.* 1995) (Table 10.1).

Table 10.1 Class, name, structure, function, and amount of the main proteins in peas (Shewry *et al.* 1995; FAOSTAT 2017).

Class	Name	Structure and molecular weight (MW)	Function	Amount (% of total pea protein content)
Globulins 49–82% of total pea protein content	Legumin	Hexamer MW: 330–400 kDa	Storage	6–24.5
	Vicilin	Trimer MW: 150–170 kDa	Storage	21–55
	Convicilin	Trimer MW: ~210 kDa	Storage	3.9–8.3
Albumins 20–30% of total pea protein content	PA1	Dimer MW: ~10 kDa	Storage	5–9
	PA2	Dimer MW: ~50 kDa	Unknown	10–20
	Lectins	Tetramer MW: ~50 kDa	Protection against endogenous proteases	2.5%
	Lipoxygenases	n/a	Oxidation of fatty acids	<1%
	Serine/trypsin protease inhibitors	Monomer MW: 10–16 kDa	Antitrypsic activity	<2%

10.3.2 Digestibility of Pea Protein

The reduced protein digestibility of leguminous plants such as peas, as compared to animal protein, partially stems from the antinutritive factors found in pulses, including proteolytic inhibitors, tannins, and phytic acid. These antinutritive compounds are destroyed or inactivated by subjecting peas to processes such as extrusion, cooking, or baking.

The true digestibility of the pea protein NUTRALYS® (Roquette) was found to be around 97.3% (Asgar *et al.* 2010) in rats. A similar product, Lysamine® GP (Roquette), was found to have digestibility of 98% in piglets (unpublished data). Other pea proteins on the market, such as Pisane® from Cosucra or Propulse™ from Nutri-Pea Ltd., also have a digestibility of 98%. The estimated Protein Digestibility Corrected Amino Acid Score (PDCAAS) of NUTRALYS® pea protein is 85% for children aged between three and 10 years and 93% for adults (World Health Organization/Food and Agriculture Organization/United Nations University 2007). Pea protein has been evaluated previously as having a quality similar to that of casein, eggs, and soy and is known to be better than other proteins from non-animal sources such as vegetables, legumes, and cereals (Asgar *et al.* 2010) (Table 10.2).

10.3.3 Amino Acid Composition of Pea Protein

The processing of peas may influence the amino acid composition of the final product. Protein quality from plant-based sources has also often been attributed to an unbalanced composition in terms of amino acids. For example, cereals are limited in lysine, while pea proteins contain limited amounts of sulfur amino acids, including methionine and cysteine, with respect to human requirements (Shewry *et al.* 1995). The value of the amino acid score for adults was found to be 96 for adults and 89 for children (3–10 years) for NUTRALYS® pea protein. Similar results are presented for the pea protein Pisane (Cosucra). Despite this limitation and due to its richness in lysine which plays an important role in growth and bone health and is a limiting amino acid in wheat, pea protein is very complementary to wheat-based products. Very balanced blends can therefore be

Table 10.2 Protein Digestibility Corrected Amino Acid Scores of pea protein in comparison to other food proteins. Pea protein PDCAAS = 93%.

Protein source	PDCAAS (%)
Casein	100
Egg white	100
Soy protein	100
Beef	92
Fruits	76
Vegetable	73
Legumes	70
Cereals	59
Whole wheat	42

prepared by, for example, mixing 80% wheat flour (amino acid score below 45) and 20% pea protein (NUTRALYS®) to reach a whole amino acid score of 100%, based on FAO 2007 recommendations (Shewry *et al.* 1995). NUTRALYS® pea protein is also rich in the branched-chain amino acids isoleucine, leucine, and valine, which are important for muscle cell composition, arginine that sustains physical effort and the efficacy of the immune system, and glutamine, a source of energy for the muscles during stress.

10.3.4 Antinutritional Factors and Toxicity

Some antinutritional factors are found in pulses, including proteolytic inhibitors, tannins, lectins, and phytic acid. These compounds can be destroyed or inactivated by subjecting peas to industrial processes including a 'clean' type of separation such as ultrafiltration without chemical solvents and possibly extrusion, cooking, or baking. The plant-based proteins that result therefore contain few antinutritional factors, few complex sugars, and a low fat content with a favourable fatty acid profile (80% unsaturated fatty acids). They also have increased bioavailability.

Internal toxicological studies conducted to ensure the food safety of NUTRALYS® pea protein (Roquette) have shown that the pea protein is safe, with no anticipated health effect. It was fully evaluated for acute toxicity (Chentouf *et al.* 2013a), genotoxic and mutagenic potential (Chentouf *et al.* 2013b), dietary safety in a 13-week feeding trial in rodents (Chentouf *et al.* 2013c) and for irritating and sensitisation effects (Chentouf *et al.* 2013d), using internationally validated methods and in accordance with OECD Guidelines (www.oecd.org/chemicalsafety/testing/oecdguidelinesforthetesting ofchemicals.htm).

10.3.5 Allergenicity

There is no available epidemiological data on the prevalence of allergy to peas. Food allergy to peas is rare and rarer than food allergy to cow's milk, wheat, soy, or peanuts. For example, in the United States, food allergies in general affect about 2% of adults and 4–8% of children; the prevalence of green pea allergy among adults in the US has been estimated to be 1% (Sell *et al.* 2005).

Pea proteins are not regarded as major food allergens. Consequently, peas are not listed among the food allergens recognised by the European Union (EU) as the 14 most significant food allergens and are not of mandatory declaration upon European or US regulations on labeling of food allergens.[1]

Some clinical data are available on sensitisation to peas (i.e. presence of pea-specific IgE antibodies in serum or positive response to standard penetration tests (SPT) and on the symptoms associated with allergic reactions to pea ingestion. Although rare, food allergy to peas has been demonstrated for some patients (Singh 1988), in most cases with mild

1 Regulation (EU) No 1169/2011 of the European Parliament and of the Council of 25 October 2011 on the provision of food information to consumers, amending Regulations (EC) No 1924/2006 and (EC) No 1925/2006 of the European Parliament and of the Council, and repealing Commission Directive 87/250/EEC, Council Directive 90/496/EEC, Commission Directive 1999/10/EC, Directive 2000/13/EC of the European Parliament and of the Council, Commission Directives 2002/67/EC and 2008/5/EC and Commission Regulation (EC) No 608/2004. Official Journal of the European Union, **L 304**, 18–63.

symptoms; however, severe anaphylaxis has been reported in at least two patients in only one study (Wensing *et al.* 2003). Immunological cross-reactivity between peas and other plants, mainly legumes (in particular peanuts (Rouge *et al.* 2010) and lentils (Bernhisel-Broadbent and Sampson 1989; Vereda *et al.* 2010)) has been demonstrated. This is due to the similarity of the pea, peanut, and lentil allergens. However, the clinical relevance of this fact,(i.e. whether this will translate in a significant proportion of individuals with reported allergic reactions after pea ingestion because of cross-reactivity with other legumes) remains controversial.

The globulins vicilin and convicilin are pea proteins which have been proposed as possibly allergenic in one study only, either directly or via cross-reactivity (Sanchez-Monge *et al.* 2004), possibly explaining the very infrequent cases observed. Proteins from the albumin fraction have also been shown to display an allergenic potential, but individual proteins have not been identified. Other allergenic proteins may include lectins such as agglutinin (which belong to the albumin fraction) or profilin (not a major protein in peas). Experimental data from one study suggest that the potential allergenicity of pea proteins may increase with the degree of maturation of the pea seed (Chentouf *et al.* 2013d). Only vicilin and convicilin are officially accepted as pea allergens according to the criteria of the IUIS Allergen Nomenclature Sub-Committee (Allergen Nomenclature 2017), under the allergen names Pis s 1 and Pis s 2, respectively.

10.4 Health Potential of Pea Proteins

Checking for the number of occurrences found in the last five years on PubMed, using research criteria oriented to health properties of plant-based proteins, indicated that rice and pea proteins were the vegetal proteins which were the most frequently studied, with a total of 16 and 10 relevant studies identified, respectively.

10.4.1 Food Intake, Satiety, and Weight Management

Related to satiety, blood glucose management has a demonstrated impact on weight control and metabolism. Indeed, one potential benefit of eating peas regularly may be related to their impact on satiety, food intake, and blood glucose (high amounts of protein and dietary fibre) (Dahl *et al.* 2012).

Adult diet-induced obese rats (Adam *et al.* 2016) (high-fat diet) supplemented for four weeks with a pea protein (Special Diet Services Ltd.), in comparison with other groups (high fat, high fat with high casein, high fat with pectin), experienced significant benefits on weight management and satiety and had decreases in body weight gain and body fat mass gain as well as decreased food intake. In another acute study (Overduin *et al.* 2015), physiological signals relevant to satiety, characterised *in vitro* via gastric digestion kinetics and *in vivo* through hormonal responses in rats indicated that pea protein (NUTRALYS®) had an intermediately fast intestinal bioavailability and an efficacy comparable to that of a dairy protein meal in triggering gastrointestinal satiety signals (anorexigenic hormonal responses). An oral preload (isovolumetric, not isocaloric) of 10 or 20 g of isolated yellow pea protein (Propulse – 82% protein on dry weight; Nutri-pea Ltd.) had a positive impact on the postprandial glycaemic response following an *ad libitum* meal taken 30 or 120 minutes later in young, healthy males

(Smith *et al.* 2012). In another study conducted by the same team (Mollard *et al.* 2014) with isocaloric meals, no significant beneficial impact of yellow pea protein on food intake, subjective appetite, and postprandial glycaemic response was observed.

10.4.2 Impact on Cholesterol

Dyslipidaemia is a major risk factor for cardiovascular disease. Managing dietary and blood cholesterol decreases the risk for heart disease. Pea protein could be part of this preventive action. The hypocholesterolaemic effect of soluble fibre and legume (pea) proteins was investigated in a study in rats fed a hypercholesterolaemic diet for 28 days, alone or in combination (Parolini *et al.* 2013) (six groups), showing that each vegetable-containing diet was associated with lower total cholesterol concentrations compared with the control. The pea diet led to lower total cholesterol and low-density lipoprotein + very low-density lipoprotein (LDL + VLDL) concentrations when compared with the control diet at both 14 and 28 days of the hypercholesterolaemic diet, with no difference between the pea and control groups for high-density lipoprotein (HDLs) cholesterol and triacylglyceride (TAG) concentrations ($P > 0{\cdot}05$).

A cross-over study (Re *et al.* 2016) in 33 human volunteers examing the short-term satiating effects of two doses (30 and 15 g) of a plant-based pea protein (NUTRALYS®) in comparison to whey protein (30 g) using a preload soup as a carrier showed that the pea protein and the whey protein reduced caloric intake equally at the *ad libitum* meal. Another team studied the effect of various plant proteins, including pea protein (Pisane®), and their combinations with soluble fibres (oat fibre or apple pectin) on plasma total and LDL-cholesterol levels in a randomised, double-blind, parallel group design (four-week period) (Sirtori *et al.* 2012). Among the seven groups tested (25 participants/group), the pea protein combined with cellulose bars induced no significant cholesterol change. However, pea protein combined with oat fibre or pea protein combined with apple pectin resulted in significant reductions of total cholesterol levels ($P = 0.05$), showing a potential clinical benefit of consuming combinations of pea protein and a soluble fibre, such as oat fibre or apple pectin.

10.4.3 Blood Pressure – Preventive Impact on Hypertension

Diet composition may affect blood pressure (BP), but the underlying mechanisms remain unclear and contradictory, particularly for plant-based proteins (Altorf-van der Kuil *et al.* 2010). Food-derived bioactive peptides are attracting research and food industry interest because of their relatively safe and multifunctional properties, including management of hypertension. Recently, peptides that inhibit angiotensin-1 converting enzyme (ACE-1) or renin or both have been discovered in a range of different plant sources, including pea (Li and Aluko 2010; Li *et al.* 2011).

For example, a team from Manitoba (Girgih *et al.* 2016) investigating short- and long-term systolic blood pressure-lowering effects in spontaneously hypertensive rats of a pea protein (Propulse™) showed that it had systolic blood pressure-lowering effects in acute conditions if heated, which were even stronger if heated and hydrolysed with the enzyme thermoase followed by filtration. Incorporation of the heated or heated, hydrolysed, and filtered pea protein had a similar effect in hypertensive rats (not in normotensive ones), with a decreasing potency in the fourth and fifth week, although systolic blood pressure values of the treated rats were still lower than the untreated controls. The antihypertensive potency of the filtered hydrolysate may have been due to the presence of easily absorbed hydrophilic peptides.

Postprandial blood pressure-related response to the ingestion of pea and animal proteins was compared in a randomised short-term study, in 48 (six groups) overweight or obese adults with untreated elevated BP (Teunissen-Beekman *et al.* 2014). This showed a significant short-term beneficial impact of pea protein on postprandial changes in blood pressure-related outcomes: diastolic blood pressure, total peripheral resistance, augmentation index, and S-nitrosothiols, and a more beneficial impact as compared to egg white protein.

10.4.4 Physical Activity, Muscle Repair, and Anabolism

Increased protein accretion and muscle synthesis happen immediately after exercise (Ogawa *et al.* 2010; Konopka and Harber 2014) as a result of a positive protein net balance (muscle catabolism versus anabolism), and protein consumption just after exercise has a positive impact. In parallel, in elderly people, maintaining lean mass by optimising protein nutrition is crucial for muscle strength and improved health (Beaudart *et al.* 2017). In this context, few studies have specifically addressed the postprandial muscle protein synthetic response to the ingestion of plant proteins, generally considered as potentially inducing a lower response than animal ones (lower digestibility, greater splanchnic extraction, relative lack of specific essential amino acids such as sulfur ones or lysine, or relative low content of leucine) (van Vliet *et al.* 2015). Pea protein is rich in branched chain amino acids, which are considered as playing an important role in muscle synthesis.

The impact of oral supplementation with 25 g pea protein (Babault *et al.* 2015) (NUTRALYS®) versus whey protein and placebo (three groups) on biceps brachii muscle thickness and strength after a 12-week resistance training programme was explored in 161 males, aged 18–35 years. In addition to appropriate training, supplementation with pea protein promoted a greater increase of muscle thickness compared to placebo in the volunteers with the lowest muscle strength at baseline. Since no difference was seen between the two protein groups, pea protein was considered as a possible alternative to whey-based dietary products.

10.5 Applications of Pea Protein in the Human Food Industry

10.5.1 Savoury and Dairy Markets: From a 'Hidden Use of Plant Proteins' to 'Plant Proteins in the Spotlight'

'Source of protein' or 'High protein' are trendy claims in the food market these days. This is especially true for the dairy and savoury markets, where approximately 20% of new product launches in the last five years have borne a protein claim (Mintel GNPD 2017).

10.5.1.1 Partial Substitution of Animal Proteins in Food Products

Convenience foods and green packaging continue to play a significant role in meat and fish markets driven by increasingly busy consumers whose hectic family, work, and social commitments lead to the search for quick answers. Moreover, consumers expect tasty and affordable products of natural origin and secured quality. The increased population, having generated limits in animal raw materials availability, has led to increased market prices for meat or dairy products. As household food expenditure share declined from 26% in 1960 to 15% in 2014, affordability has become a key purchase driver. To remain competitive and optimise their margin, food manufacturers look for cost-effective solutions, often turning to plant-based proteins. Recent consumer trends around GMO defiance and the hunt for clean labels and allergen-free products have generated interest for pea protein in this market.

The biggest challenge for plant-based proteins in partial substitution of animal protein – either meat or dairy proteins – is to mimic their behaviour. To achieve this, different grades of pea protein have been developed. Pea proteins are suitable for numerous meat and fish applications such as minced meat, fine emulsions (sausages, paté), fermented products, surimi sticks, fish nuggets or brines, and marinades.

Taking the example of processed meat products, the main pea protein functionalities are emulsification and gelation. Pea protein is thus used for binding of meat pieces, adhesion, and water/fat-binding capacity. As a meat extender, while achieving a cost reduction, it creates a consistent and attractive texture when used at concentrations up to 10% substitution, without noticeable off-flavours.

Emulsification Some pea protein grades have been developed to enhance its emulsifying properties. It is especially useful in emulsified products, like sausages, patés, and mousses. The stability and strength of cold emulsions containing pea protein give improved yield and juiciness to the end-products together with a homogeneous texture.

Other, more soluble versions have been developed to ensure the stability of hot emulsions, obtained, for example, after a heat treatment. This is particularly suited for fat-containing sauces and marinades.

Water-Binding Capacity Water-binding capacity is especially necessary in meat products such as cooked hams.

In April 2011, a pea protein was tested in the Danish Meat Research Institute (DMRI) meat processing pilot plant to investigate its functional properties in restructured 120%-yield ham (typically sandwich ham) and 140%-yield ham (typically pizza ham) in comparison to soy protein concentrate (SPC).

Vacuum packages containing 10 slices of cooked hams were used to quantify expelled liquid (purge) after three weeks' storage. The use of pea protein in both 120%- and 140%-yield hams induced a more than 70% decrease in cooking loss in cooked ham products, improving the retention of water during cooking, in the same way as SPC.

10.5.1.2 Higher Value Markets: Animal Protein Alternatives

New trends are influencing the food products market. With concerns about the environment, scarcity of natural resources, animal welfare and health, an increasing proportion of Western consumers choose vegan or flexitarian options. This provides momentum for GMO-free, organic plant alternatives, while a meat-centric diet is increasing being questioned. Furthermore, the development of social media, quickly spreading information, speeds up this trend even more. For example, in France, 34% of young consumers incorporated more vegetarian foods in their diet compared to previous years, showing a clear trend of new generations towards flexitarianism (Lightspeed/Mintel, 2017). As a consequence, global meat substitute launches grew 24% between 2011 and 2016 (IFIC Food and Health Survey, 2016).

Meat Alternatives As a protein-based meat analogue, texturised pea protein can extend the shelf-life of meat products while providing an economical, functional, and protein-rich food ingredient (65% protein approximately). Roquette has filed a patent for texturised pea proteins (WO 2007/017571 A1) based on the use of its NUTRALYS® pea protein in a granulated powder form specially designed for the extrusion process. Used for their rapid rehydration and texturising properties, texturised pea proteins can emulate the fibrous texture of meat and therefore allow more than 30% meat substitution. Beyond meat extension, texturised pea proteins offer new solutions to formulators seeking to offer food products that match current trends such as vegetarianism. Texturised pea protein can emulate the appearance and bite of meat in products including meat balls, patties, and pie fillings.

Moreover, texturised pea proteins are a unique alternative for manufacturers and consumers looking for soya-free and/or gluten-free options in a world in which most existing texturised vegetable protein is based on soy or wheat. Use of pea protein also gives manufacturers the opportunity to reduce the number of allergens that have to be listed on packaging, as pea is not a major allergen.

Dairy Alternatives The sustainability of the yellow pea crop makes it an environmentally friendly solution, while the ingredient status answers the current consumer requirement for clean-label ingredients. Furthermore, pea protein shows good technical functionalities, necessary to develop dairy-alternative products.

Let's consider the example of a fermented dessert.

- *Processing*: Pea protein makes fermentation possible.
- *Texture*: A protein gel comparable to that achieved with soy can be created.
- *Emulsion*: Pea protein helps to emulsify and stabilise the product.
- *Powder rheology*: Spray-dried pea protein grades have inherently free-flowing properties, which makes lecithin or other flow agents unnecessary. This feature is also of interest inside the manufacturing plants, providing easy dispersion, low dust, lumps, or foam properties.
- Last but not least, pea protein significantly helps decrease formulation cost.

Until now, the main restriction on the use of vegetable proteins has been their taste. This can vary slightly, depending on botanical origin, but it has always had a persistently unattractive odour and so-called 'green' (or vegetable) sensory note. The origin of this undesirable note is linked to naturally present volatile molecules in the pulse itself, such as hexanal. In recent years, pea protein manufacturers have continuously improved their extraction process in order to exclude most of these volatile compounds from raw material to isolated protein.

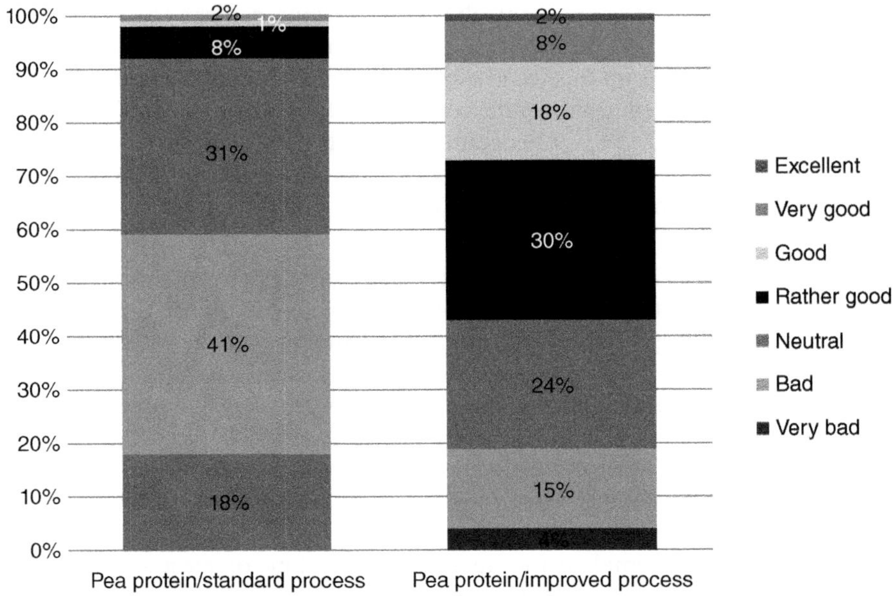

Figure 10.1 Consumer satisfaction for plant-based ice creams tested with 130 people (Roquette, internal report, 2011).

Figure 10.1 shows the results of a sensory blind test carried out using 130 French consumers tasting 100% plant-based ice creams containing 7% pea protein. One ice cream was made with a protein extracted with a typical process, and the second one was extracted via an improved process, aiming at getting rid of the volatile molecules. Due to the improved extraction process, the second ice cream was rated from 'rather good' to 'excellent' by almost 60% consumers, compared to 11% with the standard process (Roquette, internal study, 2011).

10.5.2 Baking: Driven by the Hunt for Protein Fortification and Gluten-Free Ingredients

In baking, plant proteins have always been used, driven by two main focuses: protein fortification and gluten-free requirement. Protein fortification was first achieved thanks to soy/whey, as these are the most widely consumed proteins. To reach a higher protein rate, other plant proteins are also used, including pea. While only 1% of the population have been diagnosed as coeliac (Gujral *et al.* 2012), gluten avoidance has now unexpectedly become a widespread trend. To replace both the protein content and wheat gluten technical features, pea protein has proven its relevance.

10.5.2.1 Protein Fortification: Pea Protein as a Nutritional and Technical Complement to Wheat Protein

In recent years, new cereal baked goods launches bearing a 'high-protein' claim have risen on average 30% per year (Mintel GNPD, CAGR 2010–2016). This answers a clear consumer demand for protein-fortified products in the field of baking. As an example,

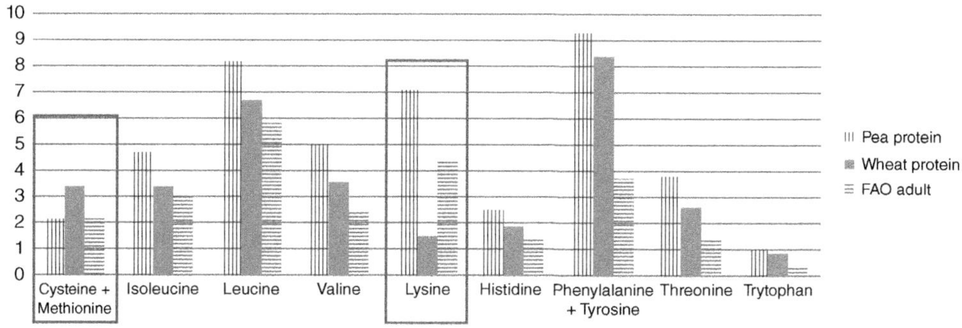

Figure 10.2 Complementarity of pea and wheat protein amino acid profiles. Orange: pea protein; red: wheat protein; brown: FAO recommendations for adults.

48% French, 51% German, 61% Spanish, and 66% Polish consumers would like to try a high-protein bread.

Achieving a 'high-protein' claim by increasing the level of gluten is not possible due to the technical characteristics of this protein. Indeed, overdose of gluten leads to texture issues (e.g. in bread, high dough consistency, high elasticity, and low volume development). The challenge is to preserve texture and cost-effectiveness, together with achieving high protein content with a balanced amino acid score.

First, pea and gluten (wheat protein) have complementary amino acid profiles, as shown in Figure 10.2. Pea, as a legume, lacks methionine while wheat, as a cereal, lacks lysine. A 75% pea protein/25% wheat protein mix enables the development of baked goods with a complete amino acid profile, in line with the FAO recommendations.

The Technical Challenge of Baking Applications: Interaction with Water In bread-like products, incorporation of ingredients with high water absorption or high interaction with water will have a negative impact on gluten hydration and therefore on gluten network formation. Biscuit doughs are very low in water and the incorporation of ingredients with high interaction with water will affect the hydration of other ingredients, especially flour and sugar. This can lead to negative impacts on texture profiles due to the extra water required in the formulation.

The Solution: Insoluble Pea Protein and Hydrolysed Gluten In order to overcome the technical hurdles, specific grades of pea protein have been developed, especially insoluble ones, to limit the interaction with water.

In bread-type products, mixing an insoluble pea protein with hydrolysed wheat protein is the best way to answer this challenge. On one hand, pea protein increases the amino acid score to high levels, while maintaining texture and volume. In the meantime, hydrolysed wheat protein maximises vital wheat gluten incorporation, preserves the texture and volume of breads, and helps to develop a cost-effective solution.

In biscuits, insoluble grades of pea protein do not interfere with the hydration of ingredients, thereby preserving texture, while allowing an increase in the amino acid score.

In other cereal products, such as snacks or extruded products, an insoluble grade of pea protein makes it possible to achieve high protein content while maintaining processibility (fluidity, water absorption, expansion, output pressure).

10.5.2.2 Gluten-Free Products

During the past five years, 34% of the new protein centric food products bore a gluten-free claim in the EU (Mintel GNPD, 2011–2016).

Gluten is a major allergen, most often associated with wheat and wheat flour. Gluten proteins in wheat flours make dough elastic and stretchy, and trap gas within baked goods, providing a light, aerated structure. These functions make it challenging to substitute, especially in bread.

The most widely used solution to replace gluten is a mix of starches and hydrocolloids (such as gums). However, these ingredients cause quick retrogradation, which leads to a weak crust and rapid staling. A functional grade of pea protein is a good answer to overcome this challenge, first, by increasing the Maillard reaction which helps to create a stronger crust and delay staling. Second, as gluten-free bakery products usually lack nutrition, it provides a high-quality protein, with an almost complete amino acid profile.

10.5.3 Specialised Nutrition

The use of plant-based proteins in the specialised nutrition market (sports nutrition, weight management, clinical nutrition) is comparable to what happened in the savoury and dairy markets. In addition, mainstreaming has developed, meaning that new consumers are now purchasing specialised nutrition products, beyond the core users. For example, baby formulas now not only concern infants but also older babies, to whom parents give growing-up milks.

10.5.3.1 Partial Substitution of Dairy Proteins

Specialised nutrition is a dynamic market, showing annual growth rates among the highest of the food and beverages industries (7% CAGR 2011–2016, Euromonitor).

Even in specialised nutrition, a market of high-value products, there is dynamic competition between producers. Manufacturers have improved their formulation costs via partial substitution of dairy proteins with plant proteins. However, not all plant proteins can be used in specialised nutrition.

Indeed, the nutritional quality and performance of the protein included in these products, from protein powders for muscle mass development to daily supplements for people recovering from surgery, must provide all the essential amino acids.

The first to be used was soy protein, due to both easy availability worldwide and good quality (PDCAAS ≈95). However, increasing concern around allergens, GMOs, and phyto-oestrogens has highlighted the relevance of pea protein in this market. With a very good digestibility and an almost complete amino acid profile, as described above, pea protein has gained success in specialised nutrition. Taking sports nutrition products as an example, manufacturers use pea protein for its very high content of branched chain amino acids and arginine, which are of interest in muscle mass development and metabolism. Overall, new high-protein product launches containing pea protein rose by 13% between 2011 and 2016, while new products containing soy protein decreased by 5% during the same period.

During this first phase of plant protein use as dairy or partial dairy substitutes, plant proteins were often 'hidden' on food labels as they were considered by both manufacturers and consumers as low-quality alternatives to their dairy counterparts.

10.5.3.2 Higher Value Plant-Based Protein Markets (Weight Management and Sports Nutrition)

These days, more and more consumers are becoming body conscious, either to look good or for health reasons. This is where diet modification and physical activity enter. This trend originated in North America, followed closely, as always, by Europe, and is now spreading into South East Asia, India, and Latin America. In these developing markets, young people are willing to buy specialised nutrition products to shape their body. These consumers are different from the core users of specialised nutrition products which include heavy bodybuilders and 'go-pro' athletes, focused on performance rather than on food ingredient origin or environmental impact. Above the body shaping effect they are seeking, these new consumers brought into sports and weight management markets trends from the general food mass market, including the interest in vegan, ethically sourced, plant-based, and free-from products. These consumers are looking for all these claims on their sports nutrition and weight management products. As an example, the 'vegan' claim has grown 77% in the EU (54% worldwide) between 2012 and 2016 (Mintel GNPD).

As a reaction from the food industry, plant proteins are now clearly highlighted on product packaging as value-added ingredients. One exception is soy. Indeed, most products bearing a 'vegan' or 'plant-based' claim also claim to be 'soy-free'. Pea protein fits into this new trend thanks to its 'clean story', and has generated great interest among specialised nutrition producers. However, this market is a challenge for pea protein. On one hand, the high protein content of the different product matrices is a real technical hurdle, and on the other hand, nutritional efficiency must be as high as dairy proteins to ensure the product accomplishes its role in the body.

Technical Challenge: Demanding Product Formulations The three most common specialised nutrition products are powders, ready-to-drink beverages, and bars. If the first one is quite easy to formulate (the different ingredients must have comparable particle sizes and densities, disperse well and create a nice smooth texture), the other two are more complicated to develop.

In the case of high-protein bars (so-called 'dough bars'), the main protein properties that influence the final texture are water-binding capacity, solubility, and particle size. All these properties refer to interactions with water. Indeed, there is an insufficient amount of water in a high-protein dough bar (≈15%) to fully hydrate all components (≈30% protein + 10% humectants + 40% carbs...), meaning that each ingredient competes for water. The solution is to mix proteins with different chemical characteristics or behaviours vis-à-vis water, both soluble and insoluble. The soluble ones will ensure a soft texture, while the insoluble ones prevent the bar from hardening over time due to too much water absorption by the soluble proteins.

In the case of ready-to-drink beverages, which are stable drinks with protein contents ranging from 6% to 10–12% protein, the main technical requirements are excellent solubility, low viscosity, and resistance to heat treatment, meaning no viscosity increase after heat treatment. Here, soluble proteins are the ones to use. Hydrolysis is one of the process steps that can enhance solubility and minimise viscosity by decreasing the molecular weight of a protein.

Nutritional Challenge: Demanding Consumers Sports people consume proteins for muscle mass development and recovery. A plant-based sports nutrition product must therefore achieve the same level of nutritional performance as a dairy counterpart. Pea protein has proved its worth in sports nutrition, as described above, and continues to gain interest from manufacturers.

Weight-conscious people buy low-calorie meal replacements and high-protein snacks to balance their daily diet in order to manage their weight. The high protein content of these products has two main missions: preserving muscle mass while losing fat, and promoting satiety to ensure good compliance with the diet and a successful weight control strategy. Pea protein has proved its effect in weight management, as described above, enhancing the good reputation this ingredient already has.

10.5.3.3 Future of Plant-Based Specialised Nutrition Products: Senior Nutrition

The last part of the specialised nutrition market that has not yet been fully explored with regard to the use of plant protein-based alternatives to dairy and soy is the area of senior and clinical nutrition. Indeed, most products targeting ageing, ill or frail people are still dairy based. However, we can already predict that a portion of this market will start to look to plant proteins in the coming years.

The worldwide population is ageing, generating a real public health issue in developed countries. 'Silver actives' are consumers wanting to remain active and maintain the same physical abilities. To achieve that, they will look for high-protein food products and as they are completely integrated in the current trends, being connected and active in the community, it is likely that some will look for plant-based alternatives to dairy products for seniors. The next five years will certainly see the development of plant-based high-protein products targeting seniors.

10.6 Conclusion

Pea protein is successful. Its growth and use in food products have been steadily double-digit for the past five years and are expected to remain so in the mid-term. Flexitarian lifestyles, health and allergen concerns about soy and wheat, rising meat costs, and better tasting vegetable protein solutions are drivers for long-term sustained pea protein growth. Pea protein answers all these new demands around clean and recognisable ingredients, environmentally friendly food products, and nutrition options.

Conflict of interest

The authors of the chapter are both employees of Roquette.

Acknowledgements

This chapter did not receive any specific grant from funding agencies in the public, commercial, or non-for-profit sectors. We thank many of our colleagues for their help and support.

References

Allergen Nomenclature (2017). WHO/IUIS Allergen Nomenclature Sub-Committee. Retrieved from: www.allergen.org/search.php?allergensource=Pisum+sativum

Altorf-van der Kuil, W., Engberink, M.F., Brink, E.J. *et al.* (2010). Dietary protein and blood pressure: a systematic review. *PLoS One* 5 (8): e12102.

Adam, C.L., Gratz, S.W., Peinado, D.I. *et al.* (2016). Effects of dietary fibre (pectin) and/or increased protein (casein or pea) on satiety, body weight, adiposity and caecal fermentation in high fat diet-induced obese rats. *PLoS One* 11 (5): e0155871.

Asgar, M.A., Fazilah, A., Huda, N. *et al.* (2010). Nonmeat protein alternatives as meat extenders and meat analogs. *Comprehensive Reviews in Food Science and Food Safety* 9: 513–529.

Babault, N., Païzis, C., Deley, G. *et al.* (2015). Pea proteins oral supplementation promotes muscle thickness gains during resistance training: a double-blind, randomized, placebo-controlled clinical trial vs whey protein. *Journal of the International Society of Sports Nutrition* 12 (1): 3–9.

Beaudart, C., Dawson, A., Shaw, S.C. *et al.* (2017). Nutrition and physical activity in the prevention and treatment of sarcopenia: systematic review. *Osteoporosis International* 28 (6): 1817–1833.

Bernhisel-Broadbent, J. and Sampson, H.A. (1989). Cross-allergenicity in the legume botanical family in children with food hypersensitivity. *Journal of Allergy and Clinical Immunology* 83 (2 Pt 1): 435–440.

Boirie, Y. (2004). Protéines « lentes », protéines « rapides ». *Nutrition Clinique et Métabolisme* 18 (1): 25–27.

Boye, J.I., Aksay, S., Roufik, S. *et al.* (2010). Comparison of the functional properties of pea, chickpea and lentil protein concentrates processed using ultrafiltration and isoelectric precipitation techniques. *Food Research International* 43: 537–546.

Chentouf, A., Looten, P., Srinivasan, M., and Srinivas, A. (2013a). Acute oral toxicity of pea protein isolate (Nutralys) in Wistar rats and Cd1 mouse. *Journal of Toxicology and Health, Photon* 103: 180–184.

Chentouf, A., Looten, P., Parvathi, M.V. *et al.* (2013b). Genotoxicological evaluation of NUTRALYS® pea protein isolate. *International Scholarly Research Notices: Toxicology* 27: 817353.

Chentouf, A., Looten, P., Srinivasan, M. *et al.* (2013c). Subchronic toxicological effects of pea protein isolate (Nutralys) on wistar rats: a ninety-day dietary. *Journal of Toxicology and Health, Photon* 103: 225–233.

Chentouf, A., Looten, P., Srinivasan, M., and Srinivas, A. (2013d). Acute skin, eye irritation responses in rabbits and skin sensitization potential in mice of pea protein isolate (Nutralys). *Journal of Toxicology and Health, Photon* 103: 219–224.

Coyne, C.J., Grusak, M.A., Razai, L. and Baik, B.-K. (2005). Variation for pea seed protein concentration in the USDA Pisum core collection. Retrieved from: http://hermes.bionet.nsc.ru/pg/37/5.htm

Dahl, W., Foster, L.M., and Tyler, R.T. (2012). Review of the health benefits of peas (*Pisum sativum* L.). *British Journal of Nutrition* 108 (S1): S3–S10.

Davis, J., Sonesson, U., Baumgartner, D.U., and Nemecek, T. (2010). Environmental impact of four meals with different protein sources: case studies in Spain and Sweden. *Food Research International* 43 (7): 1874–1884.

De Backer, C.J. and Hudders, L. (2014). From meatless Mondays to meatless Sundays: motivations for meat reduction among vegetarians and semi-vegetarians who mildly or significantly reduce their meat intake. *Ecology of Food and Nutrition* 53 (6): 639–657.

FAO (2013). *World Livestock 2013 – Changing Disease Landscapes.* Rome: Food and Agriculture Organisation of the United Nations.

FAOSTAT (2017). www.fao.org/faostat/en/#home

Fox, N. and Ward, K. (2008). Health, ethics and environment: a qualitative study of vegetarian motivations. *Appetite* 50 (2–3): 422–429.

Gausseres, N., Mahe, S., Benamouzig, R. *et al.* (1997). [15N]-Labeled pea flour protein nitrogen exhibits good ileal digestibility and postprandial retention in humans. *Journal of Nutrition* 127 (6): 1160–1165.

Girgih, A.T., Nwachukwu, I.D., Onuh, J.O. et al. (2016). Antihypertensive properties of a pea protein hydrolysate during short- and long-term oral administration to spontaneously hypertensive rats. *Journal of Food Science* 81 (5): H1281–H1287.

Gujral, N., Freeman, H.J., and Thomson, A.B.R. (2012). Celiac disease: prevalence, diagnosis, pathogenesis and treatment. *World Journal of Gastroenterology* 18 (42): 6036–6059.

Hood-Niefer, S.D., Warkentin, T.D., Chibbar, R.N. *et al.* (2012). Effect of genotype and environment on the concentrations of starch and protein in, and the physicochemical properties of starch from, field pea and fababean. *Journal of the Science of Food and Agriculture* 92 (1): 141–150.

Janzen, J.P., Brester, G.W., and Smith, V.H. (2014). *Dry Peas: Trends in Production, Trade, and PriceBriefing No. 57.* Bozeman: Agricultural Marketing Policy Center.

Konopka, A.R. and Harber, M.P. (2014). Skeletal muscle hypertrophy after aerobic exercise training. *Exercise and Sport Sciences Reviews* 42 (2): 53–61.

Li, H. and Aluko, R.E. (2010). Identification and inhibitory properties of multifunctional peptides from pea protein hydrolysate. *Journal of Agricultural and Food Chemistry* 58 (21): 11471–11476.

Li, H., Prairie, N., Udenigwe, C.C. *et al.* (2011). Blood pressure lowering effect of a pea protein hydrolysate in hypertensive rats and humans. *Journal of Agricultural and Food Chemistry* 59 (18): 9854–9860.

Mahe, S., Pelletier, X., and Tome, D. (1997). Les besoins en azote et en acides aminés et la qualité nutritionnelle des protéines alimentaires. *Les Protéines* 1 (4): 63–83.

Mollard, R.C., Luhovyy, B.L., Smith, C.E., and Anderson, G.H. (2014). Acute effects of pea protein and hull fibre alone and combined on blood glucose, appetite, and food intake in healthy young men – a randomized crossover trial. *Applied Physiology, Nutrition and Metabolism* 39 (12): 1360–1365.

Mondor, M., Aksay, S., Drolet, H. *et al.* (2009). Influence of processing on composition and anitinutritional factors of chick pea protein concentrates produced by isoelectric precipitation and ultrafiltration. *Innovative Food Science and Emerging Technologies* 10 (3): 342–347.

Nadathur, S.R., Wanasundara, J.P.D., and Scanlin, L. (2017). Introduction. In: *Sustainable Protein Sources* (ed. S.R. Nadathur, J.P.D. Wanasundara and L. Scanlin). London: Academic Press.

Overduin, J., Guerin-Deremaux, L., Wils, D., and Lambers, T.T. (2015). NUTRALYS® pea protein: characterization of *in vitro* gastric digestion and *in vivo* gastrointestinal peptide responses relevant to satiety. *Food and Nutrition Research* 59: 25622.

Ogawa, K., Sanada, K., Machida, S. *et al.* (2010). Resistance exercise training-induced muscle hypertrophy was associated with reduction of inflammatory markers in elderly women. *Mediators of Inflammation* 171023.

Parolini, C., Manzini, S., Busnelli, M. *et al.* (2013). Effect of the combinations between pea proteins and soluble fibres on cholesterolaemia and cholesterol metabolism in rats. *British Journal of Nutrition* 110 (8): 1394–1401.

Re, R., Pombo, S., Calama, W. *et al.* (2016). The satiating effect of NUTRALYS® pea protein leads to reduced energy intake in healthy humans. *Journal of Nutritional Health and Food Science* 4 (3): 1–10.

Rouge, P., Culerrier, R., Granier, C. *et al.* (2010). Characterization of IgE-binding epitopes of peanut (*Arachis hypogaea*) PNA lectin allergen cross-reacting with other structurally related legume lectins. *Molecular Immunology* 47 (14): 2359–2366.

Sanchez-Monge, R., Loez-Torrejon, G., Pascual, C. *et al.* (2004). Vicilin and convicilin are potential major allergens from pea. *Clinical and Experimental Allergy* 34 (11): 1747–1753.

Scarborough, P., Applebye, P.N., Mizdrak, A. *et al.* (2014). Dietary greenhouse gas emissions of meat-eaters, fish-eaters, vegetarians and vegans in the UK. *Climatic Change* 125 (2): 179–192.

Singh, U. (1988). Antinutritional factors of chickpea and pigeonpea and their removal by processing. *Plant Foods for Human Nutrition* 38: 251–261.

Shewry, P., Napier, J., and Tatham, A. (1995). Seed storage proteins: structures and biosynthesis. *Plant Cell* 7 (7): 945–956.

Sirtori, C.R., Tiolo, M., Boissio, R. *et al.* (2012). Hypocholesterolaemic effects of lupin protein and pea protein/fibre combinations in moderately hypercholesterolaemic individuals. *British Journal of Nutrition* 107 (8): 1176–1183.

Sell, M., Steinhart, H., and Paschke, A. (2005). Influence of maturation on the alteration of allergenicity of green pea (*Pisum sativum* L.). *Journal of Agricultural and Food Chemistry* 53 (5): 1717–1722.

Smith, C.E., Mollard, R.C., Luhovyy, B.L., and Anderson, G.H. (2012). The effect of yellow pea protein and fibre on short-term food intake, subjective appetite and glycaemic response in healthy young men. *British Journal of Nutrition* 108 (S1): S74–S80.

Teunissen-Beekman, K.F.M., Dopheide, J., Geleijnse, J.M. *et al.* (2014). Differential effects of proteins and carbohydrates on postprandial blood pressure-related responses. *British Journal of Nutrition* 112: 600–608.

Tzitzikas, E.N. 2005. Exploring variation in pea protein composition by natural selection and genetic transformation. PhD thesis, Wageningen University.

Tzitzikas, E.N., Vincken, J.P., de Groot, J. *et al.* (2006). Genetic variation in pea seed globulin composition. *Journal of Agricultural and Food Chemistry* 54 (2): 425–433.

Vereda, A., Andreae, D.A., Lin, J. *et al.* (2010). Identification of IgE sequential epitopes of lentil (Len c 1) by peptide microarray immunoassay. *Journal of Allergy and Clinical Immunology* 126 (3): 596–601.

van Vliet, S., Burd, N.A., and van Loon, L.J.C. (2015). The skeletal muscle anabolic response to plant- versus animal-based protein consumption. *Journal of Nutrition* 145 (9): 1981–1991.

Wensing, M., Knulst, A., Piersma, S. *et al.* (2003). Patients with anaphylaxis to pea can have peanut allergy caused by cross-reactive IgE to vicilin (Ara h 1). *Journal of Allergy and Clinical Immunology* 111 (2): 420–424.

WHO 2014. Factsheets 311 and 312. Retrieved from: www.who.int

World Health Organization/Food and Agriculture Organization/United Nations University, 2007. Protein and Amino Acid Requirements in Human Nutrition. Report of a Joint WHO/FAO/UNU Expert Consultation. WHO Technical Report Series No. 935. World Health Organization, Geneva.

Yang, H., Guerin-Deremaux, L., Zhou, L. *et al.* (2012). Evaluation of nutritional quality of a novel pea protein. *AgroFood Industry Hi-Tech* 23 (6): 8–10.

11

Seaweeds as a Source of Proteins for Use in Pharmaceuticals and High-Value Applications

Chigozie Louis Okolie[1,2], *Beth Mason*[1], *and Alan T. Critchley*[1]

[1] *Verschuren Centre for Sustainability in Energy and the Environment, Cape Breton University, Sydney, Nova Scotia, Canada*
[2] *Department of Plant, Food, and Environmental Sciences, Faculty of Agriculture, Dalhousie University, Truro, Canada*

11.1 Introduction

The global population is on a steady rise, with a projected addition of 2.3 billion people by the year 2050 (Bleakley and Hayes 2017). This expected spike in population makes it more than necessary to source additional, sustainable alternatives to the current sources of basic needs such as food, energy, and health. Scientific investigations into the conversion of bio-products to functional food ingredients, biofuels, plant bioactives, and plant-based drugs are ongoing, as first steps to tackling these challenges.

The oceans are the most significant natural reservoir of both living and non-living organisms (Godfray et al. 2010; Harnedy and FitzGerald 2011). A key member of the marine ecosystem, by biomass and diversity, is the group of macroscopic, photosynthetic organisms commonly referred to as seaweeds or macroalgae. Macroalgae are a diverse group of macroscopic, photosynthetic marine organisms with potentially high nutritional and pharmaceutical values. They are traditionally consumed as food (sea vegetables) and have been used as herbal medicine in the treatment of stomach ailments, such as gall stones, eczema, cancer, psoriasis, asthma, lung disease, heart disease, and others (Peng et al. 2015).

Macroalgae are classified into three major groups, largely based on their pigmentation: brown (Phaeophyta), red (Rhodophyta), and green (Chlorophyta) seaweeds (Guiry and Guiry 2017). Fucoxanthin is the major accessory pigment in brown macroalgae, phycobiliproteins are found in red macroalgae, and chlorophyll a and b in green macroalgae (Okolie et al. 2017). The macroalgae are notable for their polysaccharide and mineral content, as well as having appreciable amounts of proteins, which are known to vary across species and also within seasonal cycles. The relatively high protein content of macroalgae makes them a potential source of bioactive peptides with benefits for promoting health in animals and humans. These proteins, in addition to providing nutritional benefits, also have antibiotic, cytotoxic, antinociceptive, anti-inflammatory, anti-HIV (human immunodeficiency virus), and other beneficial effects.

Novel Proteins for Food, Pharmaceuticals, and Agriculture: Sources, Applications, and Advances, First Edition. Edited by Maria Hayes.

Exogenous bioactive peptides remain inactive in the parent protein source until released by the action of endogenous/exogenous enzymes or microbial activity (Udenigwe et al. 2017). Peptides released through these processes possess various physiological functions including antibacterial, antioxidant, antithrombotic, antihypertensive, and immunomodulatory activities, amongst others (Udenigwe and Aluko 2012). Various species of macroalgae also contain high amounts of individual amino acids (AAs), e.g. alanine, glycine, arginine, leucine, valine, and methionine. Special amino acids with health-promoting potential, differing from the regular 20 amino acids, have also been found in various seaweed species. These include D-homocysteic acid, aminobutyric acid, ornithine, citrulline, taurine, baikiain, laminine, chondrine, gigartine, pipecolic acid, demoic acid, and kainic acid (Pangestuti and Kim 2015).

Despite the many prospects for the industrial utilisation of macroalgal proteins, peptides, and amino acids, their translation into commercially viable products remains a challenge. Seasonal and geographical variability, scalable extraction processes for protein isolation, regulatory restrictions, food safety concerns, food matrix–peptide interactions and the bioavailability and bioactivity of peptides *in vivo* are some of the stumbling blocks affecting the translation of these functional ingredients into value-added food and pharmaceutical products (Bleakley and Hayes 2017; Ejike et al. 2017; Stack et al. 2017).

11.2 Macroalgal Proteins, Peptides, and Amino Acids

11.2.1 Macroalgal Proteins

Proteins are essential components of the human and animal diet and play major roles in health maintenance and promotion. *In vivo* functions include storage, transport, defence, enzyme catalysis, support, motion, and physiological regulation (Cerna 2011). With the steady rise in global population, it appears unlikely that the traditional sources of proteins (i.e. legumes, cereals, and grains) will meet the growing demand for proteins for both human and animal nutrition (Cole et al. 2015). Certain species of macroalgae provide a sustainable alternative source, since they are relatively abundant and could be made readily available for use. This is particularly true of seaweed proteins if cultivation of seaweeds at large scale is achieved.

Proteins from macroalgae contain all the essential amino acids (EAAs) and, in some cases, have been reported to possess higher quantities and qualities than most conventional sources of proteins such as cereals, eggs, fish, and soybean (Fleurence 2004; Cerna 2011; Harnedy and FitzGerald 2013). Amongst the various macroalgal groups, members of the Rhodophyta are reported to have the highest protein content, with maximum protein content of approximately 50% dry weight, followed by selected chlorophytes with a 'moderate' amount of 9–26% dry weight, followed by certain phaeophytes with the lowest at about 3–15% (Fleurence 2004; Pangestuti and Kim 2017). Table 11.1 shows the protein content of some species from the three main groups of seaweed.

The amount of protein in a particular species may be influenced by a number of factors, one of which is seasonal change (Harnedy and FitzGerald 2011). The seasonal influence on protein composition can be linked to a number of environmental conditions such as sea surface temperature (SST), salinity, availability of light, fluctuation in

Table 11.1 Protein content of some species of macroalgae.

Major group	Species	Protein content (% w/w)	Major group	Species	Protein content (% w/w)	Major group	Species	Protein content (% w/w)
	Ulva rigida	18–19		*Porphyra* sp.	30–50		*Undaria pinnatifida*	11–24
	U. fasciata	6.26		*Porphyra* sp.	24.11	Phaeophyta	*U. pinnatifida*	19.8
Chlorophyta	*U. pertusa*	20–26	Rhodophyta	*Porphyra* sp.	44		*Laminaria digitata*	15.9
	U. pertusa	17.5–26		*P. tenera*	33–47		*L. digitata*	10.7
	U. compressa	21–32		*P. umbilicalis*	29–39		*L. digitata*	8–15
	U. lactuca	8.7–32.7		*P. yezoensis*	31–44		*L. ochroleuca*	7.49
	U. lactuca	10–21		*Pyropia tenera*	28–47		*L. saccharina*	6–11
	U. lactuca	29		*Palmaria palmata*	8–35		*Laminaria* sp.	7.5
	U. lactuca	15–25		*P. palmata*	Dec-21		*Fucus vesiculosus*	3–14
	U. armoricana	18–24		*Amansia multifida*	25.6		*F. serratus*	3–10
	Ulva sp.	15–25		*Bryothamnion seaforthii*	17.3		*F. spiralis*	10.77
	Caulerpa lentillifera	9.7		*Enantiocladia duperreyi*	19.5		*Sargassum fusiforme*	11.6
	C. lentillifera	10.4		*Gelidiella acerosa*	31		*S. vulgare*	16.3
	C. lentillifera	10–13		*Gracilaria changii*	34.5		*S. polycystum*	5.4
	Caulerpa sertularoides	17–20		*Grateloupia turuturu*	22.9		*S. latissima*	5
	C. racemosa	20		*Halymenia formousa*	21.2		*S. japonica*	7–8

(*Continued*)

Table 11.1 (Continued)

Major group	Species	Protein content (% w/w)	Major group	Species	Protein content (% w/w)	Major group	Species	Protein content (% w/w)
	Codium fragile	17.8–18.4		*Laurencia filiformis*	18.3		*Ascophyllum nodosum*	5–12
	Enteromorpha intestinalis	8–11		*Plocamium brasiliense*	15.7		*Himanthalia elongata*	6–11
	E. intestinalis	10–18		*Solieria filiformis*	21.3		*Alaria esculenta*	9–20
	E. compressa	15.2					*Hizikia fusiforme*	11.6
	E. flexuosa	12.3					*Dictyota acutiloba*	12
	Cladophora glomerata	7.9					*Durvillaea antarctica*	10.4
	C. glomerata	20.4						
	Utricularia reticulata	14.1						

This table was prepared with the following references using the nomenclature as cited by: Cerna (2011); Fleurence (2004);. Harnedy and FitzGerald (2011); Pangestuti and Kim (2015).

carbohydrate levels, and types of protein present (Harnedy and FitzGerald 2011). Studies on the influence of seasonal changes on protein levels suggested that protein content was highest during the winter and spring, but lower during the summer and early autumn periods. *Palmaria palmata* (Rhodophyta) showed large variations in protein content (i.e. 9–25%), with the highest values occurring during winter and spring months (Galland-Irmouli et al. 1999). Conversely, the protein contents of *Pterocladia capillacea* and *Jania rubens* (Rhodophyta) were highest in the spring (23.72% and 12.93% respectively) and lowest in the autumn (17.35% and 9.76% respectively) (Khairy and El-Shafay 2013). Khairy and El-Shafay (2013) also reported that the highest protein content (20.12%) of *U. lactuca* occurred in the spring and was lowest in autumn (16.78%). The protein contents of the phaeophytes *Stypopodium schimperi* and *Padina pavonia* were at their highest in the spring (15.29% and 8.86%) and lowest during the summer (10.95%) and autumn periods (8.00%), respectively.

The two most prominent proteins found in various macroalgae are lectins and phycobiliproteins (Stengel et al. 2011). Lectins are naturally occurring proteins or glycoproteins of non-immune origin, with at least one catalytic binding domain, which selectively bind non-covalently to carbohydrate residues and are able to agglutinate cells/precipitate polysaccharides or glycoconjugates (Rivanor et al. 2014; Praseptiangga 2017). Lectins play a significant role in cell-to-cell or cell-to-matrix reactions as recognition molecules, and this property makes them a valuable biochemical tool in the analysis of protein–carbohydrate interactions, such as host–pathogen, induction of apoptosis, cancer metastasis, and differentiation (Harnedy and FitzGerald 2011; Praseptiangga 2017). Lectins from macroalgae are low molecular weight, monomeric, thermostable protein molecules with a high affinity for oligosaccharides or glycoproteins but not monosaccharides. Lectins are sometimes referred to as agglutinins and haemagglutinins due to their haemagglutinating activities against erythrocytes (Harnedy and FitzGerald 2011). They are applied in blood grouping, drug delivery, and as cancer biomarkers (Bleakley and Hayes 2017). Other bioactivities reported for algal lectins include antinociceptive, antibacterial, antiviral, antiadhesion, cytotoxic, and mitogenic properties (Bleakley and Hayes 2017).

Phycobiliproteins are brightly fluorescent photosynthetic, light-harvesting pigments which can be found in members of the Cyanobacteria, Cryptophyta (flagellates), and Rhodophyta (Wu et al. 2013; Johnson et al. 2014). Phycobiliproteins are water soluble and consist of multichain, holo-proteins, covalently attached to linear tetrapyrrole chromophore groups, known as bilins or phycobilins, which are similar in structure to the well-known bile pigments bilirubin and biliverdin (Harnedy and FitzGerald 2011; Johnson et al. 2014).

There are four chromophore types that are usually attached to the protein backbone of phycobiliproteins: phycocyanobilin, phycoerythrobilin, phycoviolobilin, and phycourobilin (Wu et al. 2013). Phycobiliproteins are geometrically arranged into macromolecules known as phycobilisomes, which ensure the transfer of all captured light energy to the primary, photosynthetic reaction centres (Harnedy and FitzGerald 2011; Johnson et al. 2014). The major phycobiliproteins found in macroalgae include phycocyanin, phycoerythrin, allophycocyanin, and phycoerythrocyanin. Of these, phycoerythrin is the most abundant in the Rhodophyta, constituting about 12% w/w in *P. palmata* (Chronakis 2000).

In general, phycobiliproteins have a vast potential range of applications. They are used in the biomedical field for fluorescent immunoassays, fluorescent immunohistochemistry, biomolecule labelling, and fluorescent microscopy (Harnedy and FitzGerald 2011). In foods, they are used as natural colorants in such essentials as chewing gum and dairy products. In cosmetics, they are used for essential colours in eyeliners and lipsticks. These proteins also have some pharmacological potential such as antioxidative, neuroprotective, antitumour, antiviral, hepatoprotective, and anti-inflammatory activity. They have also been applied in the treatment of arteriosclerosis, serum lipid reduction, and lipase inhibition (Sekar 2008; Samarakoon and Jeon 2012).

11.2.2 Macroalgal Peptides

Macroalgal proteins, like most food proteins, contain specific peptide sequences with corresponding and specific biological activities. These remain inactive in the parent sequence until they are released, by enzymatic hydrolysis, food processing, gastrointestinal digestion, or microbial fermentation (Udenigwe et al. 2017). Such peptides are known to positively improve physiological functions through cascades of various reactions, particularly binding to specific receptors, interacting with target cells, and inhibiting enzymatic processes (Samarakoon and Jeon 2012). In addition to being a source of bioactive peptides, various macroalgal biomasses also possess several endogenous peptides with health-promoting effects, some of the configurations of which are linear, cyclic, or peptides with their amide bonds replaced by ester bonds (Harnedy and FitzGerald 2011). Among these peptides are carnosine and glutathione, notable for their antioxidant potential, kahalalide F (a cyclic depsipeptide obtained from the green alga *Bryopsis* sp.), SECMA 1 (a mitogenic hexapeptide), galaxamide (a cyclic pentapeptide), and a dipeptide derivative, almazole D (Harnedy and FitzGerald 2011).

Recent studies have reported various pharmacological effects of macroalgal peptides, including antidiabetic (Harnedy et al. 2015), antibacterial (Beaulieu et al. 2015), antihypertensive (Fitzgerald et al. 2011; Paiva et al. 2016), antioxidative (Shiu and Lee 2005), and anticancer (Mao et al. 2017).

11.2.3 Macroalgal Amino Acids

Various macroalgal proteins contain all amino acids and meet the WHO/FAO requirements for EAA levels in dietary proteins (Cerna 2011). Varying amounts of essential AAs and non-essential amino acids (NEAAs) have been reported within different seaweed groups. Most studies reported that the highest amino acid concentrations were to be found in the red macroalgae, although there was a contrary report of higher AA levels in members of the green algae (Cerna 2011; Pangestuti and Kim 2015). Going by the statistics, the amino acid scores of proteins in red seaweeds, *Porphyra* spp. (Rhodophyta) and *Undaria* sp. (Phaeophyta), were reported to be 91 and 100, respectively, similar to those derived from animal foods and indeed higher than those of most green and other red macroalgae (Pangestuti and Kim 2015). Also, the EAAs in red seaweeds were reported to represent almost half of the AAs with an EA:AA ratio of 0.4–0.5 (Cerna 2011). In the red seaweed *P. palmata*, 26–50% of the available amino acid pool was reportedly EAAs, and this seems to be in agreement with the aforementioned data on EAA:AA ratio (Galland-Irmouli et al. 1999).

The composition of individual amino acids can also differ across the groups of macroalgae. In general, glycine, alanine, arginine, proline, glutamic, and aspartic acids make up a large proportion of amino acids, whereas cysteine, methionine, and tyrosine are found in lower amounts (Cerna 2011). Notably, aspartic and glutamic acids have been found at higher levels in many seaweeds; for example, in the brown macroalga *Fucus* sp., these two AAs represented 22–44% of the total amino acids available; in the two greens *Ulva rigida* and *U. rotundata*, the combined levels of glutamic and aspartic acids could represent up to 26% and 32% of total amino acids respectively. In some red seaweeds, for instance *Porphyra tenera*, these two amino acids constituted 15% of the total (Fleurence 2004). The relatively high amounts of these amino acids in seaweeds is a major plus, since they both participate in key physiological processes. Glutamic acid plays an important role in maintaining brain function and mental activity while aspartic acid helps to initiate important pathways such as the Krebs and urea cycles (Cerna 2011).

Various macroalgae also contain certain non-conventional amino acids, in addition to the common ones. These are reported to include homocysteic acid, aminobutyric acid, ornithine, citrulline, taurine, baikiain, laminine, chondrine, gigartine, pipecolic acid, demoic acid, and kainic acid. They are generally referred to as 'conditional essential amino acids' and have been reported to exhibit various pharmacological properties such as bile acid binding, immunomodulation, regulation of calcium levels, antihypertension, and antioxidative activity (Harnedy and FitzGerald 2011). Taurine, for instance, has been identified as a free amino-sulfonic acid, mainly acquired through the diet, that plays major role in brain development, osmoregulation, calcium homeostasis, membrane stabilisation, and neuromodulation (Cornish et al. 2017) Table 11.2 contains the amino acid profiles of some of the more common macroalgal proteins.

11.3 Extraction of Macroalgal Proteins, Peptides, and Amino Acids

Macroalgae collectively, and especially those which are cultivated, are a sustainable source of bioactive proteins, peptides, and amino acids. To access these varied bioactivities, efficient recovery (or yield) from a given biomass (feedstock) is key (Figure 11.1). Chemical extraction has been the mainstay in the extraction of proteins and has been successfully used to produce proteins from a number of different species, including *Ascophyllum nodosum* (Kadam et al. 2017), *Ulva rigida* (Fleurence et al. 1995), and *P. palmata* (Harnedy and FitzGerald 2013). Proteins may be extracted by aqueous, acidic, or alkali methods, followed by centrifugation and recovery techniques such as ultrafiltration, precipitation, or chromatography (Bleakley and Hayes 2017).

Despite the successes recorded, accessibility of seaweed proteins is still a major challenge since macroalgal cell walls are predominantly made up of highly viscous and anionic polysaccharides which are known to impede their extraction (Bleakley and Hayes 2017). Hence, a combination of physical and mechanical processes with conventional methods appeared to be the best means to enhance protein recovery. Processes such as grinding, osmotic shock, ultrasonication, and polysaccharidase-aided hydrolysis have all helped to disrupt cell walls and render their content more accessible for extraction. The protein extraction yields from *Porphyra acanthophora* var. *acanthophora* (R), *Sargassum vulgare* (P), and *Ulva fasciata* (C) were significantly improved by grinding

Table 11.2 Amino acid profiles of some macroalgal species.

	Green algae			Red algae			Brown algae		
Amino acids	*Ulva fasciata*	*Ulva pertusa*	*Caulerpa lentillifera*	*Porphyra tenera*	*Palmaria palmata*	*Chondrus crispus*	*Laminaria digitata*	*Sargassum vulgare*	*Undaria pinnatifida*
Alanine	8.5	6.1	0.9	7.4	6.7	3.8	14.4	7.2	4.8
Arginine	5.6	14.9	7.0	16.4	5.1	33.6	0.3	4.3	7.5
Aspartate	13.0	6.5	11.6	7.0	18.5	3.8	8.7	10.9	5.6
Cysteine	—	1.2	—	0.3	—	—	1.7	—	0.5
Glutamate	12.6	6.9	14.4	7.2	9.9	4.1	9.4	17.6	5.1
Glycine	6.5	5.2	0.9	7.2	13.3	3.5	4.3	5.7	4.4
Histidine	2.4	4.0	0.7	1.4	0.5	0.9	1.3	2.1	2.7
Isoleucine	3.9	3.5	5.0	4.0	3.7	1.8	2.7	4.8	2.9
Leucine	7.6	6.9	1.0	8.7	7.1	2.9	5.4	8.5	5.1
Lysine	5.1	4.5	6.6	4.5	3.3	4.9	3.7	5.4	4.3
Methionine	0.9	1.6	—	1.1	2.7	0.5	1.6	2.2	2.0
Phenylalanine	5.3	3.9	4.9	3.9	5.1	1.5	3.2	5.3	3.7
Proline	4.6	4.0	0.6	6.4	1.8	1.9	3.7	4.6	2.8
Serine	5.8	3.0	0.8	2.9	6.3	2.2	4.0	5.1	2.8
Threonine	5.1	3.1	6.4	4.0	3.6	2.2	4.4	4.8	2.4
Tryptophan	—	0.3	—	1.3	3.0	—	0.8	—	0.8
Tyrosine	3.3	1.4	3.9	2.4	3.4	1.0	1.5	2.2	1.6
Valine	5.7	4.9	0.9	6.4	6.9	—	4.2	5.8	4.1

This table was prepared with the following references: Fleurence (2004) and Pangestuti and Kim (2015).

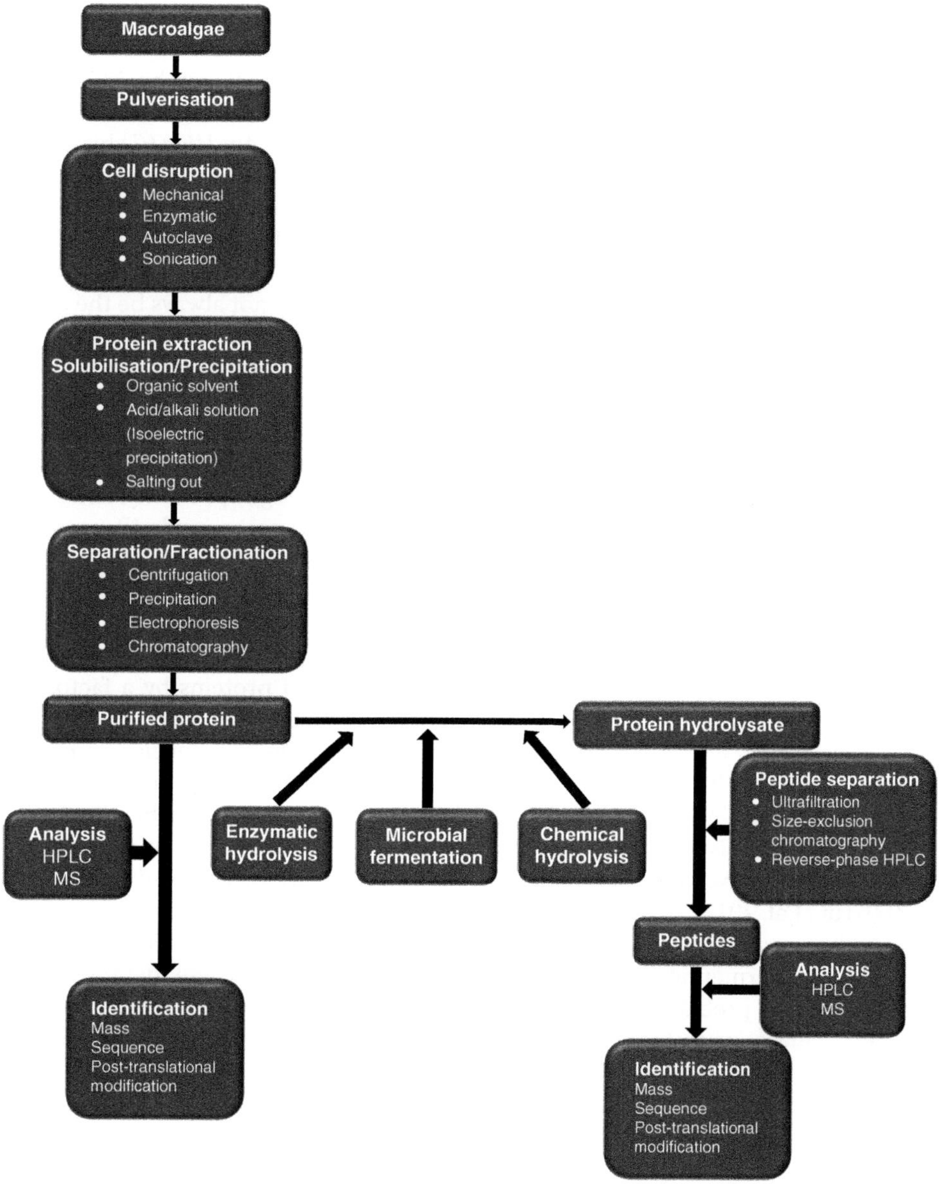

Figure 11.1 Overview of some procedures used for the extraction and analysis of macroalgae proteins and peptides. HPLC, high-performance liquid chromatography; MS, mass spectrometry.

the seaweeds before immersion in ultra-pure water (Barbarino and Louren 2005). The combination of ultrasonication and maceration in the extraction process for phycoerythrin and phycocyanin from *Gelidium pusillum* (R) resulted in high extraction efficiencies of 77% and 93% for both of the aforementioned proteins (Mittal et al. 2017). Also, combining ultrasonication and maceration significantly enhanced extraction efficiency by 31%, compared to maceration alone (Mittal et al. 2017).

Another study on the extraction of proteins from *A. nodosum* (P) using ultrasonication, in combination with acid and alkali treatments, recorded significant protein yields of 43.13% and 57.23% respectively and increased extraction efficiency by 540% and 27%,compared to acid and alkali treatments alone (Kadam et al. 2017). Pulse electric field extraction has also been used to efficiently extract proteins from *Ulva* (C) biomass, although this process also reportedly damaged some specific proteins (Polikovsky et al. 2016). Cell wall disruption with either osmotic shock or shear force resulted in protein recovery efficiency of 5.76 and 6.18 g/100 g respectively (Harnedy and FitzGerald 2013).

These studies showed significant improvement when mechanical and physical processes were combined with conventional methods, but this may not always be the case, as the application of these processes at extreme conditions may have a negative impact, depending on the structural and chemical properties of the protein of interest.

Enzymatic hydrolysis has been reported to be the best technique for the extraction of protein and peptide fractions, in the food and pharmaceutical industries (Pangestuti and Kim 2015). This method has the advantages of requiring less toxic chemicals/microbial contamination and having high catalytic efficiency, high specificity, and mild reaction conditions combined with preservation of bioactive compounds (Pangestuti and Kim 2015). Moreover, the use of cell wall-degrading polysaccharidases (e.g. cellulase, kappa-carrageenase, xylanase, beta-agarase) in disrupting the cell wall of some seaweeds has helped to render the proteins more accessible, thereby increasing protein availability (Bleakley and Hayes 2017). Enzyme-assisted extraction of proteins from *P. palmata* with xylanase and cellulase increased the yield of proteins by a factor of 10 times that of mechanical extraction alone (Joubert and Fleurence 2008). In addition, enzymatic treatment increased protein yields in certain seaweed species, such as *Sargassum muticum* (P), compared to aqueous extraction form the same biomass (Hardouin et al. 2014). Most bioactive macroalgal peptides are known to be generated via enzymatic hydrolysis. Enzymatic proteolysis enhanced the extraction of antioxidant and angiotensin-1-converting enzyme (ACE) inhibitory peptides from *Porphyra columbina* (R) (Cian et al. 2013).

Other techniques which have been used to obtain peptides from seaweed proteins include microbial fermentation and chemical hydrolysis; however, these methods have limited application, probably due to the low yields obtained. Peptides within protein hydrolysates can be separated through ultrafiltration or any of the chromatographic techniques and can be identified/quantified with high-performance liquid chromatography (HPLC) and mass spectroscopy.

11.4 Bioactivities of Macroalgal Proteins, Peptides, and Amino Acids

11.4.1 Antioxidant Properties of Macroalgal Proteins, Peptides, and Amino Acids

Elevated levels of reactive oxygen species (ROS) have been linked to a good number of pathological conditions, such as hypertension, cancer, neurodegenerative diseases, and diabetes. Oxidative stress, a condition which favours an increase in the production of free radicals and ROS, results in oxidative damage of biological macromolecules such as lipids, carbohydrates, and DNA, which can eventually lead to a multitude of health complications (Ejike et al. 2017). The effects of ROS can be cushioned by exogenous antioxidants with radical scavenging properties, thus complementing the cellular antioxidant machinery (Udenigwe et al. 2017).

Proteins, peptides, and amino acids obtained from selected seaweeds possess strong potential as natural antioxidants (Samarakoon and Jeon 2012). Phycoerythrobilin obtained from the Korean purple laver (*Porphyra* sp.) (R) demonstrated antioxidant activity by scavenging peroxyl radicals *in vitro* (Yabuta et al. 2010). Enzymatic extracts of *Ishige okamure* (P) from four proteases, alcalase, flavourzyme, kojizyme, and protamex, showed significant hydrogen peroxide radical scavenging activities of 91.62%, 93.41%, 96.27%, and 93.71% respectively (Heo and Jeon 2008). Also, a protease hydrolysate of the rhodophyte *P. palmata* had the strongest 2,2-diphenyl-1-picrylhydrazyl (DPPH) and hydroxyl radical-scavenging capacities when compared to carbohydrase hydrolysates (Wang et al. 2010). An endogenous peptide, carnosine (beta-alanyl-L-histidine), isolated from *Ancanthophora delilei* (R), showed antioxidant activity arising from its ability to chelate transition metals (Fleurence 2004; Harnedy and FitzGerald 2011).

Another antioxidant peptide isolated from a macroalgae is glutathione, which is present in relatively high proportion (Pangestuti and Kim 2015). Studies on *Ulva fasciata* showed that the inherent enzymatic (ascorbate-glutathione) and non-enzymatic (glutathione) cycles which detoxify and scavenge oxidative radicals were stimulated in response to ROS produced on exposure to UV-B radiation (Shiu and Lee 2005). Cian et al. (2013) attributed the radical scavenging activity of enzymatic extracts from *P. columbina* to the presence of low molecular weight peptides (ca. 500 Da) and polyphenols (Cian et al. 2013). Taurine, a naturally occurring beta-amino acid derived from cysteine and methionine, may be present in certain red algae at high concentrations, and also acts as an antioxidant through its protection against toxicity from various heavy metals, such as lead and cadmium (Pangestuti and Kim 2015).

Although there is no clear-cut mechanism for the antioxidative activity of macroalgal proteins and their peptides, there are suggestions that the presence of hydrophobic amino acids in the peptide sequence could be an influence (Pangestuti and Kim 2015).

11.4.2 Antihypertensive Properties of Macroalgal Proteins, Peptides, and Amino Acids

Hypertension is a major health risk for the adult population of the world and is currently the most prevalent non-communicable disease (Ejike et al. 2017). It is estimated that by 2025, 1.56 billion people will be suffering from hypertension (Samarakoon and Jeon 2012). This human condition is defined as the sustained increase in systemic arterial pressure above 140/90 mmHg (Tierney and Croft 2010). To maintain homeostasis in blood regulation, there needs to be a balance between the vasoconstrictive mechanisms (renin-angiotensin-aldosterone system (RAAS)), the sympathetic nervous system, the vasopressin system, the endothelin system, ROS-induced vasoconstriction and vasodilatory mechanisms (function of nitric oxide, the bradykinin and prostaglandin systems, the atrial natriuretic peptide system).

Hypertension can be managed by antihypertensive drugs which interfere with one or more of the aforementioned mechanisms. One of this group of drugs includes the ACE inhibitors. ACE catalyses the conversion of angiotensin 1 (ang 1) to angiotensin 2 (ang 2) in the RAAS system. High vascular concentrations of ang 2, which are responsible for regulating blood pressure, can lead to excessive blood vessel contractions, resulting in higher blood pressures (Udenigwe et al. 2017). Thus, inhibiting ACE leads to a decrease in the production of ang 2 and eventually lower blood pressure in a hypertensive condition. ACE also degrades bradykinin, an enzyme involved in the production of nitric oxide, a potent vasodilator (Udenigwe et al. 2017).

Macroalgal-derived peptides have shown interesting prospects as antihypertensive drugs through their ACE inhibitory activities. Also, some animals suffer from similar health conditions to humans, so seaweed as a source of ACE inhibitory peptides may help to improve animal health as well, thus creating a lucrative market for the application of seaweed extracts in animal feeds.

Protein extract and hydrolysates (with alcalase and trypsin) obtained from *P. columbina* exhibited ACE inhibitory activities of 24% and higher than 35%, respectively (Cian et al. 2012). Enzymatic extracts of seven brown algal species (*Ecklonia cava, E. kurome okamure, Sargassum fulvellum, S. horneri, S. coreanum, S. thunbergii*, and *Scytosiphon lomentaria*) were screened for their ACE inhibition activities. Of these species, a flavourzyme digest of *E. cava* had the highest ACE inhibition with the half maximal inhibitory concentration (IC_{50}) value of $0.3\,\mu g^{-1}$ (Athukorala and Jeon 2005). The ACE inhibition of 10 dipeptides, Tyr-His, Lys-Trp, Lys-Tyr, Lys-Phe, Phe-Tyr, Val-Trp, Val-Phe, Ile-Tyr, Ile-Trp, and Val-Tyr, has also been investigated with IC_{50} values of 5.1, 10.8, 7.7, 28.3, 3.7, 10.8, 43.7, 2.7, 12.4, and 11.3 μM respectively. Furthermore, oral administration of four of these peptides, Tyr-His, Lys-Tyr, Phe-Tyr, and Ile-Tyr, significantly reduced blood pressure in spontaneously hypertensive rats (Suetsuna et al. 2004). Laminine monocitrate, an endogenous amino acid derivative from *Laminaria angusta*, has also been reported to possess hypotensive effects (Bhakuni and Rawat 2005).

The antihypertensive activity of macroalgal peptides has been confirmed in humans and also been approved as 'food for specific health uses' in Japan (Pangestuti and Kim 2015).

11.4.3 Antiproliferative Properties of Macroalgal Proteins, Peptides and Amino Acids

Cancer is a global pandemic ranked as the sixth leading cause of death in the world (WHO 2013). In 2008, GLOBOCAN estimated that approximately 12.7 million cancer cases were reported and 7.6 million deaths were said to have occurred (Samarakoon and Jeon 2012). Cancer is characterised by the uncontrolled proliferation of cells with the potential to spread to other parts of the body. Chemotherapy has been the mainstay in management of most cancer cases.

However, proteins and peptides from macroalgae have promising prospects as anticancer agents. A natural anticancer drug, kahalalide F, which is a cyclic depsipeptide, produced by the coenocytic green alga *Bryopsis* sp. but isolated from its herbivorous grazer *Elysia rufescens*, has been reported to possess potent cytotoxic effects against cell lines from solid tumuors including colon, breast and prostate carcinomas, neuroblastoma, osteosarcoma, and chondrosarcoma (Hamann et al. 1996; Suarez et al. 2003; Harnedy and FitzGerald 2011). The antitumour potential of this putative drug have been confirmed in Phase I and Phase II clinical trial stages (Harnedy and FitzGerald 2011). SECMA 1, a mitogenic hexapeptide isolated from *Ulva* sp., has been associated with the modulation of cell proliferation molecules, glycosaminoglycans, and proteoglycans, in human foreskin fibroblasts (Harnedy and FitzGerald 2011). Galaxamide, a cyclic pentapeptide isolated from *Galaxaura filamentous* (R), has been reported to elicit antiproliferative effects against human renal cell carcinoma GRCp-I and human hepatocellular carcinoma HepG2 cell lines (Xu et al. 2008).

11.4.4 Antimicrobial Properties of Macroalgal Proteins, Peptides, and Amino Acids

Studies which investigated the antimicrobial properties of macroalgal proteins and peptides are quite limited, but the few available suggest that these peptides are quite effective in inhibiting the growth of unwanted microbes. Kahalalide A and F have shown prospects as antiviral, antifungal and antibacterial (against tuberculosis-causing mycobacterium – *Mycobacterium tuberculosis*) agents (Hamann and Scheuer 1993; Sayed et al. 2000). Almazole D, a dipeptide derivative isolated from a red alga harvested on the coast of Senegal, has also shown antibacterial effects against *Salmonella typhi* and *Serratia marcescens* (N'Diaye et al. 1996).

11.4.5 Antidiabetic Properties of Macroalgal Proteins, Peptides, and Amino Acids

Diabetes mellitus is a metabolic disorder with multiple aetiologies which is characterised by an abnormal increase in blood glucose level, resulting from lack of insulin secretion or action (Lacroix and Li-chan 2012; Harnedy et al. 2015). An estimated 377 million people will be living with diabetes by the year 2030 (Lacroix and Li-chan 2012). The use of dipeptidyl peptidase IV (DPP-IV) inhibitory drugs is one of the few ways to manage this public health menace.

Physiological postprandial blood glucose is maintained through the action of incretin hormones, glucagon-like peptide 1 (GLP-1), and glucose-dependent insulinotropic polypeptide (GIP) (Harnedy et al. 2015). DPP-IV, which is a serine protease, degrades GLP-I and GIP, leading to the loss of insulinotropic potential. Inhibition of this enzyme by DPP-IV inhibitors exerts a positive influence on glucose regulation by decreasing the inactivation of GLP-1 and GIP and, as such, enhances insulin secretion.

Peptides purified from *P. palmata* protein extracts have recently shown promising prospects as DPP-IV inhibitors. Three new peptide sequences from *P. palmata*, Ile-Leu-Ala-Pro, Leu-Leu-Ala-Pro, and Met-Ala-Gly-Val-Asp-His-Ile, were identified as DPP-IV inhibitors with IC_{50} values within the range of 43–159 μM (Harnedy et al. 2015). This result suggested that selected macroalgae could be a sustainable source of active ingredients in the management of diabetes.

Table 11.3 presents some of the reported bioactivities for macroalgal proteins, peptides, and amino acids.

11.5 Industrial Applications of Macroalgal Proteins, Peptides, and Amino Acids

Industrially, macroalgae and macroalgal proteins have a wide range of applications, especially in the food and pharmaceutical industries, cosmetics, and as animal feeds.

Table 11.3 Bioactivities of macroalgae proteins, peptides, and amino acids.

	Name/sequence	Source	Bioactivity	References
Proteins/protein hydrolysates	Lectin	*Bryothamnion seaforthii*	Wound healing	Nascimento-Neto et al. (2012)
		Griffithsia sp.	Anti-HIV	Ziółkowska and Wlodawer (2006)
		Eucheuma serra	Mitogenic activity	Kawakubo et al. (1997)
		Eucheuma serra	Cytotoxic/anticancer	Sugahara et al. (2001)
		Amansia multifida	Antinociceptive	Neves et al. (2007)
	Phycoerythrobilin	*Porphyra* sp.	Antioxidant	Yabuta et al. (2010)
	Protein hydrolysates	*Pyropia haitanensis*	Antiproliferative	Mao et al. (2017)
		Palmaria palmata	ACE inhibition/antioxidant	Beaulieu et al. (2016)
		Costaria costata	Antitumor/antioxidant/antityrosinase	Lee et al. (2005)
		Ecklonia cava	ACE inhibition/antioxidant	Cha et al. (2006); Heo et al. (2003)
		Porphyra tenera	Antioxidant, ACE inhibition, antitumor, antityrosinase	Lee et al. (2005)
		Porphyra yezoensis	Antihypertensive/antimutagenic/hypoglycaemic/hypolipidaemic/Ca^{2+} precipitation inhibition/improved hepatic function/superoxide dismutase-like activity	Suetsuna and Saito (2001)
		Codium fragile	Anticoagulant	Athukorala et al. (2006)
		Enteromorpha prolifera	ACE inhibition/antioxidant/antityrosinase	Lee et al. (2005)
		Grateloupia filicina	ACE inhibition/antioxidant/antitumor/antityrosinase	Lee et al. (2005)

Peptides	Carnosine	*Ancanthophora delilei*	Antioxidant	Fleurence (2004)
	Glutathione	*Ulva fasciata*	Antioxidant	Shiu and Lee (2005)
	Ile-Leu-Ala-Pro Leu-Leu-Ala-Pro Met-Ala-Gly-Val-Asp-His-Ile	*Palmaria palmata*	Antidiabetic activity	Harnedy et al. (2015)
	IP AFL	*Ulva rigida*	ACE inhibition	Paiva et al. (2016)
	SECMA 1	*Ulva* sp.	Modulation of cell proliferation	Shiu and Lee (2005)
	Galaxamide 1	*Galaxaura filamentosa*	Antiproliferative activity	Xu et al. (2008)
	Tyr-His Lys-Tyr Phe-Tyr Ile-Tyr	*Undaria pinnatifida*	Antihypertensive activity	Suetsuna et al. (2004)
Amino acids	Laminine monocitrate	*Laminaria angusta* *Chondria armata*	Hypotensive activity	Bhakuni and Rawat (2005)

ACE, angiotensin-converting enzyme.

Their application in the food and pharmaceutical industries stems from the many health benefits already discussed in this chapter. An additional application of macroalgal proteins in the food industry is the use of phycocyanin as colorants (blue pigment) in soft drinks, dairy products, chewing gums, confectionery, and popsicles (Spolaore et al. 2006).

In cosmetics, seaweed proteins have demonstrated substantial potential as photoprotective agents (Pangestuti and Kim 2015; Suganya et al. 2016; Guillerme et al. 2017). This function has been attributed to the presence of mycosporine-like amino acids (MAAs), abundant in red seaweeds. MAAs absorb UV radiations within the range of 310–362 nm and have been identified as a common sunscreen compound (Guillerme et al. 2017). *Corallina pilulifera* extracts have demonstrated photoprotective potential against UVA-induced oxidative stress in human dermal fibroblasts (Ryu et al. 2009). This activity may be attributed to the presence of MAAs. Also, *Fucus vesiculosus* extract has been used to reduce the appearance of circles under the eye by increasing the expression of a molecule, heme oxygenase-1 (HO-1), responsible for the elimination of heme produced under the skin (Ariede et al. 2017).

Macroalgal proteins have also found application as medical textiles due to functional properties such as elasticity, hydrophobicity, wound dressing, surgical sutures, body armor, and filters (Janarthanan and Senthil Kumar 2017). Macroalgal proteins, particularly phycobiliproteins, are also applied as natural dyes in the textile industries. Phycobiliproteins are also used in flow cytometry, fluorescent immunochemistry, fluorescent microscopy, and fluorescent labelling (Bleakley and Hayes 2017).

Proteins are essential components of animal feeds, hence macroalgae present a sustainable source of protein for both nutritional and health-promoting purposes. Most of the studies on the use of seaweed as animal feed are centred on the application of whole seaweed supplements and very little is reported on proteins extracted from seaweeds as a feed additive, although macroalgae are attractive animal feeds due to their high protein content (Makkar et al. 2016; Bleakley and Hayes 2017; Miranda and García-Vaquero 2017; Schleder et al. 2017). For instance, as fish feed, supplementation of 10% *Gracilaria chilensis* (R) significantly improved specific growth rate by 1.51% in Atlantic salmon (*Salmo solar*). The inclusion of this seaweed at 1% and 10% in the diet also exhibited antiviral effects against the infectious salmon anaemia (ISA) virus (Lozano et al. 2016). Also, hepatic function and lipid content were improved by the inclusion of 5% and 15% *P. palmata* (Wan et al. 2016). *Laminaria japonica* and *U. lactuca* have also been reported to be economical feed additives in the diets of sea cucumber *Apostichopus japonicus* with suitable digestibility and low ammonia-nitrogen production (Xia et al. 2012).

In poultry, incorporating 3% *U. lactuca* in broiler chicken diet increased breast muscle yields and decreased serum lipids, uric acid concentration and cholesterol, when compared to birds fed with a corn-only diet (Abudabos et al. 2013). Pellet binding in duck feed was improved with the inclusion of 3% red seaweed *Polysiphonia* spp. (El-Deek and Brikaa 2009). In pigs, supplementation of *L. digitata* increased body weight gain by 10% and increased iodine concentration in muscles by 45% (He et al. 2002). Also, adding *Ascophyllum nodosum* to pig diets has been reported to stimulate beneficial gut bacteria activity and increase iodine concentration (Dierick et al. 2009). Green seaweeds, including *U. lactuca*, have been suggested as a high-protein, low-energy foodstuff for sheep and goats (Bleakley and Hayes 2017). A detailed summary of the application of seaweeds as protein sources for monogastric livestock has been provided by Angell et al. (2016).

In summary, macroalgal proteins, peptides, and amino acids are attractive and sustainable starting materials for industrial applications.

11.6 Future Directions

This chapter highlights the prospects of a wide variety of macroalgae as sustainable sources of proteins, peptides, and amino acids for their health-promoting and industrial benefits. It is evident that certain seaweeds are a rich source of proteins and endogenous peptides, the more so because these macroalgae are not necessarily in short supply and are relatively easily harvested or sustainably cultivated in coastal regions of the world. Despite the proven health prospects, as yet, there seems to be very little interest in exploring this option as a sustainable source of active ingredients, probably due to the difficulties associated with the cost-effective extraction and recovery yields of these bioactives, in adequate amounts. Most reports on health benefits were at *in vitro* stage and very few have been tested in clinical trials. Future research should consider up-scaling extraction processes to obtain these bioactive proteins and peptides and possibly advance *in vitro* claims through actual clinical trials for humans or animals (companion and production), with the end-goal being the commercialisation of these readily available, active ingredients.

Acknowledgements

The authors acknowledge the support of MITACS and Acadian Seaplants Limited.

References

Abudabos, A.M. et al. (2013). Nutritional value of green seaweed (Ulva lactuca) for broiler chickens. *Italian Journal of Animal Science* 12 (2): 177–181.

Angell, A.R. et al. (2016). Seaweed as a protein source for mono-gastric livestock. *Trends in Food Science and Technology* 54: 74–84.

Ariede, M.B. et al. (2017). Cosmetic attributes of algae – a review. *Algal Research* 25 (July): 483–487.

Athukorala, Y. and Jeon, Y.-J. (2005). Screening for angiotensin-1 converting enzyme inhibitory Activity of *Ecklonia cava*. *Journal of Food Science and Nutrition* 10: 134–139.

Athukorala, Y., Kim, K.-N., and Jeon, Y.-J. (2006). Antiproliferative and antioxidant properties of an enzymatic hydrolysate from brown alga, *Ecklonia cava*. *Food and Chemical Toxicology* 44: 1065–1074.

Bleakley, S. and Hayes, M. (2017). Algae proteins: extraction, application, and challenges concerning production. *Foods* 6 (33): 1–35.

Barbarino, E. and Louren, S.O. (2005). An evaluation of methods for extraction and quantification of protein from marine macro- and microalgae. *Journal of Applied Phycology* 17: 447–460.

Beaulieu, L. et al. (2015). Characterization of antibacterial activity from protein hydrolysates of the macroalga *Saccharina longicruris* and identification of peptides implied in bioactivity. *Journal of Functional Foods* 17: 685–697.

Beaulieu, L., Sirois, M., and Tamigneaux, É. (2016). Evaluation of the *in vitro* biological activity of protein hydrolysates of the edible red alga, *Palmaria palmata* (dulse) harvested from the Gaspe coast and cultivated in tanks. *Journal of Applied Phycology* 28 (5): 3101–3115.

Bhakuni, D.S. and Rawat, D.S. (2005). Bioactive metabolites of marine algae, Fungi and Bacteria. In: *Bioactive Marine Natural Products*, 1–25. New York: Springer.

Cerna, M. (2011). Seaweed proteins and amino acids as nutraceuticals. In: *Advances in Food and Nutrition Research* (ed. S. Kim and S. Taylor), 297–312. St Louis: Elsevier.

Cha, S.-H. et al. (2006). Screening of extracts from marine green and brown algae in Jeju for potential marine angiotensin-I converting enzyme (ACE) inhibitory activity. *Journal of the Korean Society of Food Science and Nutrition* 35 (3): 307–314.

Chronakis, I.S. (2000). Biosolar proteins from aquatic algae. In: *Novel Macromolecules in Food Systems* (ed. G. Doxastakis and V. Kiosseoglou), 39–75. Amsterdam: Elsevier.

Cian, R.E., Martínez-augustin, O., and Drago, S.R. (2012). Bioactive properties of peptides obtained by enzymatic hydrolysis from protein byproducts of *Porphyra columbina. FRIN* 49 (1): 364–372.

Cian, R.E. et al. (2013). Enzyme proteolysis enhanced extraction of ACE inhibitory and antioxidant compounds (peptides and polyphenols) from *Porphyra columbina* residual cake. *Journal of Applied Phycology* 25 (4): 1197–1206.

Cole, A.J., de Nys, R., and Paul, N.A. (2015). Biorecovery of nutrient waste as protein in freshwater macroalgae. *Algal Research* 7: 58–65.

Cornish, M.L., Critchley, A.T., and Mouritsen, O.G. (2017). Consumption of seaweeds and the human brain. *Journal of Applied Phycology* 29 (5): 2377–2398.

Dierick, N., Ovyn, A., and de Smet, S. (2009). Effect of feeding intact brown seaweed *Ascophyllum nodosum* on some digestive parameters and on iodine content in edible tissues in pigs. *Journal of the Science of Food and Agriculture* 89 (4): 584–594.

Ejike, C.E.C.C. et al. (2017). Prospects of microalgae proteins in producing peptide-based functional foods for promoting cardiovascular health. *Trends in Food Science and Technology* 59: 30–36.

El Sayed, K.A. et al. (2000). Marine natural products as antituberculosis agents. *Tetrahedron* 56: 949–953.

El-Deek, A.A. and Brikaa, M.A. (2009). Nutritional and biological evaluation of marine seaweed as a feedstuff and as a pellet binder in poultry diet. *International Journal of Poultry Science* 8 (9): 875–881.

Fitzgerald, C. et al. (2011). Heart health peptides from macroalgae and their potential use in functional foods. *Journal of Agricultural and Food Chemistry* 59: 6829–6836.

Fleurence, J. et al. (1995). Comparison of different extractive procedures for proteins from the edible seaweeds *Ulva rigida* and *Ulva rotundata*. *Journal of Applied Phycology* 7: 577–582.

Fleurence, J. (2004). Seaweed proteins. In: *Proteins in Food Processing* (ed. R.Y. Yada), 198–213. Cambridge: Woodhead Publishing.

Galland-Irmouli, A. et al. (1999). Nutritional value of proteins from edible seaweed *Palmaria palmata* (Dulse). *Journal of Nutritional Biochemistry* 10: 353–359.

Godfray, H.C.J. et al. (2010). Food security: the challenge of feeding 9 billion people. *Science* 327: 812–819.

Guillerme, J.-B., Couteau, C., and Coiffard, L. (2017). Applications for marine resources in cosmetics. *Cosmetics* 4 (3): 35.

Guiry, M.D. & Guiry, G.M. 2017. *AlgaeBase*. Available at: www.algaebase.org
Hamann, M.T. and Scheuer, P.J. (1993). Kahalalide F: a bioactive depsipeptide from the sacoglossan mollusk *Elysia rufescens* and its algae diet *Bryopsis* sp. *Journal of the American Chemical Society* 115: 5825–5826.
Hamann, M.T. et al. (1996). Kahalalides F: a bioactive peptides from a marine Mollusk *Elysia rufescens* and its algal diet Bryopsis sp. *Journal of Organic Chemistry* 3263 (9): 6594–6600.
Hardouin, K. et al. (2014). Biochemical and antiviral activities of enzymatic hydrolysates from different invasive French seaweeds. *Journal of Applied Phycology* 26: 1029–1042.
Harnedy, P.A. and FitzGerald, R.J. (2011). Bioactive proteins, peptides, and amino acids from macroalgae. *Journal of Phycology* 47: 218–232.
Harnedy, P.A. and FitzGerald, R.J. (2013). Extraction of protein from the macroalga *Palmaria palmata. LWT – Food Science and Technology* 51 (1): 375–382.
Harnedy, P.A., Keeffe, M.B.O., and Fitzgerald, R.J. (2015). Purification and identification of dipeptidyl peptidase (DPP) IV inhibitory peptides from the macroalga *Palmaria palmata. Food Chemistry* 172: 400–406.
He, M.L., Hollwich, W., and Rambeck, W.A. (2002). Supplementation of algae to the diet of pigs: a new possibility to improve the iodine content in the meat. *Journal of Animal Physiology and Animal Nutrition* 86 (3–4): 97–104.
Heo, S. and Jeon, Y. (2008). Radical scavenging capacity and cytoprotective effect of enzymatic digests of *Ishige okamurae. Journal of Applied Phycology* 20: 1087–1095.
Heo, S. et al. (2003). Antioxidant effect of enzymatic hydrolyzate from a kelp, *Ecklonia cava. Algae* 18 (4): 341–347.
Janarthanan, M. and Senthil Kumar, M. (2017). The properties of bioactive substances obtained from seaweeds and their applications in textile industries. *Journal of Industrial Textiles* Retrieved from: http://journals.sagepub.com/doi/10.1177/1528083717692596.
Johnson, E.M. et al. (2014). Physicochemical parameters optimization, and purification of phycobiliproteins from the isolated *Nostoc* sp. *Bioresource Technology* 166: 541–547.
Joubert, Y. and Fleurence, J. (2008). Simultaneous extraction of proteins and DNA by an enzymatic treatment of the cell wall of *Palmaria palmata* (Rhodophyta). *Journal of Applied Phycology* 20: 55–61.
Kadam, S.U. et al. (2017). Extraction and characterization of protein from Irish brown seaweed *Ascophyllum nodosum. Food Research International* 99 (Part 3): 1021–1027.
Kawakubo, A. et al. (1997). The marine red alga *Eucheuma serra* J. Agardh, a high yielding source of two isolectins. *Journal of Applied Phycology* 9: 331–338.
Khairy, H.M. and El-Shafay, S.M. (2013). Seasonal variations in the biochemical composition of some common seaweed species from the coast of Abu Qir Bay, Alexandria. *Oceanologia* 55 (2): 435–452.
Lacroix, I.M.E. and Li-chan, E.C.Y. (2012). Evaluation of the potential of dietary proteins as precursors of dipeptidyl peptidase (DPP)-IV inhibitors by an in silico approach. *Journal of Functional Foods* 4 (2): 403–422.
Lee, J.M., You, S.G., and Kim, S.M. (2005). Functional activities of low molecular weight peptides purified from enzymatic hydrolysates of seaweeds. *Journal of the Korean Society of Food Science and Nutrition* 34 (8): 1124–1129.
Lozano, I. et al. (2016). Red macroalgae *Pyropia columbina* and *Gracilaria chilensis*: sustainable feed additive in the Salmo salar diet and the evaluation of potential antiviral activity against infectious salmon anemia virus. *Journal of Applied Phycology* 28 (2): 1343–1351.

Makkar, H.P.S. et al. (2016). Seaweeds for livestock diets: a review. *Animal Feed Science and Technology* 212: 1–17.

Mao, X. et al. (2017). Anti-proliferation peptides from protein hydrolysates of *Pyropia haitanensis*. *Journal of Applied Phycology* 29: 1623–1633.

Miranda, M. and García-Vaquero, M. (2017). Macroalgae for functional feed development: applications in aquaculture, ruminant and swine feed industries. In: *Seaweeds: Biodiversity, Environmental Chemistry and Ecological Impacts* (ed. P. Newton). New York: Nova Science Publishers.

Mittal, R. et al. (2017). Ultrasound assisted methods for enhanced extraction of phycobiliproteins from marine macro-algae, *Gelidium pusillum* (Rhodophyta). *Ultrasonics Sonochemistry* 38: 92–103.

N'Diaye, I. et al. (1996). Almazole D, a new type of antibacterial 2, 5-Disubstituted Oxazolic dipeptide from a red alga of the coast of Senegal. *Tetrahedron Letters* 37 (17): 3049–3050.

Nascimento-Neto, L.G. et al. (2012). Characterization of isoforms of the lectin isolated from the red algae *Bryothamnion seaforthii* and its pro-healing effect. *Marine Drugs* 10: 1936–1954.

Neves, S.A. et al. (2007). Antinociceptive properties in mice of a lectin isolated from the marine alga *Amansia multifida* Lamouroux. *Brazilian Journal of Medical and Biological Research* 40: 127–134.

Okolie, C.L. et al. (2017). Prospects of brown seaweed polysaccharides (BSP) as prebiotics and potential Immunomodulators. *Journal of Food Biochemistry* 41 (5): e12392.

Paiva, L. et al. (2016). Isolation and characterization of angiotensin I-converting enzyme (ACE) inhibitory peptides from *Ulva rigida* C. Agardh protein hydrolysate. *Journal of Functional Foods* 26: 65–76.

Pangestuti, R. and Kim, S.-K. (2015). Seaweed proteins, peptides, and amino acids. In: *Seaweed Sustainability: Food and Non-food Applications* (ed. B.K. Tiwari and D. Troy), 125–139. London: Academic Press.

Pangestuti, R. and Kim, S. (2017). Bioactive peptide of marine origin for the prevention and treatment of non-communicable diseases. *Marine Drugs* 15 (3): ii.

Peng, Y. et al. (2015). Chemical composition of seaweeds. In: *Seaweed Sustainability: Food and Non-food Applications* (ed. B.K. Tiwari and D. Troy), 79–123. London.: Academic Press.

Polikovsky, M. et al. (2016). Towards marine biorefineries: selective proteins extractions from marine macroalgae *Ulva* with pulsed electric field. *Innovative Food Science and Emerging Technologies* 37: 194–200.

Praseptiangga, D. 2017. Development of seaweed-based biopolymers for edible films and Lectins. International Conference on Food Science and Engineering, Indonesia: 1–5.

Rivanor, R.L.d.C. et al. (2014). A lectin from the green seaweed Caulerpa cupressoides reduces mechanical hyper-nociception and inflammation in the rat temporomandibular joint during zymosan-induced arthritis. *International Immunopharmacology* 21 (1): 34–43.

Ryu, B.M. et al. (2009). Anti-photoaging activity and inhibition of matrix metalloproteinase (MMP) by marine red alga, *Corallina pilulifera* methanol extract. *Radiation Physics and Chemistry* 78 (2): 98–105.

Samarakoon, K. and Jeon, Y. (2012). Bio-functionalities of proteins derived from marine algae — a review. *Food Research International* 48 (2): 948–960.

Schleder, D.D. et al. (2017). Brown seaweeds as feed additive for white-leg shrimp: effects on thermal stress resistance, midgut microbiology, and immunology. *Journal of Applied Phycology* 29 (5): 2471–2477.

Sekar, S. (2008). Phycobiliproteins as a commodity: trends in applied research, patents and commercialization. *Journal of Applied Phycology* 20: 113–136.

Shiu, C. and Lee, T. (2005). Ultraviolet-B-induced oxidative stress and responses of the ascorbate – glutathione cycle in a marine macroalga *Ulva fasciata*. *Journal of Experimental Botany* 56 (421): 2851–2865.

Stengel, D.B., Connan, S., and Popper, Z.A. (2011). Algal chemodiversity and bioactivity: sources of natural variability and implications for commercial application. *Biotechnology Advances* 29 (5): 483–501.

Spolaore, P. et al. (2006). Commercial applications of microalgae. *Journal of Bioscience and Bioengineering* 101 (2): 87–96.

Stack, J. et al. (2017). Seasonal variation in nitrogenous components and bioactivity of protein hydrolysates from *Porphyra dioica*. *Journal of Applied Phycology* 29 (5): 2439–2450.

Suarez, Y. et al. (2003). Kahalalide F, a new marine-derived compound, induces oncosis in human prostate and breast cancer cells. *Molecular Cancer Therapeutics* 2: 863–872.

Suetsuna, K. & Saito, M. 2001. Enzyme-decomposed materials of laver and uses therof. US Patent No. 6217879B1.

Suetsuna, K., Maekawa, K., and Chen, J. (2004). Antihypertensive effects of *Undaria pinnatifida* (wakame) peptide on blood pressure in spontaneously hypertensive rats. *Journal of Nutrititional Biochemistry* 15: 267–272.

Sugahara, T. et al. (2001). The cytotoxic effect of Eucheuma serra agglutinin (ESA) on cancer cells and its application to molecular probe for drug delivery system using lipid vesicles. *Cytotechnology* 36: 93–99.

Suganya, T. et al. (2016). Macroalgae and microalgae as a potential source for commercial applications along with biofuels production: a biorefinery approach. *Renewable and Sustainable Energy Reviews* 55: 909–941.

Tierney, M.S. and Croft, A.K. (2010). A review of antihypertensive and antioxidant activities in macroalgae. *Botanica Marina* 53: 387–408.

Udenigwe, C.C. and Aluko, R.E. (2012). Food protein-derived bioactive peptides : production, processing, and potential health benefits. *Journal of Food Science* 71 (1): 11–24.

Udenigwe, C.C. et al. (2017). Ribulose-1,5-bisphosphate carboxylase as a sustainable and promising plant source of bioactive peptides for food applications. *Trends in Food Science and Technology* 69 (A, 74): –82.

Wan, A.H.L. et al. (2016). The inclusion of *Palmaria palmata* macroalgae in Atlantic salmon (Salmo salar) diets: effects on growth, haematology, immunity and liver function. *Journal of Applied Phycology* 28 (5): 3091–3100.

Wang, T. et al. (2010). Enzyme-enhanced extraction of antioxidant ingredients from red algae *Palmaria palmata*. *LWT – Food Science and Technology* 43 (9): 1387–1393.

WHO. 2013. Global health estimates summary tables: projection of deaths by cause, age and sex: 2011–2013. Retrieved from: www.who.int/healthinfo/global_burden_disease/estimates/en/index1.html.

Wu, X. et al. (2013). Modular generation of fluorescent phycobiliproteins. *Photochemical and Photobiological Sciences* 12: 1036–1040.

Xia, S. et al. (2012). Effects of different seaweed diets on growth, digestibility, and ammonia-nitrogen production of the sea cucumber *Apostichopus japonicus* (Selenka). *Aquaculture* 338–341: 304–308.

Xu, W. et al. (2008). Isolation, structure determination, and synthesis of galaxamide, a rare cytotoxic cyclic pentapeptide from a marine alga *Galaxaura filamentosa*. *Organic Letters* 10 (20): 4569–4572.

Yabuta, Y. et al. (2010). Antioxidant activity of the phycoerythrobilin compound formed from a dried Korean purple laver (*Porphyra* sp.) during in vitro digestion. *Food Science and Technology Research* 16 (4): 347–351.

Ziółkowska, N.E. and Wlodawer, A. (2006). Structural studies of algal lectins with anti-HIV activity. *Acta Biochimica Polonica* 53 (4): 617–626.

12

Microalgal Bioactive Compounds Including Protein, Peptides, and Pigments: Applications, Opportunities, and Challenges During Biorefinery Processes

Maria Hayes[1*], *Leen Bastiaens*[2], *Luisa Gouveia*[3], *Spyros Gkelis*[4], *Hanne Skomedal*[5], *Kari Skjanes*[5], *Patrick Murray*[6], *Marco García-Vaquero*[7], *Muge Isleten Hosoglu*[8], *John Dodd*[9], *Despoina Konstantinou*[4], *Ivo Safarik*[10], *Graziella Chini Zittelli*[11], *Vytas Rimkus*[12], *Victória del Pino*[13], *Koenraad Muylaert*[14], *Christine Edwards*[15], *Morten Laake*[16], *Joana Gabriela Laranjeira da Silva*[17], *Hugo Pereira*[17], *and Joana Abelho*[17]

[1] *Teagasc Ashtown Food Research Centre, Food Biosciences Department, Dublin, Ireland*
[2] *VITO, Mol, Belgium*
[3] *LNEG, National Laboratory of Energy and Geology, Bioenergy Unit, Lisbon, Portugal*
[4] *School of Biology, Aristotle University of Thessaloniki, Department of Botany, Thessaloniki, Greece*
[5] *Norwegian Institute of Bioeconomy Research – NIBIO, Ås, Norway*
[6] *Limerick Institute of Technology (LIT), Limerick City, Ireland*
[7] *School of Veterinary Medicine, University College Dublin (UCD), Dublin, Ireland*
[8] *Canakkale Onsekiz Mart University, Canakkale, Turkey*
[9] *AlgaeCytes Ltd., Discovery Parkhouse, Sandwich, UK*
[10] *Department of Nanobiotechnology, Biology Centre, ISB, Academy of Sciences, Ceske Budejovice, Czech Republic*
[11] *National Research Council, Institute of Ecosystem Study, Florence, Italy*
[12] *Spila, UAB, Vilnius, Lithuania*
[13] *Necton S.A., Olhão, Portugal*
[14] *KU, Leuven campus Kortrijk, Laboratory Aquatic Biology, Kortrijk, Belgium*
[15] *School of Pharmacy and Life Sciences, Robert Gordon University, Aberdeen, UK*
[16] *Sigtun Innovation AS, HUB Lillehammer Regional Innovation Center, Lillehammer, Norway*
[17] *Allmicroalgae, Lisbon, Portugal*

12.1 Introduction

Proteins are necessary to maintain good human nutrition and bioactive peptides which can be derived from proteins have demonstrated biological activities that enhance health and can be applied in the prevention, management, and treatment of health conditions including cancer, cardiovascular diseases (CVDs), diabetes, and others (Cicero et al. 2017). Bioactive peptides are sequences of between 2 and 30 amino acids in length that impart a health benefit to the consumer that goes above and beyond basic human nutrition (Hayes 2014). They can be generated from any protein source using hydrolysis with commercial proteolytic enzymes, lactic acid bacteria (LAB), or high pressure. Their use in cosmetics and functional foods is a developing area of research and several food protein sources, including egg, milk, meat, soy, and others, have been identified as rich in bioactive peptides.

*Corresponding Author

Novel Proteins for Food, Pharmaceuticals, and Agriculture: Sources, Applications, and Advances, First Edition. Edited by Maria Hayes.

However, the use of primary human food proteins as precursors for bioactive peptide generation does not appear sustainable (Henchion et al. 2017). There is therefore a need to identify sustainable proteins that may be used as source materials for the generation of bioactive peptides for improved functional food product development and health. These protein sources may include algae, such as seaweeds and microalgae, insects, and other sources such as duckweed, marine rest-raw materials (van der Spiegel et al. 2013) and by-products from meat processing and the fifth quarter, that is, blood (Lafarga et al. 2015).

Microalgae are ubiquitous and generally contain 40–70% protein. In the past, microalgae, including *Nostoc* sp., were consumed in China as a source of nutrition and today, commercially cultivated microalgae include *Chlorella*, *Arthrospira* (*Spirulina*), *Dunaliella*, *Nannochloropsis*, *Nitzschia*, *Crypthecodinium*, *Schizochytrium*, and *Skeletonema* species although only *Chlorella* and *Spirulina* sp. are approved by the European Food Safety Authority (EFSA) for sale in the European Union (EU). Microalgae-derived protein can be a nutritious protein source as microalgae can contain all 20 essential amino acids and indeed, microalgal protein has been incorporated into food products previously due to its technofunctional attributes. However, the amount of protein in microalgae depends on the species type, growth phase, and light quality where the alga is cultivated and can be changed by nutrient adjustment and environmental stress (Safafar et al. 2016).

Furthermore, microalgae can produce other metabolites that can be used in the prevention and management of diseases including cancer, microbial infection, oxidative stress, diabetes, and hyperlipidaemia. In addition, microalgae can be associated with negative health effects. This chapter will focus only on bioactive peptides derived from microalgal proteins and will present the current state of the art regarding microalgal biorefinery.

12.2 Cultivation of Microalgae

Microalgae provide the basis of the aquatic food web. Today, microalgae are cultivated for food and feed, and there is substantial research on their use in biofuel generation. Due to limited control of cultivation conditions and contamination, the use of open cultivation units is restricted to a relatively small number of 'robust' microalgae species (e.g. *Chlorella*, *Scenedesmus*, and *Nannochloropsis*) that grow rapidly or under very selective conditions (e.g. *Arthrospira*, *Dunaliella*). However, as closed photo-bioreactors (PBRs) support a controlled environment, potentially free of contaminants, a much wider selection of strains can be produced.

Cultivation can occur indoors in specially designed bioreactors as well as in PBR systems (tubular, planar, inclined, and vertical) that are enclosed but outside where they can use solar energy. Mass production methods include open pond systems, PBRs, and immobilised culture systems. Open pond cultivation is less expensive than the other two methods but is susceptible to contamination of the culture, low productivity, water loss, and high harvesting costs. Use of immobilised culture systems is is limited by expensive material costs and lack of variability in strains that are suitable for use in this system (Tafreshi and Shariati 2009). Energy-efficient cultivation and harvesting methods are necessary to make the process economically viable. Suspended culture cultivation and harvesting methods are energy intensive as they usually involve centrifugation, pumping or electrophoresis steps. Medium modification by inclusion of high salt or nutrient deprivation is often used to enhance production of specific chemicals, mainly lipid or carbohydrates.

Biofilm and biofouling of microalgae have been studied as methods for large-scale production of biomass and have advantages for harvesting as they use less water and are more energy efficient (Tafreshi and Shariati 2009). Recently, Estime et al. (2017) developed a tris-acetate-phosphate-pluronic (TAPP) medium with thermo-reversible sol–gel transition properties for energy-efficient cultivation and harvesting of microalgae. This system worked well with the alga *Chlamydomonas reinhardtii*.

Methods used to eliminate unwanted micro-organisms from natural water used to culture microalgae – including unwanted microalgae, protozoans, moulds, and most bacteria – include filtration, pasteurisation, and bleaching (chlorination–dechlorination). Existing industrial applications of microalgal culture mainly depend upon high-value, extractable chemical components; large raceway systems growing the halophilic chlorophyte *Dunaliella* sp. supply a lot of the demand for beta-carotene along with that derived synthetically.

Cultivation of microalgae is a very technically skilled process and it is likely that advances will be species specific and relate to the ability of a given microalga to produce a unique and useful biochemical compound. The dominant microalgal species are *Arthrospira* (*Spirulina*) and *Chlorella*.

12.3 Biorefinery of Microalgae

Microalgae accumulate a variety of compounds that form the basis for potential marketable products as whole cell biomass or as fractions after an extraction process. A cascading biorefinery approach aims at recovering and separating multiple algae compounds to valorise the complete biomass and improve the economics of the value chain (Figure 12.1). Despite the various publications on theoretical algal biorefinery concepts, the literature describing experimental data supporting the feasibility of the concepts is scarce and mostly related to small-scale experiments.

Biorefinery approaches can be grouped based on the value of the main target compound (Harisko and Posten 2014), resulting in three categories. The first category focuses on the production of low-value compounds combined with valorisation of the residual biomass. The targeted compounds can be lipids which form the basis for biofuels, carbohydrates which can be converted into bioethanol or pyrolysis end-products. The remaining de-oiled fractions of algae biomass contain mainly carbohydrates and proteins and are marketable for many applications such as energy carriers, algae meal for animal and fisheries feed development, biosorbents to remove dyes and heavy metals, bioplastics or as compost after anaerobic digestion (Bryant 1982).

The second category focuses on the extraction of medium-value bulk products, including proteins with prices ranging from only a few €/kg up to 100 €/kg.

The third biorefinery category, focusing on high-value products coupled with use of residual biomass, is currently economically the most attractive. Examples of high-value products (>100 €/kg) include carotenoids like astaxanthin and beta-carotene, produced by *Haematococcus pluvialis* (1.5–4%) and *Dunaliella salina* respectively, phycobiliproteins like phycocyanin and phycoerythrin, produced by *Spirulina* and *Porphyridium cruentum* respectively (Bryant 1982), highly purified compounds like essential lipids such as omega-3 polyunsaturated fatty acid (PUFAs), eicosapentaenoic acid (EPA) and docosahexaenoic

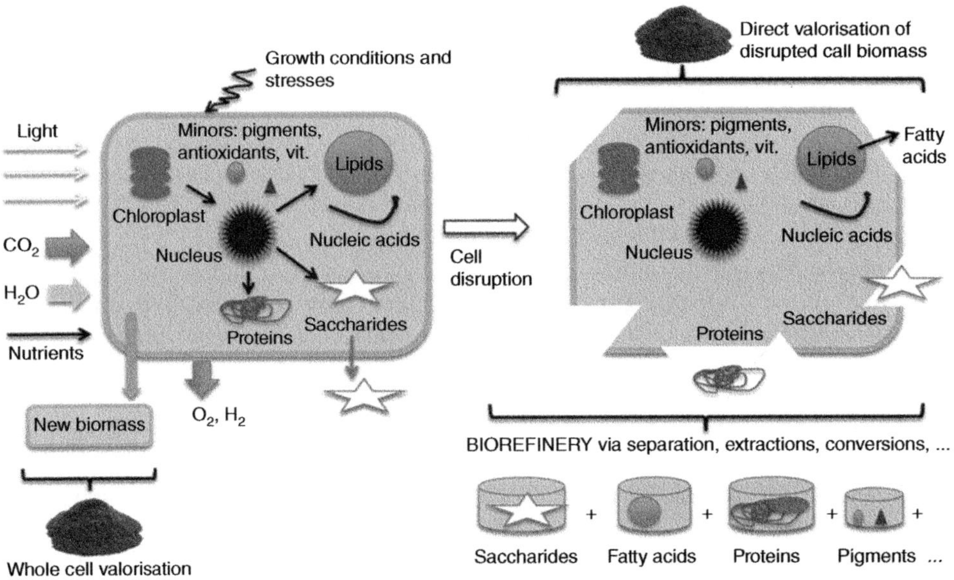

Figure 12.1 Schematic representation of algae biomass valorisation options.

acid (DHA), polypeptides along with other compounds including sterols and vitamins. The target high-value markets are often associated with nutraceuticals, cosmetics, and pharmaceutics.

Any biorefinery concept aims at maximum valorisation of the algal biomass, where increasing concentrations of target compounds are more favourable. The latter can be achieved by proper selection of the algal strain, taking into account its cell content and growth characteristics, as well as by stimulating the production of the main target compounds in the selected algae cell during cultivation. An overview of biorefinery studies with experimental data is given in Table 12.1 while different biorefinery approaches are detailed in Figure 12.2.

Table 12.1 Overview of bio-refinery studies carried out to date.

Algae	Targeted compounds	Reference
Dunaliella tertiolecta	Pyrolysis products bio-oil and char from defatted algae biomass, lipids (beta-carotene, phytosterol, fatty acids)	Soyler et al. (2017)
Nannochloropsis gaditana	Biodiesel, carotenoids, proteins	Li et al. (2015)
Nannochloropsis sp.	Enriched lipid fractions (fatty acids or carotenoids/EPA), biohydrogen	Nobre et al. (2013)
Isochrysis galbana	Fucoxanthin-carotenoids, polar lipids	Gilbert-Lopez et al. (2015)
Scenedesmus acutus	Bioethanol (from hydrolysed sugars), oil	Miranda et al. (2012)
Scenedesmus sp.	Biogas, amino acids	Posada et al. (2016)

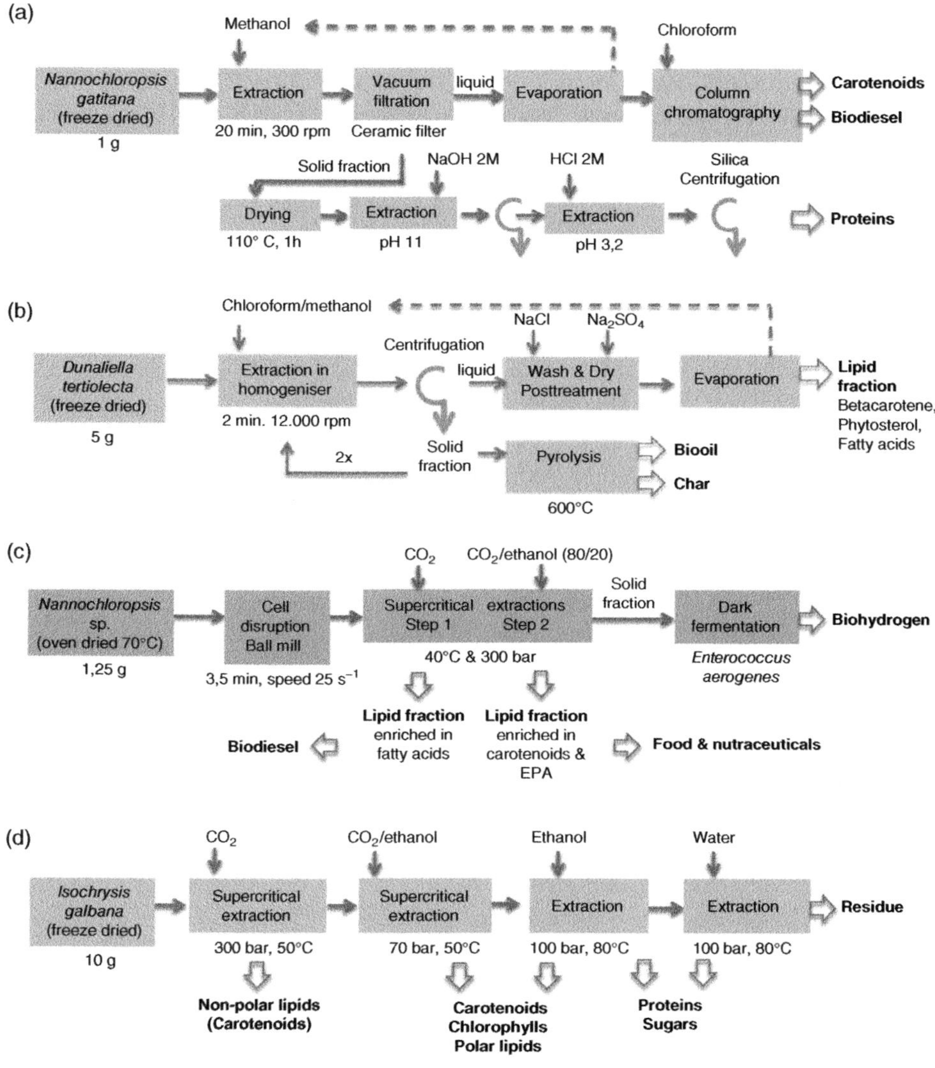

Figure 12.2 Different bio-refinery approaches.

Recently, a mild biorefinery process was investigated on the microalga *Nannochloropsis gaditana* to obtain an enriched fraction of water-soluble proteins free from chlorophyll. A $100\,g\,L^{-1}$ solution of cells was subjected to cell disruption using high-pressure or enzymatic treatment and both fractions were then subjected to ultrafiltration and dia-filtration. The main objective of the study was to carry out a mild biorefinery process with concentrated microalgae to obtain a protein fraction in the filtrate after testing two cell disruption methods (Safi et al. 2017). The study found that homogenisation was more efficient for this purpose than enzymatic treatment and that filtration revealed higher efficiency for the protease-treated samples with a 25% protein yield obtained in the filtrate of the 300 kDa membrane.

12.4 Microalgae as a Source of Protein

Marine microalgae are currently considered as a source of alternative protein for the generation of protein isolates and hydrolysates for the functional foods industry. Microalgal species rich in protein include *Chlorella*, *Arthrospira* (*Spirulina*), *Scenedesmus*, *Micractinium*, *Dunaliella*, *Chlamydomonas*, and *Euglena* and protein content in these species can be as high as 50% dry weight. In addition, the quality of protein is good and often contains all the essential amino acids, including the branched chain amino acids and lysine, valine, isoleucine, and leucine which are required for various functions of the body, especially in relation to muscle protein health (Lubitz 1963; Bleakley and Hayes 2017).

Bioactive peptides also play an important role in human health and nutrition. The freshwater and marine algae *Chlorella vulgaris*, *Arthrospira* (*Spirulina*) *platensis*, *Navicula incerta*, and *Pavlova lutheri* are a few species that can be used to generate bioactive peptides and a number of antioxidant and antimicrobial peptides have been isolated to date, including antioxidant peptides derived from hydrolysed *P. lutheri* biomass and a microbial hydrolysis of *Chlamydomonas* sp. that was shown to suppress *Helicobacter pylori*-induced carcinogenesis. Other angiotensin-1-converting enzyme (ACE-i) inhibitory peptides are detailed in Table 12.2.

12.5 Microalgae as a Source of Pigments

Marine microalgae have been studied for years as a source of pigments. Carotenoids including beta-carotene, astaxanthin, canthaxanthin, lutein, and violaxanthin are produced by freshwater and marine microalgae. There are over 700 different types of carotenoids and several can exert beneficial effects on human health. Beta-carotene is known to enhance immunity through protection of the cell from oxidative stress as well as facilitating monocyte function to increase the number of surface molecules expressed, and it is thought to play a role in protection against conditions including cancers,

Table 12.2 Heart health beneficial (angiotensin-converting enzyme (ACE)-1 inhibitory and antioxidant) peptides from microalgae.

Name	Sequence	Chem. mass	Bioactivity
ACE inhibitor from red algae	MKY	440.5420	ACE inhibitor
ACE inhibitor from red algae	AKYSY	630.6730	ACE inhibitor
ACE inhibitor from microalgae	IAPG	356.4080	ACE inhibitor
ACE inhibitor from microalgae	FAL	349.4160	ACE inhibitor
ACE inhibitor from microalgae	IAE	331.3550	ACE inhibitor
Anticancer peptide from algae protein waste (*Chlorella vulgaris*)	VECYGPNRPQF	1326.3050	Multifunctional
Antioxidative peptide derived from algae protein waste (*Chlorella vulgaris*)	VECYGPNRPQF	1326.3050	Antioxidative

Table 12.3 Pigments sourced from marine microalgae.

Pigment	Colour	Microalgal source	Application	Reference
Beta-carotene	Orange/red	*Dunaliella salina* *Nannochloropsis gaditana* *Synechococcus* spp.	Antioxidant Provitamin A source Anticancer agent	Guedes et al. (2011)
Chlorophyll a	Green	*Nannochloropsis gaditana*	Anticarcinogenic Antioxidant Antimutagenic	Tomolo et al. (2012)
Violaxanthin	Orange	*Dunaliella tertiolecta*	Antiproliferative	Pasquet et al. (2011), Berry et al. (2003)
Fucoxanthin	Green/yellow	*Phaeodactylum tricornutum*	Reduced cholesterol Antidiabetic Anticancer agent	Kim et al. (2012)

rheumatoid arthritis, neurodegenerative disease, and CVD. Sources of beta-carotene include the green alga *D. salina* which is reported to contain up to 14% beta-carotene in some cultures. It produces beta-carotene in harsh environmental conditions and is commercially cultured for this purpose. Zeaxanthin is found in the microalga *Synechococcus* spp. and *Nannochloropsis gaditana* is a source of pigments including chlorophyll a, violaxanthin, and vaucheriaxanthin. Table 12.3 details pigments from microalgae and associated health benefits/applications.

12.6 Legislation Governing Use of Microalgae in Europe

Marketing of proteins and peptides as 'bioactives with health benefits' now requires substantial scientific evidence in Europe and it is necessary to provide evidence of any benefit before any health claim can be made for a functional food product. European regulations on nutrition and health claims became effective on 19 January 2007, and all EU member states now have a common regulation allowing health claims on foods, which may be considered the 'gold standard'. The use of bioactive components in food is regulated under the General Food Law (178/2002/EC), which assigns primary legal responsibility for the safety of products on the market to the business operator. The EFSA is in charge of the regulation. Article 2 of the General Food Law states that 'Food' means any substance or product, whether processed, partially processed, or unprocessed, intended to be or reasonably expected to be ingested by humans, excluding (a) feed, (b) live animals, (c) plants prior to harvesting, (d) medicinal products within Directive 2004/27/EC, (e) cosmetics within Directive 76/768/EC, tobacco, narcotics or residues and contaminants (Regulation (EC) No. 178/2002, 2002).

If algae are to be categorised as food or feed ingredients according to Article 2, they must be characterised and further defined as one of the following.

a) Novel food (Directive 258/1997/EC) – DHA-rich algal oils have been granted a novel foods claim
b) Foods for particular nutritional use, covering dietetic foods (Directive 89/398/EC)
c) Food additives (Directive 89/107/EC)
d) Flavourings (Directive 91/71/EC completing Directive 88/388/EC)

12.6.1 Nutrition Claims

There are a number of foods on the European market with labels bearing nutrition and health claims. These labels are regulated by Regulation 1924/2006/EC of the European Parliament and of the Council of December 2006 on nutrition and health claims made on foods. The main objective of this regulation is to ensure high consumer protection against fraudulent marketing and fair competition within the food industry.

Nutrition claims apply to all commercial communications including labelling, advertising of food and promotional campaigns, trademarks, brand names, and any information that may be interpreted as a nutritional claim. This relates to any claim that states, suggests, or implies that a food has a particular beneficial nutritional property due to: (i) energy and (ii) the nutrients or other substances it contains or contains in reduced or increased portions or does not contain.

12.6.2 Health Claims

This relates to any claim that 'states, suggests or implies that a relationship exists between a food category, a food or one of its constituents and health.' Specific conditions and requirements for the use of health claims can be found on the EFSA nutrition and health claims register: (http://ec.europa.eu/food/safety/labelling_nutrition/claims/register/public/?event=register.home).

Briefly, Article 13 (1) deals with health claims describing or referring to:

a) the role of a nutrient or other substance in growth, development, and the functions of the body
b) behavioural and psychological functions
c) weight control or reduction in the sense of hunger or increased satiety or a reduction in available energy from the diet.

These claims should be supported by generally accepted scientific evidence as assessed by the EFSA panel of experts and they should be well understood by the consumer. Article 13 (5) relates to claims that are made 'based on newly developed scientific evidence and/or which include a request for the protection of proprietary data.' Applications made using this claim require an extensive dossier. The EFSA provides its judgement on a product or functional food component within five months following the request and a further two months is required by the Commission following consultation with the member states.

Article 14 relates to reduction of disease risk claims and claims referring to children's development and health. Claims submitted according to Article 14 undergo rigorous authorisation procedures and an extensive dossier must be submitted. The EFSA has published guidelines for dossier submission under Article 14. An extensive literature review and studies must be submitted to support a claim made under Article 14. Submissions are made to the competent authority of the member state in question for further handling by the EFSA and the Commission. Following a review of the EFSA nutrition and health claims register, no microalgal-derived products have obtained health claims to date.

12.6.3 Additive Claims

12.6.3.1 Feed

The EFSA evaluates the safety and/or efficacy of additives, products, or substances used in animal feed before they can be authorised for use in the EU. The European Commission decides whether or not to authorise the feed additive application following the EFSA evaluation. This claim is governed by Regulation EC 1831/2003. The procedure for feed additive applications requires submission of an application to the European Commission, a technical dossier to the EFSA (electronic format only), and three reference samples of the feed additive to the European Union Reference Laboratory.

12.6.3.2 Food

Food additives include colours which are used to add or restore colour in a food. A recent review by Lehto et al. (2017) describes the EU and US regulations applicable to food colours. It found that the overall approach is similar, based on risk assessment procedures and risk management measures, but differences affecting free movement of goods can be found in the details and implementation of regulations. Additives being approved only in the US or in the EU means that producers aiming to export need to adjust their product composition to the export market. Failure to comply may give rise to claims of adulteration, misbranding or non-compliance, and rejection at the border or recall from the market.

Differences between the EU and USA in terms of regulation include the following. The EU sets maximum limits for four heavy metal contaminants in almost all colour additives: arsenic, lead, cadmium, and mercury. The US does not regulate the content of cadmium. The use of colour additives in the EU is limited to defined food categories and the addition of food colourants to unprocessed food, food labelled as organic, baby food and, for example, unflavoured milk, bread, and oil is not permitted at all. Use of colour additives is banned in some EU traditional foods too. Other differences are outlined by Lehto et al. (2017).

- Preservatives – these are added to prolong the shelf-life of foods by protecting them against micro-organisms.
- Antioxidants – substances which prolong the shelf-life of foods by protecting them against oxidation (i.e. fat rancidity and colour changes).
- Flour treatment agents – added to flour or to dough to improve its baking quality.

The EFSA Panel on Food Additives and Nutrient Sources added to food (ANS) assesses the safety of food additives and colours. The Panel's safety evaluations of food colours and other food additives involve a review of all available, relevant scientific studies as well as data on toxicity and human exposure, from which the Panel draws conclusions regarding the safety of the substance. Consideration of whether certain food colours are likely to trigger adverse allergic reactions is carried out by the EFSA Panel on Dietetic Products, Nutrition, and Allergies.

The list of approved food additives for use in the EU took effect from 1 June 2013. Regulation (EC) No. 1331/2008 is relevant to food additives.

12.6.4 Novel Food Claims

On 10 November 2016, EFSA published two guidance documents on novel food and on traditional food following the new European regulation on novel food. The regulation came into effect in January 2018, introducing a centralised authorisation and assessment procedure.

Arthrospira platensis (Spirulina (DE) (SL), blue-green algae (EN), Spirulina (PT) (ES), spirulina (PL) (FI) (CZ), sini-rohevetikas (ET), Spirulina alga (HU), σπιρουλίνα (EL), Spirulīna, zilaļģe (LV)) has a novel food claim as it was on the market as a food or food ingredient and consumed to a significant degree before 15 May 1997. Thus its access to the market is not subject to the Novel Food Regulation (EC) No. 258/97. However, other specific legislation may restrict the placing on the market of this product as a food or food ingredient in some member states. Therefore, it is recommended to check with the national competent authorities. In addition, *Chlorella luteoviridis, C. pyrenoidosa,* and *C. vulgaris* have novel food claims for the same reasons. DHA- and EPA-rich algal oil from *Schizochytrium* sp. was granted a claim as a novel food ingredient in the context of Regulation (EC) No. 258/97. In addition, beta-carotene from *Dunaliella* sp. and DHA from *Crypthecodinium cohnii* are also already approved as food ingredients by the EFSA and *Chlorella* and *Spirulina* are commonly sold as food supplements in the EU and USA. The novel food catalogue can be accessed at: http://ec.europa.eu/food/safety/novel_food/catalogue/search/public/index.cfm.

12.7 Advantages of Microalgal Use

Microalgae are an extremely diverse group of organisms comprising eukaryotic photoautotrophic protists and prokaryotic cyanobacteria. The genetic and phenotypic diversity of microalgae makes their distribution almost ubiquitous in the biosphere. Microalgae are an essential component of the food web in aquatic environments, with different degrees of salinity, light, and temperature conditions. In addition, they also occur in terrestrial habitats, alone or associated with fungus such as lichens (Singh and Saxena 2015), and this variability makes them an extremely attractive source of a wide range of biomolecules (Wijffels et al. 2010; Borowitzka 2013). Many microalgae species have yet to be explored globally and within Europe. Moreover, the possible strain improvements offered by genetic engineering make microalgae an appealing biomass to scale up the production of biodiesel and high-value compounds such as lipids and proteins (Wijffels et al. 2010).

Oil and protein demands are increasing globally due to the rising population. The major imported sources for lipids and proteins in Europe are palm oil and soybean products respectively. Both products are derived from terrestrial crops which lead to deforestation and contribute to water scarcity in their regions of origin (Patel et al. 2017). Soybean imports in Europe reached 14.5 million tonnes in 2016, meaning 5.8 million tonnes of protein from beans with 40% of protein content. The same amount could be achieved by the microalgae industry producing 19.3 million tonnes of dried algae with 30% protein content. Imports of palm oil into Europe reached 6.7 million tonnes in 2016, which could be produced from 16.7 million tonnes of microalgae biomass containing 40% lipids. The approximately 19 million tonnes of dried algae needed as a protein source will also cover the demands of lipids if both products are efficiently separated. With production of 50 tonnes per hectare per year, 19 million tonnes of algae could be produced in approximately 7 million hectares, representing only 14% of the agricultural land in use in Spain in 2016.

These figures show that the cultivation of microalgae in Europe could be a good alternative to proteins and lipids from terrestrial crops (Vanthoor-Koopmans et al. 2013). Furthermore, microalgae production systems offer certain advantages with respect to terrestrial crops since they can be placed on land which is non-productive for conventional feedstocks, mitigating the competition for land, and have low requirements for water, which does not need to be potable, as well as trade-offs between bioenergy production and food security. In addition, harvesting could be done in a daily basis, avoiding storage.

A multidisciplinary research approach is needed to achieve the full potential of the microalgae industry, including fundamental and system biology, metabolomics, strain development, engineering (bioreactor design and downstream processing), scale-up, biorefineries, integrated production chain and whole-system design with logistics (García-Vaquero and Hayes 2016).

12.7.1 Potential Applications and Uses in the Vegetarian and Vegan Foods Arena

The incorporation of algae into food-based products allowed the microalgae industry to diversify and grow outside the field of health food supplements, a market that currently uses 75% of the biomass for the production of powders, tablets, and capsules. For the successful inclusion of microalgae as novel ingredients, their chemical composition and functional properties need to be determined. Features such as water activity, water- and oil-holding capacity, colour, solubility, foaming, and emulsifying properties should be comparable or better than traditional ingredients as these properties will contribute to the taste, texture, and consumer acceptance of the final food products (García-Vaquero et al. 2017).

The incorporation of microalgae into traditional foods (e.g. breakfast cereals, spreads, breads, cookies, brownies, energy bars, mayonnaises, gelled desserts, pastas, and beverages), which are largely consumed on a daily basis, may be beneficial whilst avoiding the potential problems associated with changing food habits and therefore solving possible resistance to the introduction of different products into the diet. This strategy has many advantages from both the economic and health hazard points of view and has proved to be efficient in the long term, whilst not presenting the drawbacks of traditional therapeutic actions based on medicines which generally have a short-term impact.

Several microalgae have already been tested in different food products, such as *Chlorella vulgaris* (green and carotenogenic) for pigmentation purposes (egg yolks, rainbow trout, seabream, ornamental fishes, emulsions such as mayonnaises, biscuits, puddings, and pastas) (Gouveia et al. 2008). *Haematococcus pluvialis* and *Arthrospira (Spirulina) maxima* were used for ornamental fish pigmentation and food products such as puddings and pastas (Fradique et al. 2013). *Isochrysis galbana* and *Diacronema vlkianum* have been investigated as a potential source of PUFAs and have been tested in puddings and pastas (Fradique et al. 2013). *Chlorella vulgaris* and *Spirulina maxima* were incorporated in pasta products, at inclusion percentages between 0.5% and 2%, resulting in increased firmness of pasta with green and orange colours that would be positively received in the market when evaluated by a sensory panel (Fradique et al. 2013). *Spirulina* biomass was also included in pasta products at 5% and 10% without affecting the technological quality of the pasta as measured by its texture and cooking properties, resulting in a final product with increased protein, phenols, and antioxidant properties (Enzing et al. 2014). The inclusion of *Isochrysis galbana*, *Tetraselmis suecica*, *Scenedesmus almeriensis*, and *Nannochloropsis gaditana* at 1.5% in bread resulted in breads with modified colouration (increased browning and green–yellow tonalities), but had no significant effect on the main physicochemical properties, texture, and porosity of the products (Garcia-Segovia et al. 2017). Recently, Batista et al. (2017) incorporated *Arthrospira* (*Spirulina*), *Chlorella vulgaris*, *Tetraselmis suecica*, and *Phaeodactylum tricornutum* in butter cookies at 2–6%, resulting in products rich in phenolic compounds, with high colour and texture stability. The addition of microalgae also resulted in interesting rheological properties (Batista et al. 2011).

Although the incorporation of full microalgal biomass into traditional products could be promising, the strong green colour could limit the number of microalgae used in certain foods. Moreover, the acceptability of microalgal-enriched products seems to depend on the targeted population or market (Liang et al. 2004). In China and other Asian countries in which traditional cuisine includes algae, the incorporation of microalgae is considered a favourable trait appreciated by consumers (Liang et al. 2004) while the opposite situation seems to occur in Western countries.

To avoid these potential negative effects, the extraction of valuable compounds, such as lipids and proteins, for inclusion in foods in substitution of traditional ingredients is becoming a new trend to create novel food products. Meat, milk, eggs, fish, and seafood products are considered the main animal sources for proteins and lipids while a wide variety of protein and oil crops, fruits, and vegetables are the most important sources of vegetable proteins and oils (Draaisma et al. 2013).

Proteins extracted from *Chlorella vulgaris* showed high content of essential amino acids (38%) and emulsifying properties comparable to or even higher than other commercial ingredients such as soybean proteins or sodium caseinate from milk (Ursu et al. 2014). Furthermore, protein isolates from *Arthrospira* (*Spirulina*) showed good foaming and emulsifying properties together with the capacity to develop films that could allow the development of edible coatings for different food applications (Enzing et al. 2014).

12.7.2 Microalgal Products on the Market

Enzing et al. (2014) reviewed microalgal-based products for food and feed production. Table 12.4 details food, feed, and cosmetic products currently available. Over 75% of microalgae produced globally are used for dietary supplement applications. There is great potential to use microalgal-derived proteins and peptides in foods for health. Indeed, algal-derived DHA is currently found in 99% of infant and baby food in the USA. Although Table 12.4 lists a number of microalgal species that are used in the manufacture of food, feed, and cosmetic products, two algae – *Arthrospira* (*Spirulina*)

Table 12.4 Microalgal-derived products in food, feed, and cosmetics.

	Product	Microalga	Company
Food			
DHA/EPA	Dietary supplement	*Spirulina/Chlorella* sp.	Seambiotic Israel EID Perry India
Algalin Flour (protein additive)	Flour replacer/ food additive	*Spirulina/Chlorella* sp.	Solazyme-Roquette Nutraceuticals partnership (which has dissolved)
Phycocyanin	Colourant	*Spirulina/Chlorella* sp.	EID Perry India Inner Mongolia Biomedical Eng. (Mongolia)
Feed			
DHA/EPA	Animal feed	*Spirulina/Chlorella* sp.	Seambiotic Israel
Cosmetics			
Whole algae	Personal skincare products	*Spirulina/Chlorella* sp.	Soliance (France) LVMH (France)
Polysaccharides/ alguronic acid	Personal skincare products	*Spirulina/Chlorella* sp.	Solazyme/Algenist
Porphyridium cruentum	Hydrating skincare products	*Porphyridium cruentum*	Codif (France)
Anti-inflammatory peptide	Anti-inflammatory skincare	*Phaeodactylum tricornutum*	Soliance (France)
Slimming product	Slimming products		Soliance (France)
	Fluorescent protein marker	Unknown	Martek/DSM
	Suncare product	Unknown	Cyanotech

DHA, docosahexaenoic acid; EPA, eicosapentaenoic acid.

and *Chlorella* sp. – still dominate the market and are produced in large quantities for DHA/EPA, beta-carotene, and astaxanthin.

China dominates the production of *Arthrospira* (*Spirulina*) and *Chlorella* sp. bulk products. The cost of microalgal products varies worldwide. For example, US-grown *Spirulina* sp. (Earthrise LLC, California, and Cyanotech, Hawaii) command a premium price of 20–28 €/kg (bulk price). Chinese growers can sell *Spirulina* sp. for 6–9 €/kg (bulk price). Quality issues including nutrient content, microbiological contamination, irradiation, pesticide, and heavy metals content are important factors in terms of product safety and suitability for different applications. EU production costs are closer to those found in the USA and to remain viable, production costs must be reduced by using novel processes for harvesting, dewatering, and extraction of bioactive compounds, including proteins/peptides. Despite the high costs, currently microalgal production continues in countries including Greece, Portugal, Spain, France, Germany, Italy, The Netherlands, Bulgaria, and others.

The algal market refers to initial bulk algae grown and harvested. Data presented in many studies are retail market size data including B2B (business to business) data. B2B should be attributed to ingredients extracted from algae. Microalgae production market volume and market size in general are still relatively small compared to other commodities but they have grown since 1999 when global production volumes of microalgae were estimated at only 1000 tonnes dry weight. A fivefold increase in growth was observed in 2004 with 5000 tonnes dry weight reported. In 2011, the total production volume had risen to 9000 tonnes dry weight (Enzing et al. 2014). This represents an 11-fold increase over a period of 12 years. This market by volume is dominated primarily by *Spirulina* and *Chlorella* sp. Algae commodity market value estimates vary by sources but currently stand at approximately ~€120 million. The ingredients market where algae like *Dunaliella* sp. and *Haematococcus* sp. are used for production of astaxanthin, beta-carotene, and algae oils is valued at €200 million. The algae industry has a limited number of growers/farmers which together occupy about 1000 ha of growing area. This is a small fraction of the area designated for agriculture production of proteins and food products. For comparison, 13.4 billion ha of land is used in crop production (arable land and land under permanent crops). The potential of biomass value creation in the algae business is very high.

12.8 Conclusion

It is possible to foresee a huge increase in the demand for cultured algae, in terms of both quantity and diversity. The depletion of natural resources, the increasing world population and the general perception that natural products are healthier and more environmentally friendly continuously drive the necessity to produce alternative sources of food ingredients, cosmetics, and new compounds for drug manufacture. The current market for microalgal biomass is estimated to be around 20 000 tonnes per year; the price ranges between 30 and 300 €/kg.

Biorefinery is the only strategy that can efficiently contribute to the economic sustainability of the microalgae sector. Mass cultures of microalgae have traditionally been cultivated in open ponds which are much cheaper and easier to operate than PBRs. More efficient methods to harvest biomass and extract products are under study. There is great scope to extend the number and type of protein/peptide products derived from microalgae but key challenges exist, including but not limited to reduction of biomass production costs; regulatory approval for strains other than *Arthrospira* and *Chlorella*, especially in the EU; thorough food safety assessments; and improved algal strain stability. In addition, novel and innovative technologies may play a role in the future in improving extraction of protein and other metabolites.

Novel processing technologies, including magnetically responsive nano- and microparticles which have already been used in many biotechnology applications (Safarik and Safarikova 2009, 2014; Borlido et al. 2014; Garcia et al. 2015) and which may be employed for magnetic flocculation of microalgae cells from cultivation media, may be used for efficient harvest of microalgae (Safarik et al. 2016). Extraction of protein from microalgae using novel processing technologies, including pressurised liquid extraction, microwave-assisted extraction, sonication, the pH shift extraction method, and others, is also currently under investigation.

References

Batista, A.P., Nunes, M.C., Fradinho, P. et al. (2011). Novel foods with microalgal ingredients – effect of gel setting conditions on the linear viscoelasticity of *Spirulina and Haematococcus* gels. *Journal of Food Engineering* (Special Issue 110 (2): 182–189.

Batista, A.P., Niccolai, A., Fradinho, P. et al. (2017). Microalgae biomass as an alternative ingredient in cookies: sensory, physical and chemical properties, antioxidant activity and in vitro digestibility. *Algal Research* 26: 161–171.

Bleakley, S. and Hayes, M. (2017). Extraction, application and challenges concerning production. *Foods* 6 (5): 33.

Borowitzka, M.A. (2013). High value products from microalgae – their development and commercialisation. *Journal of Applied Phycology* 25 (3): 743–756.

Bryant, D.A. (1982). Phycoerythrocyanin and Phycoerythrin: properties and occurrence in cyanobacteria. *Journal of General Microbiology* 128: 835–844.

Cicero, A.F.G., Fogacci, F., and Colletti, A. (2017). Potential role of bioactive peptides in prevention and treatment of chronic disease: a narrative review. *British Journal of Pharmacology* 174 (11): 1378–1394.

Draaisma, R.B., Wijffels, R.H., Slegers, P.M. et al. (2013). Food commodities from microalgae. *Current Opinion in Biotechnology* 24 (2): 169–177.

Enzing, C., Ploeg, M., Barbosa, M., and Sijtsma, L. (2014). Microalgae-based products for the food and feed sector: an outlook for Europe. *JRC Scientific and Policy Reports* 1–82.

Estime, B., Ren, D., and Sureshkumar, R. (2017). Cultivation and energy efficient harvesting of microalgae using thermoreversible sol–gel transition. *Scientific Reports* 7: 40725. doi: 10.1038/srep40725.

Fradique, M., Batista, A.P., Nunes, M.C. et al. (2013). *Diacronema vlkianum* and *Isochrysis galbana* biomass incorporation in pasta products. *LWT – Food Science and Technology* 50: 312–319.

Garcia, P.F., Brammen, M., Wolf, M. et al. (2015). High-gradient magnetic separation for technical scale protein recovery using low cost magnetic nanoparticles. *Separation and Purification Technology* 150: 29–36.

Garcia-Segovia, P., Pagan-Moreno, M.J., Lara, I.F., and Martinez-Monzo, J. (2017). Effect of microalgae incorporation on physicochemical and textural properties in wheat bread formulation. *Food Science and Technology International* 23 (5): 437–447.

García-Vaquero, M. and Hayes, M. (2016). Red and green macroalgae for fish and animal feed and human functional food development. *Food Reviews International* 32 (1): 15–45.

Gilbert-Lopez, B., Mendiola, J.A., Fontecha, J. et al. (2015). Downstream processing of *Isochrysis galbana*: a step towards microalgal biorefinery. *Green Chemistry* 17 (9): 4599–4609.

Gouveia, L., Coutinho, C., Mendonça, E. et al. (2008). Functional biscuits with PUFA from *Isochrysis galbana*. *Journal of Science of Food and Agriculture* 88: 891–896.

Guedes, A.C., Amaro, H.M., and Malcata, F.X. (2011). Microalgae as sources of carotenoids. *Marine Drugs* 9: 625–644.

Harisko, I. and Posten, C. (2014). Biorefinery of microalgae – opportunities and constraints for different production scenarios. *Biotechnology Journal* 9: 739–752.

Hayes, M. (2014). Bioactive peptides and their potential use for the prevention of diseases associated with Alzheimer's disease and mental health disorders: food for thought? *Annals of Psychiatry and Mental Health* 2 (3): 1017–1026.

Henchion, M., Hayes, M., Mullen, A.M. et al. (2017). Future protein supply and demand: strategies and factors influencing a sustainable equilibrium. *Foods* 6 (7): 53.

Lafarga, T., Rai, D.K., O'Connor, P., and Hayes, M. (2015). A bovine fibrinogen enriched fraction as a source of peptides with in vitro renin and angiotensin-converting enzyme inhibitory activities. *Journal of Agricultural and Food Chemistry* 63 (39): 8876–8684.

Lehto, S., Bachweitz, M., Klimm, A. et al. (2017). Comparison of food colour regulations in the EU and the US: a review of current provisions. *Food Additives and Contaminants. Part A, Chemistry, Analysis, Control, Exposure and Risk Assessment* 34 (3): 335–355.

Li, J., Liu, Y., Cheng, J.J. et al. (2015). Biological potential of microalgae in China for bio-refinery based production of biofuels and high value compounds. *New Biotechnology* 32 (6): 588–596.

Liang, S., Liu, X., Chen, F., and Chen, Z. (2004). Current microalgal health food R & D activities in China. *Hydrobiologia* 512 (1): 45–48.

Lubitz, J.A. (1963). The protein quality, digestibility and composition of algae, Chlorella 71105. *Journal of Food Science* 28 (2): 229–232.

Nobre, B.P., Villalobos, F., Barragán, B.E. et al. (2013). A biorefinery from Nannochloropsis sp. microalga: extraction of oils and pigments. Production of biohydrogen from the leftover biomass. *Bioresource Technology* 135: 128–136.

Patel, A., Gami, B., Patel, P., and Patel, B. (2017). Microalgae: antiquity to era of integrated technology. *Renewable and Sustainable Energy Reviews* 71: 535–547.

Posada, J.A., Brentner, L.B., Ramirez, A., and Patel, M.K. (2016). Conceptual design of sustainable integrated microalgae biorefineries. Parametric analysis of energy use, greenhouse gas emissions and techno-economics. *Algal Research* 17: 113–131.

Regulation (EC) No 178/2002. 2002. Regulation (EC) No 178/2002) of the European Parliament and of the Council of 28th Jan 2002. Official Journal of the European Communities, L31/1.

Safafar, H., Nørregaard, P.U., Ljubic, A. et al. (2016). Enhancement of protein and pigment content in two *Chlorella* species cultivated on industrial process water. *Journal of Marine Science and Engineering* 4: 84.

Safarik, I. and Safarikova, M. (2009). Magnetic nano- and microparticles in biotechnology. *Chemical Papers* 63: 497–505.

Safarik, I. and Safarikova, M. (2014). One-step magnetic modification of non-magnetic solid materials. *International Journal of Materials Research* 105 (1): 104–107.

Safarik, I., Prochazkova, G., Pospiskova, K., and Branyik, T. (2016). Magnetically modified microalgae and their applications. *Critical Reviews in Biotechnology* 36 (5): 931–941.

Safi, C., Olivieri, G., Campos, R.P. et al. (2017). Biorefinery of microalgal soluble proteins by sequential processing and membrane filtration. *Bioresource Technology* 225: 151–158.

Söyler, N., Goldfarb, J.L., Ceylan, S., and Saçan, M.T. (2017). Renewable fuels from pyrolysis of *Dunaliella tertiolecta*: an alternative approach to biochemical conversion of microalgae. *Energy* 120: 907–914.

Tafreshi, A.H. and Shariati, M. (2009). Dunaliella biotechnology: methods and applications. *Journal of Applied Microbiology* 107 (1): 14–35.

Ursu, A.-V., Marcati, A., Sayd, T. et al. (2014). Extraction, fractionation and functional properties of proteins from the microalgae *Chlorella vulgaris*. *Bioresource Technology* 157: 134–139.

van der Spiegel, M., Noordam, M.Y., and van der Fels-Klerx, H.J. (2013). Safety of novel protein sources (insects, microalgae, seaweed, duckweed and rapeseed) and legislative aspects for their application in food and feed products. *Comprehensive Reviews in Food Science and Food Safety* 12 (6): 662–678.

Vanthoor-Koopmans, M., Wijffels, R.H., Barbosa, M.J., and Eppink, M.H. (2013). Biorefinery of microalgae for food and fuel. *Bioresource Technology* 135: 142–149.

Wijffels, R.H., Barbosa, M.J., and Eppink, M.H.M. (2010). Microalgae for the production of bulk chemicals and biofuels. *Biofuels, Bioproducts and Biorefining* 4: 287–295.

13

Current and Future Trends in Protein Use and Consumption

Maria Hayes

Teagasc Food Research Centre, Food BioSciences Department, Dublin, Ireland

13.1 Introduction

Improvements in food security will probably lead to improvements in global social, economic, and ecological situations as food and feed production and consumption are inextricably linked to nutrition, health, poverty, climate change, and resource depletion and cannot be tackled in isolation. Nitrogen and protein are linked to issues of sustainability and have been shaped by evolution in the fields of agriculture and industry (Aiking 2011). Production of food for the growing world population, in particular protein, is a challenge but is achievable. However, doing so without compromising sustainability is a greater challenge (Tilman et al. 2009). Protein is essential for the well-being of the human population as nitrogen is an indispensable constituent of DNA, RNA, and proteins. Growth in incomes and continued urbanisation of societies have led to dietary transformations and increased growth in the consumption of animal proteins. However, there has also been an increase in utilisation of plant protein and research into the use of novel protein sources, including insects and algae, to support the production of protein-rich foods and to help reduce the strain of intensive animal protein production (Day 2013).

13.2 Land-Based Plant Proteins

Plant proteins are considered cheap and more abundant than animal proteins and direct consumption of proteins from plants is conventional for humans but has a number of limiting factors. Indeed, most plant-produced proteins are currently used in animal feed for the production of animal-based proteins despite the fact that conversion of plant proteins to animal proteins is inefficient, with less than 15% of the plant proteins obtained from feed crops turned into animal proteins for human consumption (Pimentel and Pimentel 2003; Aiking 2011; Day 2013). Production of plant proteins has several advantages over production of animal proteins. For example, 100 times more water is required

Novel Proteins for Food, Pharmaceuticals, and Agriculture: Sources, Applications, and Advances,
First Edition. Edited by Maria Hayes.

to produce animal protein compared to plant proteins (Day 2013). However, plant proteins often do not have the complete amino acid profile of an ideal protein, compared to animal protein sources. In addition, their functional attributes and increased costs associated with processing of plant proteins compared to animal protein sources are challenges concerning their use (de Boer and Aiking 2011).

13.3 Cereal Proteins

Cereal seed proteins may be classified into groups on the basis of their extraction and solubility in a series of solvents: water (albumins), dilute aqueous salts (globulins), alcohol/water mixtures (prolamins), and dilute acids or alkalis (glutelins). This has provided a framework for the modern study of cereal seed proteins and, indeed, for the classification of proteins in general (Osborne 1924; Shewry et al. 1986).

Wheat may be considered a very important food crop and is eaten by over one billion people (USDA-FAS 2012). It can contain between 8% and 15% protein, depending on the variety of grain used, and the primary storage proteins in wheat include gluten which imparts unique functional properties to leavened products (Wrigley 1996). In Western society, it is used to make baked goods including breakfast cereals, pasta, noodles, and to fortify flour with protein. Gluten is also used in non-baked foods as a meat replacement in vegetarian foods. Texturised wheat protein is used increasingly to replicate the look and texture of meat products (Day 2013). Gluten has disadvantages, however, and consumption of gluten is thought to cause coeliac disease in susceptible individuals. Gluten is found in three cereals: wheat, barley, and rye, and is also found in many processed foods. Coeliac disease is now known to be a common condition that affects approximately 1 in 100 people. Women are two to three times more likely to develop coeliac disease than men, with cases of coeliac disease diagnosed in people of all ages.

Wheat protein products on the market include vital wheat gluten (VWG), isolated wheat protein (IWP), texturised wheat proteins, and enzyme hydrolysed proteins; products contain between 75% and 90% protein and are manufactured by major suppliers including Cargill, MGP Ingredients, and Manildra (Day 2013). The processes that are normally applied to wheat grains for extraction and processing into food ingredients are shown in Figure 13.1.

In addition to providing energy and amino acids essential for growth, it is increasingly recognised that some dietary proteins contain biologically active peptides that can impart a beneficial physiological effect. Such bioactive peptides are typically 2–30 amino acids in length and can exert their physiological response as a result of their hormone-like properties or by acting as antagonistic receptor inhibitors. Bioactive peptides are inactive within the precursor protein and can be released through hydrolysis by digestive enzymes or fermentation. The function of these peptides covers a variety of activities, including antihypertensive, hypoglycaemic, antiamnestic, antimicrobial, antithrombotic, antioxidative, hypocholesterolaemic, gastrointestinal absorption modulation, appetite suppression, opioid, immunomodulation, and cytomodulation. In addition to its nutritional benefits, wheat is also a promising source of bioactive peptides, with several already identified (Cavazos and Gonzalez de Mejia 2013; Gangopadhyay et al. 2016).

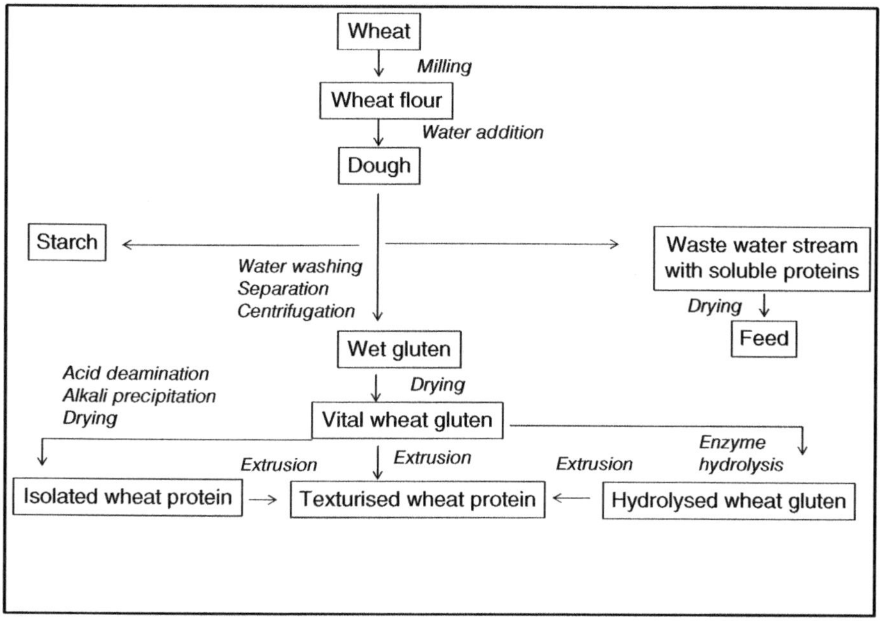

Figure 13.1 Protein extraction and processing from wheat.

13.4 Rice, Corn, and Sorghum Proteins

Rice has the lowest protein content of all cereals and contains 7–9% protein by weight. Rice protein is, in fact, a co-product produced as a result of rice starch processing from bran and broken rice kernels (Fabian and Ju 2011). The gluten-free food products market uses a lot of rice protein powders due to its hypoallergenic nature (Shih 2003).

Corn or maize is the most important food crop in the USA and has a protein content of between 9% and 12%. It is used primarily as animal feed or for ethanol production. Corn flour is used in several food products including baby foods, meat products, and fermented products.

Barley protein contains 10–17% protein and was historically the most important food grain but today two-thirds of global production is used as animal feed while the majority of the rest is used in malting. Human consumption is largely limited to northern Africa and western Asia. Barley is rich in beta-glucans which have a positive health claim for lowering blood cholesterol (Baik and Ullrich 2008; Day 2013).

Sorghum is a very resistant plant and suitable for use in climates where drought conditions prevail (Rooney 2000). It has a protein content of 9–17% and is used as coarse flour. Protein digestibility of sorghum flour after cooking decreases due to cross-linking of disulfide bonds and digestibility of sorghum proteins can be increased by extrusion processing (Correia et al. 2011). Sorghum is considered the fifth most important cereal.

13.5 Soy Protein

Soybean is rich in protein and contains about 35–40% protein on a dry weight basis compared to 17–26% protein for flaxseed and 7–9% for rice (Day 2013). Soybean is a dominant oil seed crop and global production is estimated at 271 million metric tonnes per annum (USDA-FAS 2012). Defatted soybean meal is largely used for animal feed production and a small portion is consumed by humans, particularly in East Asia where there is a tradition of soybean consumption. Soy flour, soy protein concentrate, soy protein isolate, and hydrolysed soy protein are largely produced in the USA using different processing methods including use of alkali extractions (soy protein isolate) and extrusion (textured soy protein) (Day 2013). Soy protein has several applications including in infant formula and as an ingredient in formulated foods, where it is used as a texturising agent or as a water and fat binder or for increasing the emulsification, foaming, and gelation properties of a product.

Currently, approximately 2–3% of soybean production is used for human food with the remainder going for animal feed (Johnson et al. 1992, 2008). Soy protein is known to have potential health benefits, including alleviation of chronic diseases (Baier and McClements 2005).

13.6 Pulses

Pulses include peas, lupin, chickpea, beans, and other seeds from the pods of legume plants. They are an important protein source and pea protein has been used extensively as an important commercial protein, as outlined in Chapter 10. Peas are easily grown and easy to process and can contain 25% protein (Boye et al. 2010). Chickpea is consumed largely in India. Chickpea flour has excellent emulsification properties and can be used as an extender in emulsified meat products (Withana-Gamage et al. 2009; Day 2013). Lupin seeds can contain approximately 40% protein, are non-starch leguminous seeds and have high fibre content (Evans et al. 2003). Yellow and white lupins are grown in Europe and South America and can be used for animal feed with white lupins consumed in areas of the Mediterranean for direct human food use. Western Australia produces 85% of the world lupin crop. Challenges including the alkaloid content of lupin, which limits its use as a human food ingredient, and lupin protein flour must be free of alkaloids for use as a functional ingredient (Kohajdová et al. 2011).

13.7 Nut and Tuber Proteins

Tuber plants such as potatoes can contain up to 1.5% protein but are mainly utilised for their starch content. However, by-products from starch manufacture may be used as a substrate for protein isolation (Camire et al. 2009). Nuts also contain proteins but may have limited applications due to the highly allergic nature of the proteins. Many nut allergies are triggered when the immune system recognises specific proteins in the food and releases the antibody immunoglobulin E (IgE) to latch on to the allergen, thereby causing reactions from mild itching to life-threatening anaphylaxis, a whole-body reaction that may include an itchy rash, throat swelling, and low blood pressure. Recently, scientists have adapted cashew proteins so that IgE cannot recognise them. Nut protein-containing products include almond milk and peanut protein (Mattison et al. 2016).

13.8 Insect Protein

Entomophagy is the eating of insects and is practised in 113 countries in the world where over 2000 insect species are considered edible (MacEvilly 2000). Insects have higher feed conversion efficiency compared to livestock and have a higher fecundity (Nakagaki and Defoliart 1991) and can be grown on organic waste (Rumpold and Schlüter 2013). However, much research is required to develop this novel protein source in terms of cost-effective growing and harvesting processes. Insects are reported as an excellent source of protein in the literature, with consumption of 100 g of caterpillars able to provide 76% of the daily required amount of proteins (Agbidye et al. 2009). In addition, insect protein, with the exception of palm weevil larvae which is oil fat, often has a high-quality amino acid profile with high contents of phenylalanine and tyrosine and generally meets the requirements for good-quality protein with the exception of the amino acid methionine. In addition, not all insects are safe to eat. In Nigeria, for example, *Anaphe venata* is responsible for a seasonal ataxic syndrome (Nishimune et al. 2000). Furthermore, cases of botulism and food poisoning have been reported due to consumption of insects (Schabel 2010). Work concerning the safety and consumer acceptance of insect proteins should be carried out if this protein source is to become mainstream in the future.

13.9 Fungal and Microbial Protein

Mushrooms and truffles are a rich protein source and proteins from fungal and microbial sources have considerable promise and indeed have been commercialised. Quorn protein is commercially produced for human food consumption and is widely available (Wiebe 2004; Boland et al. 2013).

13.10 Algal Proteins

13.10.1 Microalgae Protein

Algae include both microalgae and seaweeds (macroalgae). Microalgae, which include some cyanobacteria, have potential for use in human and animal nutrition as they can be cultivated without competing for land, water, or even with crops and can be grown using wastewaters with high concentrations of nitrogen (N_2) and phosphorus (P). They have a favourable nutritional profile, having a protein fraction with an equilibrated amino acid profile and a lipid fraction in which essential polyunsaturated fatty acids (PUFAs) are present. Microalgae have also been used in the bioremediation of highly contaminated wastewaters, due to unique characteristics including the ability to grow in media with high concentrations of N and P; photosynthetic activity; ability to fix carbon dioxide (CO_2) and liberate oxygen (O_2); and reduction of external aeration needs (Munoz and Guieysse 2006). Indeed, contaminants including polycyclic, aromatic hydrocarbons, phenolic compounds, and organic solvents can act as nutrients for microalgae and thus bioremediation of contaminants could be used as a strategy to produce protein-rich biomass while at the same time treating pollution.

Scenedesmus almeriensis (Chlorophyceae) is an industrially important microalgal strain that is usually colonial, forming four- or eight-celled coenobia arranged linearly. Cell size ranges from 3–78 long × 2–10 μm width, and the cell wall consists of hemicellulosic and sporopolleninic layers. This makes *S. almeriensis* especially resistant to several cell disruption techniques. They naturally occur in open treatment ponds. *Nannochloropsis gaditana* (Eustigmatophyceae) is a small spherical marine microalga of between 2 and 4 μm in diameter that is widely used for pigment extraction and in the biodiesel research field due to its ability to accumulate lipids, especially the essential PUFA eicosapentaenoic acid (EPA). *Isochrysis T-Iso* (Coccolithophyceae) is a marine species with great potential for lipid production due to its fatty acid composition (Roleda et al. 2013). This microalga is also used as a feed constituent in hatcheries, principally for molluscs but also for young fish and crustaceans, since it has a crude protein average content of above 40% (Hayes et al. 2015). In contrast, *Spirulina platensis* (Cyanophyceae) is a gram-negative cyanobacterium, with soft cell walls that consist of complex sugars and protein. The main photosynthetic pigment found in *S. platensis* is phycocyanin, which is blue in colour, but this microalga also contains chlorophyll and carotenoids. Their main advantage is that they are edible and have perceived health beneficial properties which have been known since ancient times.

Increasing interest in the health qualities of algal proteins, as well as a strong financial need to valorise all fractions of the cultivated biomass, has encouraged the development of alternative, safe, and scalable technologies to access quality algal proteins. Pre-extraction of proteins from algal biomass is being explored globally at present. Recently, Munoz et al. (2015) developed a procedure for extraction of soluble proteins under alkaline conditions, prior to the removal of lipids for biodiesel production. This method can be scaled by industry and others utilising microalgae for extracting useable proteins, augmenting efficiencies of follow-on solvent-based lipid extractions or thermal hydrolysis, and for removal of components detrimental to process efficiencies. Protein fractions, although convertible, are generally undesirable as protein-rich biomass creates bio-oils generally rich in nitrogen, creating excessive NO when fuel is burned (Wang et al. 2013; Belotti et al. 2014). These fractions can be recovered for food and non-food applications.

An appropriate cell disruption process must be developed in order to maximise the yield and the value of microalgal proteins. Cell disruption techniques that may disintegrate the microalgal cell precisely without chemical contamination or degradation of the target protein are required. For large-scale production, it is also important that the disintegration process can be scaled up and that it is rapid. In addition, integration of the cell disruption method into the downstream processing steps has to be easy and it should not have a negative impact on subsequent processing steps.

13.10.2 Macroalgae (Seaweed) Protein

Seaweeds have higher protein yield per unit area (4–15 tonnes/Ha/year) compared to terrestrial crops, such as soybean, pulse legumes, and wheat (0.6–1.2 tonnes/Ha/year, 1–2 tonnes/Ha/year, and 1.1 tonnes/Ha/year, respectively). Terrestrial agriculture already requires approximately 75% of total global fresh water, with animal protein production in particular requiring 100 times more water than if the equivalent amount of protein was produced from plant sources. Marine algae do not require fresh water or arable land to grow, maximising resources that can be used for additional food production or other cash crops. Furthermore, due to their harsh environment and phototropic life, algae are often exposed to high oxidative and free radical stresses. This has led to the evolution of natural protective systems, such as the production of pigments (e.g. carotenes, chlorophylls, and phycobiliproteins) and polyphenols (e.g. catechins, flavonols, and phlorotannins), which can impart health benefits to the consumer (Bleakley and Hayes 2017).

Macroalgae have for centuries been consumed whole among the populations of China, Korea, and Japan. Due to the environment in which they grow, macroalgae produce unique and interesting biologically active compounds. Protein can account for up to 47% of the dry weight of macroalgae, depending on species and time of cultivation and harvest. Peptides derived from macroalgae are proven to have hypotensive effects in the human circulatory system. Hypertension is one of the major, yet controllable, risk factors in cardiovascular disease (CVD). CVD is the main cause of death in Europe, accounting for over 4.3 million deaths each year. In the United States it affects one in three individuals. Hypotensive peptides derived from marine and other sources have already been incorporated into functional foods such as beverages and soups (Fitzgerald et al. 2012).

The red seaweed *Porphyra* species is traditionally used in Wales to make a food item known as laver bread. Bakery and pasta products, being the most widely consumed food products in the world, are some of the best potential vehicles for delivery of bioactive compounds. Previously, wakame (*Undaria pinnatifida*) was incorporated into pasta to improve its biofunctional properties and found that incorporation of up to 20% wakame enhanced interactions between starch granules and the protein matrix (Fitzgerald et al. 2012).

13.11 Proteins from Animals and Animal By-Products

Ruminants play a key role in converting fibre-containing materials into protein with a high nutritional value and meat can contain 20–25% protein depending on the source and fat content. Proteins from animal meat sources have several advantages over plant proteins. They are nutritionally complete, containing all the essential amino acids which are bioavailable and easily digested. Meat protein also contains vitamins and minerals including vitamins A, B9, B12, D, and E (Berner et al. 1985). There are also negative attributes associated with meat proteins, including high fat content and significant contribution to climate change and land use change (Henchion et al. 2017). Increased efficiencies in production practices are required in order to reduce the impact of meat production on climate change and the environment.

One approach that could be adopted is total utilisation of the by-products from primary meat processing; that is, utilisation of co-products such as blood, offal, and others. According to the European Union, the term *meat* refers to the edible parts removed from the carcase of domestic animals, poultry, and wild game. Meat is a valuable livestock product and is the first-choice source of animal protein for many consumers. By-products represent environmental and economic problems for processors and include carcases, skins, bones, meat trimmings, blood, fatty tissues, horns, feet, hoofs, and internal organs (di Bernardini et al. 2011). By-products can be used as an excellent protein source and indeed a source of bioactive peptides. Certain meat proteins have important physiological activities. For example, collagen has a positive influence on the delivery and bioactivity of bone morphogenic protein 2 and ectopic bone formation, enhancing bone healing (Bhakta 2013). Furthermore, bioactive peptides may be generated from meat proteins by hydrolysis, cooking, and fermentation (Lafarga and Hayes 2014).

Proteins from meat can be broadly classified as water soluble, insoluble, or soluble in high or low concentrations of salt. Water-soluble proteins are sarcoplasmic proteins. Myofibrillar proteins are soluble in salt solutions and contain stromal proteins or proteins from connective tissue such as collagen and elastin. Myofibrillar proteins account for about 55% of muscle proteins and contain actin, titin, and myosin (Herzog 2014). Blood is a common by-product in proteins and can be divided into the cellular elements that contain platelets, white, and red blood cells and plasma that contains albumin, fibrinogen, and the globulins (Bjelanović et al. 2015).

Recently, Lafarga and Hayes (2014) have looked at the potential of blood and bovine meat processing by-products (or what should more correctly be termed co-products) as a source of bioactive peptides with health benefits, including dipeptidyl peptidase IV inhibitory, angiotensin-1 converting enzyme (ACE-1) inhibitory, and antioxidant activities. Total utilisation of by-products could help to reduce the impact of processing on the environment and reduce pollution and costs for processors as well as contributing to a reduction of the impact of greenhouse gases (GHG)s and negative impacts on the environment.

Dairy proteins are highly developed products and have myriad functional and health benefits supported by science. The market for dairy is complex, multifaceted, and driven by evolving markets and dairy protein ingredients are made ordinarily as powder products that can be conveniently transported. By-products from primary milk production such as whey are used in performance nutrition and beverages such as infant formula and new products have been generated through use of technologies including microfiltration.

13.12 Future Protein Demands

It is clear that the way in which protein is produced and consumed has an impact on our global environment and on human and animal health (Henchion et al. 2017). Since the 1960s, research has been carried out to develop methods to reduce the impact of existing protein sources, especially animal proteins, on GHG emissions, land, energy, and water use and biodiversity as well as health impacts. Breeding programmes and nutrition efforts have led to significant increases in feed conversion efficiency per animal yields and a decrease in GHG emissions per kg of animal product. There is a need for further adoption of existing technologies that may alleviate climate change and bring both economic and environmental benefits.

Many different scenarios are possible regarding future protein demand and balancing driving and opposing forces will result in a world where more protein is required but is produced in a sustainable way. The concerns regarding traditional protein sources, including meat proteins, are well documented. Also, problems including use of antibiotics in meat and milk proteins are an additional concern. Indeed, moderate consumption of meat protein is more pragmatic and better for the environment compared to a vegetarian or vegan diet (Henchion et al. 2017).

Boland et al. (2013) suggest that three simultaneous changes will need to be made to meet future animal-derived protein demand: shifting protein sources up the supply chain, using plant-based substitutes or extenders for animal-derived protein foods, and use of novel sources for both animal and human nutrition. However, Henchion et al. (2017) found that different factors influence the potential of existing and novel protein sources. Existing protein sources are primarily hindered by their negative environmental impacts, with some concerns around health. The positives associated with traditional protein sources include social economic benefits and consumer acceptance. Novel protein sources do not currently enjoy the same level of consumer acceptance and new value chains need to be developed. In addition, positive environmental benefits cannot be assumed with novel protein sources and care must be taken to ensure the environment will not suffer due to their development.

13.13 Conclusion

Consumers are now demanding increased transparency across food chains and are consistently looking for access to company information, nutritional, and health benefits of food products, including novel proteins as well as sustainably sourced ingredients. Sustainable nutrition in line with the United Nations' 2030 sustainable development goals is becoming important in terms of processing food products, including proteins, as are sustainable use of natural resources and reduction of global waste, in line with the UN target for 2030 of reducing global waste at the retail level by half globally. Consumption of traditional and novel protein sources will continue to be influenced by the current 'on-the-go' nutrition trend as well as a reduction in portion sizes and easily accessible product formats. These innovations themselves may help to tackle issues such as increasing obesity and food waste in the future.

References

Agbidye, F.S., Ofuya, T.I., and Akindele, S.O. (2009). Marketability and nutritional qualities of some edible forest insects in Benue State, Nigeria. *Pakistan Journal of Nutrition* 8: 917–922.

Aiking, H. (2011). Future protein supply. *Trends in Food Science and Technology* 22: 112–120.

Baier, S.K. and McClements, D.J. (2005). Influence of cosolvent systems on the gelation mechanism of globular protein: thermodynamic, kinetic and structural aspects of globular protein gelation. *Comprehensive Reviews in Food Science and Food Safety* 4: 43–54.

Baik, B.K. and Ullrich, S.E. (2008). Barley for food: characteristics, improvement, and renewed interest. *Journal of Cereal Science* 48: 233–242.

Belotti, G., de Caprariis, B., de Filippis, P. et al. (2014). Effect of *Chlorella vulgaris* growing conditions on bio-oil production via fast pyrolysis. *Biomass and Bioenergy* 61: 687–695.

Berner, L.K., Miller, D.D., and Vancampen, D. (1985). Availability to rats of iron in ferric hydroxide polymers. *Journal of Nutrition* 115: 1042–1045.

Bhakta, G. (2013). The influence of collagen and hyaluronan matrices on the delivery of bioactivity of bone morphogenetic protein-2 and ectopic bone formation. *Acta Biomaterialia* 9 (11): 9098–9106.

Bjelanović, M., Grabež, V., Vučić, G. et al. (2015). Effects of different production systems on carcass and meat quality of sheep and lamb from Western Balkans and Norway. *Biotechnology in Animal Husbandry* 31 (2): 203–221.

Bleakley, S. and Hayes, M. (2017). Algal proteins: extraction and application, and challenges concerning production. *Foods* 6 (5): 33.

de Boer, J. and Aiking, H. (2011). On the merits of plant-based proteins for global food security: marrying macro and micro perspectives. *Ecological Economics* 70: 1259–1265.

Boland, M.J., Allan, N.R., Vereijken, J.M. et al. (2013). The future supply of animal-derived protein for human consumption. *Trends in Food Science and Technology* 29: 62–73.

Boye, J.I., Aksay, S., Roufik, S. et al. (2010). Comparison of the functional properties of pea, chickpea and lentil protein concentrates processed using ultrafiltration and isoelectric precipitation techniques. *Food Research International* 43: 537–546.

Camire, M.E., Kubow, S., and Donnelly, D.J. (2009). Potatoes and human health. *Critical Reviews in Food Science and Nutrition* 49 (10): 823–840.

Cavazos, A. and Gonzalez de Mejia, E. (2013). Identification of bioactive peptides from cereal storage proteins and their potential role in prevention of chronic diseases. *Comprehensive Reviews in Food Science and Food Safety* 12: 364–380.

Correia, I., Nunes, A., Saraiva, J.A. et al. (2011). High pressure treatments largely avoid/revert decrease of cooked sorghum protein digestibility when applied before/after cooking. *LWT – Food Science and Technology* 44: 1245–1249.

Day, L. (2013). Proteins from land plants – potential resources for human nutrition and food security. *Trends in Food Science and Technology* 32: 25–42.

Di Bernardini, R., Rai, D.K., Bolton, D. et al. (2011). Isolation, purification and characterisation of antioxidant peptidic fractions from a bovine liver sarcoplasmic protein thermolysin hydrolysate. *Peptides* 32 (2): 388–400.

Evans, J., Scott, G., Lemerle, D. et al. (2003). Impact of legume 'break' crops on the residual amount and distribution of soil mineral nitrogen. *Australian Journal of Agricultural Research* 54: 763–776.

Fabian, C. and Ju, Y.H. (2011). A review on rice bran protein: its properties and extraction methods. *Critical Reviews in Food Science and Nutrition* 51: 816–827.

Fitzgerald, C., Mora-Soler, L., Gallagher, E. et al. (2012). Isolation and characterisation of bioactive pro-peptides with in vitro renin inhibitory activities from the macroalga *Palmaria palmata*. *Journal of Agricultural and Food Chemistry* 60 (30): 7421–7427.

Gangopadhyay, W., Wynne, K., O'Connor, P.M. et al. (2016). *In silico* and *in vitro* analyses of the angiotensin-I-converting enzyme inhibitory activity of hydrolysates generated from crude barley (*Hordeum vulgare*) protein concentrates. *Food Chemistry* 203: 367–374.

Hayes, M., García-García, M., Fitzgerald, C., and Lafarga, T. (2015). Seaweed and milk derived bioactive peptides and small molecules in functional foods and cosmeceuticals. In: *Biotechnology of Bioactive Compounds: Sources and Applications* (ed. V.K. Gupta and M.G. Tuohy). Chichester: Wiley.

Henchion, M., Hayes, M., Mullen, A.M. et al. (2017). Future protein supply and demand: strategies and factors influencing a sustainable equilibrium. *Foods* 6: 53.

Herzog, W. (2014). The role of titin in eccentric muscle contraction. *Journal of Experimental Biology* 15 (217): 2825–2833.

Johnson, L.A., Myers, D.J., and Burden, D.J. (1992). Soy protein's history, prospects in food, feed. *Inform* 3: 429–430.

Johnson, L.A., White, P.J., and Galloway, R. (2008). *Soybeans. Chemistry, Production, Processing and Utilisation.* Urbana: AOCS Press.

Kohajdová, Z., Karovičova, J., and Schmidt, Š. (2011). Lupin composition and possible use in bakery: a review. *Czech Journal of Food Sciences* 29 (3): 203–211.

Lafarga, T. and Hayes, M. (2014). Bioactive peptides from meat muscle and by-products: generation, functionality and application as functional ingredients. *Meat Science* 98: 227–239.

MacEvilly, C. (2000). Bugs in the system. *Nutrition Bulletin* 25: 267–268.

Mattison, C.P., Bren-Mattison, Y., Vant-Hull, B. et al. (2016). Heat induced alterations in cashew allergen solubility and IgE binding. *Toxicology Reports* 3: 244–281.

Munoz, R. and Guieysse, B. (2006). Algal–bacterial processes for the treatment of hazardous contaminants: a review. *Water Research* 40 (15): 2799–2815.

Munoz, R., Navia, R., Ciudad, G. et al. (2015). Preliminary bio-refinery process proposal for protein and biofuel recovery from microalgae. *Fuel* 150: 425–433.

Nakagaki, B.J. and Defoliart, G.R. (1991). Comparison of diets for mass-rearing *Acheta domesticus* (*Orthoptera, Gryllidae*) as a novelty food and comparison of food conversion efficiency with values reported for livestock. *Journal of Economic Entomology* 84: 891–896.

Nishimune, T., Watanabe, Y., Okazaki, H., and Akai, H. (2000). Thiamin is decomposed due to *Anaphe* spp. entomophagy in seasonal ataxia patients in Nigeria. *Journal of Nutrition* 130: 1625–1628.

Osborne, T.B. (1924). *The Vegetable Proteins*, 2e. London: Longmans Green.

Pimentel, D. and Pimentel, M. (2003). Sustainability of meat-based and plant-based diets and the environment. *American Journal of Clinical Nutrition* 78: 660S–663S.

Roleda, M.Y., Slocombe, S.P., Leakey, R.J.G. et al. (2013). Effects of temperature and nutrient regimes on biomass and lipid production by six oleaginous microalgae in batch culture employing a two-phase cultivation strategy. *Bioresource Technology* 129: 439–449.

Rooney, L.W. (2000). Sorghum. In: *Handbook of Cereal Science and Technology*, 2e (ed. K. Kulp and J.G. Ponte), 149–175. New York: Marcel Dekker.

Rumpold, B.A. and Schluter, O.K. (2013). Nutritional composition and safety aspects of edible insects. *Molecular Nutrition and Food Research* 57 (5): 802–823.

Schabel, H.G. (2010). Forest insects as food: a global review. In: *Forest Insects as Food: Humans Bite Back* (ed. P.B. Durst, D.V. Johnson, R.N. Leslie and K. Shono), 37–64. Bangkok: FAO.

Shewry, P.R., Tatham, A.S., Forde, J. et al. (1986). The classification and nomenclature of wheat gluten proteins – a re-assessment. *Journal of Cereal Science* 4: 97–106.

Shih, F.F. (2003). An update on processing of high-protein rice products. *Nahrung-Food* 47: 420–424.

Tilman, D., Socolow, R., Foley, J.A. et al. (2009). Beneficial biofuels – the food, energy and environment trilemma. *Science* 325: 270–271.

USDA-FAS. 2012. Current world production, markets and trade reports. Retrieved from: www.fas.usda.gov.

Wang, K., Brown, R.C., Homsy, S. et al. (2013). Fast pyrolysis of microalgae remnants in a fluidized bed reactor for bio-oil and biochar production. *Bioresource Technology* 127: 494–499.

Wiebe, M.G. (2004). Quorn™ Myco-protein – overview of successful fungal product. *Mycologist* 18 (1): 17–20.

Withana-Gamage, T.S., Wanasundara, J.P.D., Pietrasik, Z., and Shand, P.J. (2009). Characterisation of chickpea (*Cicer arietinum L.*) flours and application in low fat pork Bologna as a model system. *Food Research International* 43 (2): 617–625.

Wrigley, C.W. (1996). Biopolymers – giant proteins with flour power. *Nature* 381: 738–739.

14

Allergenicity of Food Proteins

Maria Hayes

Teagasc Food Research Centre, Food BioSciences Department, Dublin, Ireland

14.1 Introduction

Food allergens are almost always proteins, but not all food proteins are allergens (Bannon 2004). A food allergen can first elicit an immunoglobulin E (IgE) response, and then, on subsequent exposure it can elicit a clinical response to the same or similar protein. In order to elicit this response, proteins that cause an allergic reaction must have specific biochemical characteristics that allow them to survive food processing, escape the digestive and microbial enzymes that are found in the gastrointestinal (GI) tract and then subsequently interact with the immune system (Bannon 2004; Verhoeckx et al. 2015). Both the protein and the immune system of people allergic to the selected protein are involved in the development of allergy. The 'big eight' food protein allergens include proteins from cow milk, hen eggs, peanuts, tree nuts, soy, wheat, shellfish, and fish (Young et al. 1994; Hefle et al. 1996). This list was expanded to 14 recently and now includes proteins from cereals containing gluten, crustaceans, molluscs, eggs, fish, peanuts, tree nuts, soybeans, milk, celery, mustard, sesame, lupin, and sulfur dioxide (Commission-Directive 2006/142/EC). Food-processing methods and other factors can affect protein allergenicity.

14.2 What is Human Allergy to Protein?

Protein allergy refers to a physical effect in which immunological mechanisms (allergic reactions) can occur in sensitised subjects following dietary exposure to certain proteins in food and it is an important human health problem (Sampson 2004). IgE-mediated allergy is an immunoreaction that occurs when the immune system reacts improperly to otherwise innocuous proteins, designated allergens (Jiang et al. 2007a).

Protein allergy has two stages; in the first stage, susceptible consumers are immunologically primed to specific proteins, resulting in sensitisation. Sensitisation can also occur via non-oral routes, for example via inhalation or skin contact (Verhoeckx et al. 2015). The second stage involves elicitation of an allergic reaction when a sensitised

Novel Proteins for Food, Pharmaceuticals, and Agriculture: Sources, Applications, and Advances, First Edition. Edited by Maria Hayes.

person encounters sufficient levels of the protein. Symptoms range from mild to systemic anaphylaxis, which can be fatal (Perry and Pesek 2013; Sicherer and Sampson 2014). Binding of an allergen to IgE attached to mast cells and basophils stimulates the release of inflammatory substances, such as histamine, leading to disease reactions (Figure 14.1). Common symptoms include pins and needles, swelling of the oral cavity and alimentary canal, and respiratory tract reactions (Jiang et al. 2007). This phenomenon is known as cross-reactivity.

Manifestations occur not only in the GI tract but also in other organ systems. Allergic sensitisation to food allergens develops early in childhood. Mechanisms of pathogenesis include cross-linking of mast cell- and basophil-bound IgE and immediate release of inflammatory mediators, as well as late-phase and chronic allergic inflammation, resulting from T-cell, basophil, and eosinophil activation (Valenta et al. 2015).

More than 700 allergen sequences have been identified from food and non-food sources. However, despite increasing knowledge of the structure and amino acid sequences of the identified allergens, only a few biochemical characteristics can be associated with food allergens. Food allergen characteristics, including abundance of the protein in the food, multiple, linear IgE binding epitopes, resistance of the protein to digestion and processing, and allergen structure, have been studied (Bannon 2004). *In silico* methodologies can help to identify whether a novel protein is an existing allergen or whether it has potential to cross-react with an existing allergen. However, they

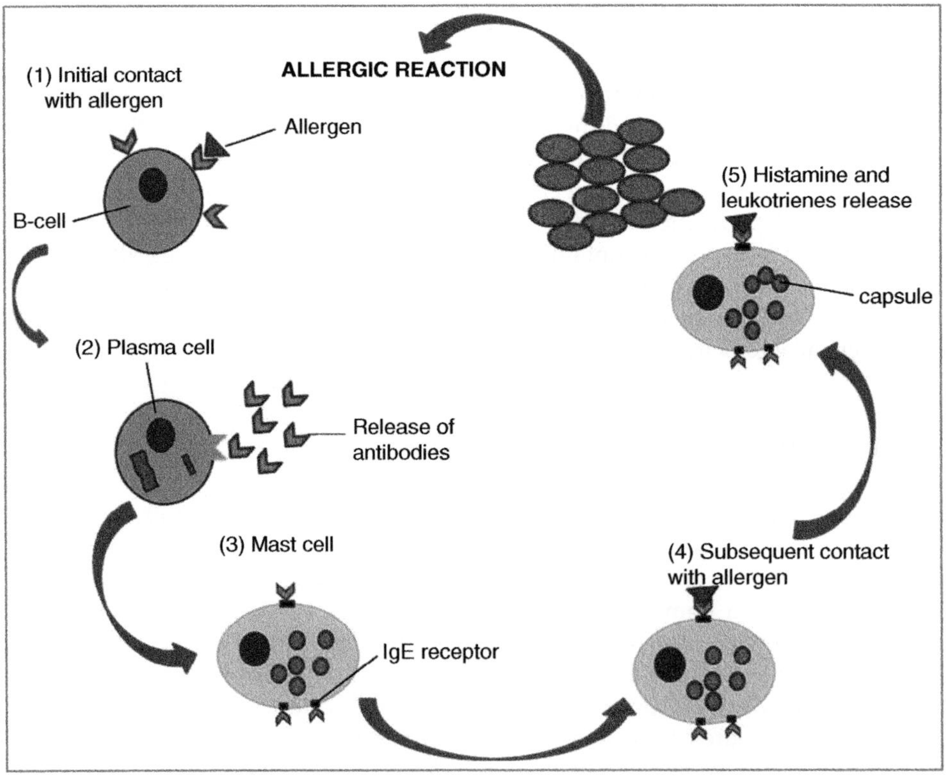

Figure 14.1 The steps leading to allergic reaction.

cannot identify whether a novel protein will 'become' an allergen (Hefle 2002; Hayes et al. 2015).

In silico methodologies and tools such as databases and comparison software play an important role in the assessment of protein allergenicity and allergenicity mechanisms. AllergenOnline (www.allergenonline.org/) is the tool currently used for the safety assessment of novel proteins, but other tools are also available, including the Structural Database of Allergenic Proteins (SDAP) and AllerTOP. PeptideRanker (http://bioware.ucd.ie/~compass/biowareweb/Server_pages/peptideranker.php) and the new, international allergen database known as the Comprehensive Protein Allergen Resource (COMPARE), as well as the Hydrophobic Cluster Analysis (HCA) method used for identifying IgE-binding epitopes in food allergens, are also useful and were reviewed recently (Hayes et al. 2015). Table 14.1 shows *in silico* methods and databases that can be used to assess proteins for the potential to cause human allergy.

14.3 Types of Food Protein Allergens

Allergens can be separated into two groups based on their pathway of entrance leading to sensitisation. Non-food allergens mainly enter by the airborne pathway. Ingested allergens can be dubbed alimentary canal sensitised (ACS), while others can be dubbed non-alimentary canal sensitised (NACS). Jiang et al. (2007) identified dissimilarities between food allergens of the ACS type and non-allergens. A bifurcation of food allergens into

Table 14.1 *In silico* methods for predicting potential human allergy to proteins.

Web tool	Web tool access address	Advantages of method	Reference
AllergenOnline	www.allergenonline.org/	• Methodology currently used for the allergenicity assessment of novel proteins • Peer-reviewed allergen list (by independent scientists and clinicians) and sequence searchable tool (FASTA, exact match search) • Used for the identification of proteins that may present a potential risk of allergenic cross-reactivity • Also coeliac disease protein database risk assessment tool • Hosted by the University of Nebraska, USA	Goodman et al. 2008
AllerHunter	http://tiger.dbs.nus.edu.sg/AllerHunter	• Cross-reactive allergen prediction program that uses a combination of Support Vector Machine (SVM) and pairwise sequence similarity • Hosted by the University of Singapore, Singapore	

(*Continued*)

Table 14.1 (Continued)

Web tool	Web tool access address	Advantages of method	Reference
PREAL	http://gmobl.sjtu.edu.cn/PREAL/index.php	• Built on a combination of SVM and protein features • Uses AllFam, UIS and Allergome allergen databases and ProAP webtool • Integrates protein biochemical and physicochemical properties (molecular weight, secondary structure propensity, hydrophobicity, polarisability, solvent accessibility, normalised van der Waals volume, polarity, and length) • Integrates sequential features and subcellular locations • mRMR and IFS used to identify allergenicity features • Hosted in the Shanghai Jiao Tong University, China	Wang et al. 2013
AllerTOP 1.0	-www.pharmfac.net/allertop/	• Based on physicochemical protein properties • Uses a protein sequence mining method (autocross covariance transformation of protein sequences into uniform equal-length vectors).The proteins are classified by k-nearest neighbour algorithm (kNN, k = 3) based on training set containing 2210 known allergens from different species and 2210 non-allergens from the same species • Hosted by Sofia University, Bulgaria	Dimitrov et al. 2013
SDAP	http://fermi.utmb.edu/SDAP/	• Investigation of the cross-reactivity between known allergens and in predicting the IgE-binding potential of food proteins • 3-D searches • Possibility to retrieve information related to an allergen from the most common protein sequence and structure databases (SwissProt, PIR, NCBI, PDB), to find sequence and structural neighbours for an allergen, and to search for the presence of an epitope from the whole collection of allergens • Various computational tools that can assist structural biology studies related to allergens • Hosted by the University of Texas, USA	

Table 14.1 (Continued)

Web tool	Web tool access address	Advantages of method	Reference
AlgPred	www.imtech.res.in/raghava/algpred/	• Allows prediction of allergen (and its position) based on similarity with known IgE epitopes • Uses several tools (SVM, MEM/MAST, BLASTBLAST, 2890 allergen-representative peptides) and combined approaches • Hosted by the CSIR-Institute of Microbial Technology, India	Saha et al. 2005
BIOPEP	www.uwm.edu.pl/biochemia	• Contains data on allergenic proteins including names, sequence, sequences of experimental/predicted epitopes • Includes AllFam allergen family and epitopes • Hosted by the University of Warmia and Mazury, Poland	Minkiewicz et al. 2015 Dziuba et al. 2013
Pôle Bioinformatique Lyonnais (PBIL)	http://pbil.univ-lyon1.fr/	• Presents information concerning peptide sequence bioactivities on predicted and known allergenic proteins	Perriére et al. 2003
FLAPs	ulfh@slv.se	• Structure prediction of proteins • Filter-length adjusted allergen peptides (FLAPSs) database	Björklund et al. 2005
BepiPred 2.0	www.cbs.dtu.dk/services/BepiPred/	• Predicts the location of linear B-cell epitopes using both a hidden Markov model and a propensity scale method • Hosted by the Technical University of Denmark	Jespersen et al. 2017
ABCpred	http://omictools.com/abcpred-s6519.html	• Predicts B-cell epitopes in an antigen sequence, using artificial neural network • Is able to predict epitopes with 65.93% accuracy using recurrent neural network • Hosted by the CSIR-Institute of Microbial Technology, India	Saha et al. 2005
Bpredictor	https://code.google.com/p/my-project-bpredictor/	• Prediction of conformational B-cell epitopes from 3-D structures by random forests with a distance-based feature • Limited update: last update in 2011	Zhang et al. 2011
Epitopia	http://epitopia.tau.ac.il/	• Detection of immunogenic regions in protein structures or sequences (PDB and FASTA) • Machine learning scheme (i.e. naive Bayes classifier) to rank individual amino acids in the protein, according to their potential for eliciting a humoural immune response • Identifies B-cell epitopes (physicochemical and structural-geometrical properties) • Hosted by Tel Aviv University, Israel	Rubinstein et al. 2009

ACS and NACS proteins was required to clearly demonstrate this pattern, since the latter group did not exhibit appreciable protease resistance in an *in silico* digestion model.

14.3.1 Plant Food Allergens

Certain nutritious proteins from wholesome foods can act as harmful allergens, and sometimes deadly poisons, to people who possess an allergy to them (Mills and Shewry 2004). Plants are major sources of allergens which elicit an allergenic response by IgE-mediated mechanisms. Plant protein allergens can result in responses such as pruritus and swelling of the lips, tongue, and soft palate, mild laryngeal symptoms as a sensation of tightness, itching, cough, gastrointestinal symptoms, rhinitis, asthma, cutaneous reactions, and more severe systemic anaphylaxis (Sinha et al. 2014).

Allergenicity of plant protein is due to the structural motifs present on the allergens which act as the allergenic determinants or the epitopes that are responsible for binding to B- and T-cells. B-cell epitopes are mostly discontinuous motifs forming conformational epitopes whereas T-cell epitopes are linear and continuous (Sinha et al. 2014). Plant pathogenesis-related (PR) proteins are generally induced by various types of pathogens such as viruses, bacteria, and fungi (Sinha et al. 2014). Their size, stability, and protease resistance, hydrolytic and membrane-permeabilising ability make them ideal candidates to elicit an allergenic response. A review of PR causing plant protein allergens concluded that there are few similarities among the different families of PR proteins in terms of their sequence identities or structures but similarity in the amino acid sequences of PR proteins from diverse plants within the same family can result in cross-reactivity (Sinha et al. 2014).

14.3.2 Milk Protein Allergic Reactions

Cow milk protein allergy (CMPA) is an adverse immune response to dietary antigen in cow milk (McWilliams and Collins 2014; Rangel et al. 2016). Caseins, which make up approximately 80% of cow milk proteins, are the primary causative agents. There are four caseins found in cow milk: the alpha S1 and S2 groups, beta, and kappa caseins. It is thought that binding of beta-casomorphins with the A1 allele of beta-casein can cause allergies and this has led to much research into A2 type milk production and products. CMPA usually affects children and is not prevalent in adults (Rangel et al. 2016). Predisposing factors for development of CMPA in newborns include neonatal antecedents that alter the formation of intestinal flora, prematurity, antibiotic therapy, and a history of atopy in first-degree relatives (Rangel et al. 2016). Breastfeeding a newborn seems to prevent the development of this allergic response. However, it is not just caseins that are potential protein allergens. In fact, all proteins found in milk are potential allergens. Table 14.2 shows the sensitisation percentage of protein fractions.

Beta-casein is the second most abundant protein in milk and is composed of 209 amino acids divided into 12 variants: A1, A2, A3, B, C, D, E, F, H1, H2, I, and G, with A1 and A2 variants commonly found in cow milk, and which differ from each other by one nucleotide at position 67 of the chain where A1 type beta-casein contains a histidine and A1 type beta-casein contains a proline residue (Vercesi Filho 2011). This nucleotide change results in these beta-caseins being digested differently where A2 beta-casein is not hydrolysed by the gut enzymes, resulting in the production of beta-casomorphin-9

Table 14.2 Sensitisation percentage of protein fractions.

Protein fraction	Percentage of sensitive individuals
Beta-lactoglobulin	66–72
Casein	43–60
Alpha-lactalbumin	41–53
Bovine serum globulin	27
Bovine serum albumin	18

Source: Griesmeier et al. (2009).

(BMC-9). Digestion of beta-casein A1 results in the production of beta-casomorphin-7 (BMC-7) which is an opioid peptide (Sodhi et al. 2012). BMC-7 has been linked to the development of allergic reactions but also to diseases including diabetes type 1 (Thorsdottir et al. 2000) and even sudden infant death syndrome and autism, although there is some controversy surrounding these studies (Woodford 2008; Rangel et al. 2016), with some researchers claiming that the A2 allele has no connection with these health problems (Kaminski et al. 2007). Dairy breeds including Norwegian Red cows (Nilsen et al. 2009) and Jersey and Guernesy breeds have a favourable association with A2 allele which is also associated favourably with the production of BCM-9 but not BCM-7 and is considered non-allergenic (Ivanciuc et al. 2009).

14.3.3 Shellfish Allergy

In shellfish, including crustaceans and molluscs, the protein tropomyosin is recognised as the major allergen provoking ingestion-related allergic reactions (Rahman et al. 2009). This protein is 281 amino acids in length with a molecular weight ranging between 38 and 41 kDa with a highly conserved amino acid sequence which is thought to be responsible for its identification as a pan-allergen for cross-reactivity between crustaceans, insects, arachnids, and different mollusc classes. Other shellfish allergens include arginine kinase, which is 40 kDa in size and is novel to shrimp (Yu et al. 2003). Sarcoplasmic reticulum calcium binding protein (SCP) is an invertebrate calcium buffering protein and an immune-reactive protein found in black tiger shrimp (Shiomi et al. 2008).

14.3.4 Fish Allergy

Atlantic cod was first used for studying fish allergens and a 12 kDa glycoprotein called Gad c1, which is parvalbumin and has a higher natural abundance in the lower invertebrates, is the major clinical cross-reactive fish allergen. It contains heat-resistant linear epitopes (Beale et al. 2009). Furthermore, collagen and gelatine isolated from fish skin and muscle tissues and fish hormones such as vitellogenin, which is found in caviar, have been reported (Lopata and Lehrer 2009).

14.3.5 Peanut Allergy

Peanut allergy is a severe food-related allergic reaction that often presents early in life, and affected individuals generally do not outgrow it. In highly sensitised people, trace quantities can induce an allergic reaction. Allergic reaction provoked by peanuts is strictly an IgE-mediated type 1 hypersensitivity reaction and seven peanut allergy-causing proteins have been identified to date (Al-Muhsen et al. 2003).

14.4 Protein Processing and Allergy

The impact of protein processing, including heat and non-heat treatment methods, on the allergenic and antigenic integrity (IgE binding and IgG binding, respectively) and allergenicity of food proteins was reviewed by Verhoeckx et al. (2015). The overall conclusion drawn from this work was that processing may influence, but does not abolish, the allergenic potential of proteins. Indeed, it was concluded that only enzymatic and acid hydrolysis or microbial fermentation have the potential to reduce allergenic integrity and allergenicity to such an extent that reactions will not be elicited.

Food processing often changes the secondary and tertiary structure of proteins and can help to reduce, in some instances, the response of the immune system by reducing intramolecular disulfide bonds of the proteins and epitope accessibility through destruction or masking. However, some protein allergens remain stable after processing. For both established and new protein sources, establishing their safety and threshold levels in food products is important to consumers, food manufacturers, and public health and regulatory agencies (Okolie et al. 2018). Efforts are required to provide food composition information with declared allergenic ingredients or the presence of traces of allergens due to contamination or which are formed during food processing, present allowable limits in the products, and determine if a manufactured food is safe for consumption (Okolie et al. 2018).

Food processing can affect two aspects of the allergenic properties of proteins:

- the integrity of epitopes recognised by IgG antibodies or IgE antibodies which may influence the ability of antibodies to bind to the modified secondary or tertiary structure of the protein
- processing could induce allergic sensitisation by affecting the ability of a protein to stimulate the production of IgE antibody (Verhoeckx et al. 2015).

It is commonly assumed that protein allergens are relatively stable proteins and are resistant to proteolysis by pepsin in simulated gastric fluid. It is possible, however, that processing and also the food matrix in which the protein is placed may have a significant impact on the digestibility of a protein by altering susceptibility to gastrointestinal enzymes. This was demonstrated with egg ovalbumin (Takagi et al. 2003). Therefore, both processing applied to the protein and the digestion of the food protein in the food matrix must be taken into account when determining the potential allergenicity of a protein.

Zukiewicz-Sobczak et al. concluded that allergens may be divided into two groups: resistant and non-resistant to heat treatment. This relates to the order of amino acids in the protein segment and to allergic reactions caused by fish allergens such as those in cod, milk lactoglobulin, and egg white ovalbumin which are all resistant to heat inactivation (Zukiewicz-Sobczak et al. 2013). Heat can cause changes in the spatial structure of the protein in allergens that are not resistant to heat treatment, including fruit and vegetable allergens, and meat where sensitisation can be lessened by boiling.

14.5 Management of Protein Allergy

Management of protein allergy really relies on avoidance of known allergenic proteins in manufactured products, restaurants, during travel, and at work or school (Sicherer and Sampson 2018). Recently, research was carried out as part of the Learning Early About Peanut (LEAP) trial to develop a prevention approach to peanut allergy development. Infants aged 4–11 months who were at high risk of developing peanut allergy, due to severe eczema and egg allergy, were fed peanuts. Those who were sensitised to peanut had a 10.6% rate of peanut allergy compared to 35.3% in the avoidance group (du Toit et al. 2018; Sicherer and Sampson 2018).

The potential of allergy prevention through early introduction of allergens remains uncertain due to limited studies in this area. Other research is ongoing concerning the use of probiotics as a preventive strategy (Berni Canani et al. 2017). Administration of *Lactobacillus rhamnosus* to infants with cow milk allergy appeared to reduce risk in those with additional atopic diseases including eczema, asthma, rhinoconjunctivitis, and other allergies.

14.6 Conclusion

Nutritional allergies are associated with better diet and nutritional standards and access to different types of foods as multicultural societies, travel, and new dietary food trends affect the diversity of foods consumed. Foods tolerated in one region of the world may be considered allergenic to other populations. Indeed, allergies disproportionately affect people from industrialised/westernised societies. Furthermore, protein allergy was observed in some studies to be more prevalent in different races and ethnicities (Greenhawt et al. 2013; Fox et al. 2015) while in others they were not (Taylor-Black et al. 2014).

In summary, various genetic and environmental factors contribute to protein allergy. Much research has been carried out and several clinical trials are under way where Food and Drug Administration (FDA)-approved therapies are being used. However, we still do not understand the reason why food allergy is increasing and there is still no best diagnostic approach for food allergy or best novel therapeutic option. A personalised medicine approach to both these points is probably required but remains elusive (Sicherer and Sampson 2018).

References

Al-Muhsen, S., Clarke, A.E., and Kagan, R.S. (2003). Peanut allergy: an overview. *Canadian Medical Association Journal* 13 (168): 1279–1285.

Bannon, G.A. (2004). What makes a food protein an allergen? *Current Allergy and Asthma Reports* 4 (1): 43–46.

Beale, J.E., Jeebhay, M.F., and Lopata, A.L. (2009). Characterisation of purified paralbumin from five fish species and nucleotide sequencing of this major allergen from Pacific pilchard, Sardinops sagax. *Molecular Immunology* 46 (15): 2985–2993.

Berni Canani, R., di Costanzo, M., Bedogni, G., Amoroso, A., Cosenza, L. and di Scala, C. 2017. Extensively hydrolysed casein formula containing *Lactobacillus rhamnosus* GG reduces the occurrence of other allergic manifestations in children with cow's milk allergy: 3 year randomised controlled trial. *Journal of Allergy Clinical Immunology*, 139: 1906–1913.

Björklund, A.K., Soeria-Atmadja, D., Zorzet, A. et al. (2005). Supervised identification of allergen-representative peptides for in silico detection of potentially allergenic proteins. *Bioinformatics* 21 (1): 39–50.

Commission Directive 2006/142/EC (2006). Amending Annex IIIa of Directive 2000/13/EC of the European Parliament and of the Council listing the ingredients which must under all circumstances appear on the labelling of foodstuffs. *Official Journal of the European Union* 110–111.

Dimitrov, I., Flower, D.R., and Doytchinova, I. (2013). AllerTOP – a server for *in silico* prediction of allergens. *BMC Bioinformatics* 14 (6): S4.

Du Toit, G., Sampson, H.A., Plaut, M. et al. (2018). Food allergy: update on prevention and tolerance. *Journal of Allergy and Clinical Immunology* 141: 30–40.

Dziuba, M., Minkiewicz, P., and Dąbek, M. (2013). Peptides, specific proteolysis products as molecular markers of allergenic proteins – *in silico* studies. *Acta Scientiarum Polonorum Technologia Alimentaria* 12: 101–112.

Fox, A.T., Kaymakcalan, H., Perkin, M. et al. (2015). Changes in peanut allergy prevalence in different ethnic groups in 2 time periods. *Journal of Allergy and Clinical Immunology* 135: 580–582.

Goodman, R.E., Vieths, S., Sampson, H.A. et al. (2008). Allergenicity assessment of genetically modified crops – what makes sense? *Nature Biotechnology* 26: 73–81.

Greenhawt, M., Weiss, C., Conte, M.L. et al. (2013). Racial and ethnic disparity in food allergy in the United States: a systematic review. *Journal of Allergy and Clinical Immunology* 1: 378–386.

Griesmeier, U., Vazquez-Cortes, S., Bublin, M. et al. (2009). Expression levels of parvalbumins determine allergenicity of fish species. *Allergy* 65: 2.

Hayes, M., Rougé, P., Barre, A. et al. (2015). In silico tools for exploring potential human allergy to proteins. *Drug Discovery Today: Disease Models* 17–18: 3–11.

Hefle, S.L. (2002). Bioinformatic methods for allergenicity assessment using a comprehensive allergen database. *International Archives of Allergy and Immunology* 128: 280–291.

Hefle, S.L., Nordlee, J.A., and Taylor, S.L. (1996). Allergenic foods. *Critical Reviews in Food Science and Nutrition* 36: S69–S89.

Ivanciuc, O., Garcia, T., Torres, M. et al. (2009). Characteristic motifs for families of allergenic proteins. *Molecular Immunology* 46 (4): 559–568.

Jespersen, M.C., Peters, B., Nielsen, M., and Marcatili, P. (2017). BepiPred-2.0: improving sequence-based B-cell epitope prediction using conformational epitopes. *Nucleic Acids Research* 3 (45): W25–W29.

Jiang, B., Qu, H., Hu, Y. et al. (2007). Computational analysis of the relationship between allergenicity and digestibility of allergenic proteins in simulated gastric fluid. *BMC Bioinformatics* 8: 375.

Kaminski, S., Cieslinska, A., and Kostyra, E. (2007). Polymorphism of bovine beta-casein and its potential effect on human health. *Journal of Applied Genetics* 48 (3): 189–198.

Lopata, A.L. and Lehrer, S.B. (2009). New insights into seafood allergy. *Current Opinion in Allergy and Clinical Immunology* 9 (3): 270–277.

McWilliams, L.M. and Collins, A.D. (2014). Cow's milk allergy. In: *Encyclopedia of Medical Immunology* (ed. I.R. Mackay, N.R. Rose, D.K. Ledford and R.F. Lockey). New York: Springer.

Mills, E.N.C. and Shewry, P.R. (2004). *Plant Food Allergens*, 42–56. Oxford: Blackwell Science.

Minkiewicz, P., Sokolowska, J., and Darewicz, M. (2015). The occurrence of sequence identical with epitopes from the allergen Pen a 1.0102 among food and non-food proteins. *Polish Journal of Food and Nutrition Sciences* 65 (1): 21–29.

Nilsen, H., Olsen, H.G., Hayes, B. et al. (2009). Casein haplotypes and their associations with milk production traits in Norwegian Red cattle. *Genetics, Selection, Evolution* 41 (24): 1–12.

Okolie, C., Aryee, A. and Udenigwe, C. 2018. Detection and deactivation of allergens in food. In: Yada, R. Y. (ed.), *Proteins in Food Processing*, 2nd edn. Woodhead/Elsevier, Cambridge.

Perriére, G., Combet, C., Penel, S. et al. (2003). Integrated databanks access and sequence/structure analysis services at the PBIL. *Nucleic Acids Research* 31 (13): 3393–3394.

Perry, T.T. and Pesek, R.D. (2013). Clinical manifestations of food allergy. *Pediatric Annals* 42: 96–101.

Rahman, A. M., Helleur, R. J., Jeebhay, M. F. and Lopata, A. L. 2009. Characterization of seafood proteins causing allergic diseases. Open access article. doi: 10.5772/25316.

Rangel, A.H., Sales, D.C., Urbano, A.A. et al. (2016). Lactose intolerance and cow's milk protein allergy. *Food Science and Technology* 36 (2): 179–187.

Rubinstein, N.D., Mayrose, I., Martz, E., and Pupko, T. (2009). Epitopia: a web-server for predicting B-cell epitopes. *BMC Bioinformatics* 10: 287–292.

Saha, G., Bhasin, M., and Raghawa, S.P.S. (2005). Bcipep: a database of B-cell epitopes. *BMC Genomics* 6: 1–7.

Sampson, H.A. (2004). Update on food allergy. *Journal of Allergy and Clinical Immunology* 113 (5): 805–819.

Shiomi, K., Sato, Y., Hamamoto, S. et al. (2008). Sarcoplasmic calcium-binding protein: identification as a new allergen of the black tiger shrimp Penaeus monodon. *International Archives of Allergy and Immunology* 146 (2): 91–98.

Sinha, M., Singh, R. P., Kushwaha, G. S., et al.. 2014. Current overview of allergens of plant pathogenesis related protein families. *Scientific World Journal*, article ID 543195, 1–19.

Sicherer, S.H. and Sampson, H.A. (2014). Food allergy: epidemiology, pathogenesis, diagnosis and treatment. *Journal of Allergy and Clinical Immunology* 133: 291–307.

Sicherer, S.H. and Sampson, H.A. (2018). Food allergy: a review and update on epidemiology, pathogenesis, diagnosis, prevention, and management. *Journal of Allergy and Clinical Immunology* 141 (1): 41–58.

Sodhi, M., Mukesh, M., Kishore, A. et al. (2012). Screening of taurine and crossbred breeding bulls for A1/A2 variants of beta casein gene. *Indian Journal of Animal Sciences* 82 (1): 2–9.

Takagi, K., Teshima, R., Okunuki, H., and Sawada, J.I. (2003). Comparative study of *in vitro* digestibility of food proteins and effect of pre-heating on the digestion. *Biological and Pharmaceutical Bulletin* 26: 969–973.

Taylor-Black, S.A., Mehta, H., Weiderpass, E. et al. (2014). Prevalence of food allergy in New York City school children. *Annals of Allergy, Asthma and Immunology* 112: 554–556.

Thorsdottir, I., Birgisdottir, B.E., Johannsdottir, I.M. et al. (2000). Different beta-casein fractions in Icelandic versus Scandinavian cow's milk may influence diabetogenicity of cow's milk in infancy and explain low incidence of insulin dependent diabetes mellitus in Iceland. *Pediatrics* 106 (4): 719–724.

Valenta, R., Hochwallner, H., Linhart, B., and Pahr, S. (2015). Food allergies: the basics. *Gastroenterology* 148 (6): 1120.e4–1131.e4.

Vercesi Filho, A. and E. (2011). Identificacao de alelos A1 e A2 para o gene da beta-caseins na raca Gir Leiteiro. *Pesquisa e Tecnologia* 8 (11): 91–96.

Verhoeckx, K.C.M., Vissers, Y.M., Baumert, J.L. et al. (2015). Food processing and allergenicity. *Food and Chemical Toxicology* 80: 223–240.

Wang, J., Zhang, P., and Li, J. (2013). PREAL: predicition of allergenic protein by maximum reference minimum redundancy (mRMR) feature selection. *BMC Systems Biology* 7: S9–S15.

Woodford, K. 2008. A1 beta-casein, type 1 diabetes and links to other modern illnesses. International Diabetes Federation Western Pacific Congress, Wellington, New Zealand.

Young, E., Stoneham, M.D., Petruckevitch, A. et al. (1994). A population study of food intolerance. *Lancet* 343 (8906): 1127–1130.

Yu, C.J., Lin, Y.F., Chiang, B.L., and Chow, L.P. (2003). Proteomics and immunological analysis of a novel shrimp allergen, Pen m 2. *Journal of Immunology* 170: 445–453.

Zhang, W., Xiong, Y., Zhao, M. et al. (2011). Prediction of conformational B-cell epitopes from SD structures by random forests with a distance-based feature. *BMC Bioinformatics* 17 (12): 341–345.

Zukiewicz-Sobczak, W., Wroblewska, P., Adamczuk, P., and Kopczynski, P. (2013). Causes, symptoms and prevention of food allergy. *Postepy Dermatologii I Alergologii* 2: 113–116.

15

Industrial Processing of Proteins

Maria Hayes

Teagasc Food Research Centre, Food BioSciences Department, Dublin, Ireland

15.1 Introduction

Protein ingredients include mammalian and fish-derived gelatine and collagen, dairy proteins and plant-derived proteins – such as those from wheat, soy, rice, and pea – and more recently protein from cultivated algae. Cutting-edge processes for separation and purification of specific protein components from their original native sources have been the driver behind many of these ingredients. Regardless of their source, all proteins are composed of a sequence of amino acids and the order of the amino acids within the protein sequence and their ratio determines a protein's physical and biological properties, including molecular size (shape) and charge, solubility, and isoelectric point (IEP). A protein's IEP is the pH at which the molecule's charge is neutral and is generally no longer soluble in water-based solutions.

Protein extraction methods include enzymatic hydrolysis, physical processes, and chemical extraction and novel methods such as ultrasound-assisted extraction (UAE), pulsed electric field (PEF), and microwave-assisted extraction (MAE). Existing protein enrichment methods used in the dairy industry and the potential of these methods to generate high-value ingredients such as bioactive peptides and functional ingredients are discussed.

15.2 Processing of Dairy Proteins

Use of dairy ingredients in different food applications can be tailored by altering the structure of the proteins. Processes that may be used for this include physical modifications using heat treatment, acidification, addition of mineral salts, homogenization, and shear; enzymatic modifications such as renneting, use of enzymes and enzymatic or acid hydrolysis, and transglutamination; and chemical modifications using chemical agents and the Maillard reaction (Augustin and Udabage 2007a). Novel processing technologies such as high pressure and ultrasound can also be applied to dairy proteins. Using dairy ingredients for the delivery of nutrients and bioactive components, while

Novel Proteins for Food, Pharmaceuticals, and Agriculture: Sources, Applications, and Advances, First Edition. Edited by Maria Hayes.

maintaining physical functionality, presents sensory, bioactivity, and processing challenges and it is important to examine the various stresses for further differentiation of milk products and for the delivery of ingredients with consistent quality for target applications (Augustin and Udabage 2007a).

15.3 Membrane Technologies

Membrane technology has become an important separation technology in recent decades. Features associated with this technology make it suitable for industrial and, in particular, food applications: no chemicals required, relatively low use of energy, low processing costs, and easy scale-up (Muro et al. 2013).

A membrane is a selective barrier separating two adjacent phases that allows the selective transport of particular species between the two phases in a specific or non-specific manner. Synthetic membranes can be made from organic materials (polymers) and inorganic materials including ceramic, metals, and their associated alloys. Membrane chemistry determines binding affinity for solutes and particles and biocompatibility. Membrane technology can be used to process new ingredients from food protein sources and is considered a green technology (Dutournié et al. 2017).

Ceramic membranes are commonly used in dairy production and fruit juice clarification. Polymer membranes currently dominate the wastewater treatment sector. Due to their lower purchase price, systems using polymer membranes are widespread in the wastewater filtration market. Polymer membranes are cheaper than ceramic membranes but require more frequent filter replacements. Ceramics can operate at pH values from 0 to 14 but polymer membranes can only withstand a much narrower pH range; they can be tailored to resist neutral, acid or basic pH values, but generally not all three ranges with the same material. Ceramic membranes are also better for high-temperature applications, can be sterilised and are stronger and more rigid.

Ceramic membranes are made from inorganic materials including alumina, zirconia, titania, and silica and they exhibit high resistance to aggressive media such as acids, alkalis, and strong solvents (Galanakis et al. 2016). They are usually thermally stable and although the cost of production is higher than with polymer membranes, they are ecologically friendly, durable, and have a longer life-span.

15.4 Pressure-Driven Membrane Processes

Pressure-driven membrane processes include nanofiltration (NF), microfiltration (MF), ultrafiltration (UF), and reverse osmosis (RO) (Dutournié et al. 2017). Transmembrane pressure is the main driving force for separation of these processes, that is, the pressure difference between the retentate side and permeate side. The membrane pore size generally decreases from microfiltration to reverse osmosis. The pore size, charge of the molecules in the solute, and the membrane composition and affinity of the solute for this material are important factors that must be taken into account.

Microfiltration can be used for separation of suspended particles and microbes from the solute. UF can be used to recover soluble proteins and peptides and NF can be used as a demineralisation strategy and for concentration of protein products. Pressures applied for MF rarely exceed 3 bar while those used in UF range from 3 to 7 bar and up to 30 bar for NF and RO procedures. In milk filtration processes, microfiltration (1.4–1.8 m) is used to recover somatic cells, fat globules and bacteria and spores. Microfiltration (0.1–0.2 m) is also used to recover casein micelles while UF is used to recover whey proteins free from bacteria, NF is used to recover non-protein nitrogen (NPN), lactose and polyvalent minerals, and RO is used to recover monovalent minerals.

Membrane filtration can be used to recover retentate fractions enriched with micellar caseins and sterile permeate containing native whey proteins. UF can be used to produce a dairy product called milk protein isolate that is similar to the original milk in terms of whey and casein proteins but has a much lower fat content. Milk protein isolate finds applications in food products that require superior functionality, including infant foods, meat products, soups and sauces, and confectionery and baked products.

15.5 Ultrafiltration and Diafiltration in the Production of Pharmaceutical Proteins

Ultrafiltration and diafiltration (UF/DF) procedures are often used to produce pharmaceutical protein formulations at high concentrations. The formulation buffer salt and other excipients such as cryoprotectants required for achieving the desired pharmaceutical protein formulations are often introduced during the UF/DF process. One of the challenges associated with delivering pharmaceutical protein formulations at high concentrations is that, when protein is concentrated during the UF operation, significant volume exclusion and charge–charge interaction between the proteins and the charged buffer components can result in preferential distribution of charged buffer components into the permeate. The latter is the so-called 'Donnan' effect which causes a substantially lower level of charged buffer components in the product containing retentate than that in the DF buffer. To compensate for this loss, a DF buffer with a somewhat higher level of charged buffer components than the desired concentration in the final composition needs to be used in order to meet the desired formulation composition. Thus, the Donnan effect causes an increase in the *ab initio* calculated amount of buffer components used in protein-based pharmaceutical products (Wang et al. 2012).

15.6 Extraction of Proteins from Algae

Protein extraction methods used on algae to date have been limited to laboratory environments for the most part and have not been applied for commercial use. Conventional, mechanical, and enzymatic methods for protein extraction may also affect the integrity of extracted algal protein due to the release of proteases from cytosolic vacuoles. Furthermore, these methods are laborious and time consuming. Improved methods of cell disruption and extraction are therefore required. Pretreatment with cell disruption techniques aid the breakdown of the tough algal cell wall, increasing the availability of proteins and other high-value components for later extraction. Some examples of novel protein extraction methods include Ultrasound assisted extraction (UAE), Pulse electric field (PEF), and Microwave assisted extraction (MAE) (Bleakley and Hayes 2017).

Seaweed protein digestibility in its raw, unprocessed form is poor and for this reason emphasis has been placed on developing improved methods for algal protein extraction in order to improve bioavailability. Algal proteins are conventionally extracted by aqueous, acidic, and alkaline methods, followed by several rounds of centrifugation and recovery using techniques such as UF, precipitation, or chromatography. Chemical extraction methods, such as two-phase acid and alkali treatments, have been especially efficient for extracting proteins from the brown seaweed *Ascophyllum nodosum* as well as green species including *Ulva* (Bleakley and Hayes 2017). The availability of protein can be substantially hindered by the high viscosity and anionic cell wall polysaccharides, including alginates in brown seaweed and carrageenans in red seaweed. Cell disruption methods and selected chemical reagents are therefore used to improve the efficiency of algal protein extraction. Methods include mechanical grinding, osmotic shock, ultrasonic treatment, and polysaccharidase-aided hydrolysis (Bleakley and Hayes 2017).

15.7 Enzyme Use for Protein Extraction from Algae

Seaweed is rich in several types of polysaccharides, including cellulose, galactans, xylans, fucoidan, laminarin, alginates, carrageenans, and floridean starch, which can reduce the availability of algal proteins and decrease protein extraction efficiency (Barbarino and Lourenco 2005). Enzymes such as polysaccharidases can be used to disrupt the cell wall and increase protein extraction. Polysaccharidases commonly used include kappa-carrageenase, beta-agarase, xylanase, and cellulase and these have been applied to the seaweeds *Chondrus crispus, Gracilaria verrucosa and Palmaria palmata* during protein extractions previously (Fleurence et al. 1995).

15.8 Novel Extraction Methods

Up-scaled production of algal proteins is limited. Conventional mechanical and enzymatic methods for protein extraction may also affect the integrity of extracted algal proteins due to the release of proteases from cytosolic vacuoles (Bleakley and Hayes 2017). Novel protein extraction methods include UAE, PEF, and MAE (Kadam et al. 2013).

15.8.1 Pulsed Electric Field

Pulsed electric field involves applying high electric currents in order to perforate a cell wall or cell membrane, causing reversible or irreversible electroporation. It is a fast and green technology that may be used for food protection (inactivation of microorganisms) by irreversible electroporation and aiding the release of intracellular contents of plant cells (Postma et al. 2016). Previously, protein yield from the microalgae *Chlorella* sp. and *Spirulina* sp. was reported to increase by 27% and 13%, respectively, following PEF treatment at $15\,kV\,cm^{-1}$ and $100\,kJ\,kg^{-1}$ (Töpfl 2006). PEF has not been used on seaweeds to date but holds potential as a laboratory method for cell disruption in seaweeds.

15.8.2 Microwave-Assisted Extraction

Microwave-assisted extraction includes heating a material, causing moisture to evaporate, and subsequent creation of bubbles under high pressure which can then rupture to disrupt cell contents. It has been used to extract fucoidan from marine algae (Mussatto 2015). The method involves a closed MAE system, ultrapure water as extraction solvent, and suitable conditions of time, pressure, and algal biomass/water ratio (Mussatto 2015). It has not been used to date to extract proteins from seaweeds but could be useful in disrupting the cell wall and subsequent release of proteins from seaweeds.

15.8.3 Ultrasound Treatment

Ultrasound may be used as a pretreatment to enhance protein extraction from seaweeds. Acoustic cavitation causes the formation, growth, and implosion of bubbles which can cause cell rupture. Ultrasound pretreatment has been used to increase protein extraction of *Ascophyllum nodosum* with acid and alkaline treatment alone by 540% and 27%, respectively, as well as reducing processing time from 60 to 10 minutes (Kadam et al. 2017). It has also been applied successfully to microalgae for the extraction of protein, as reported by Bleakley and Hayes (2017).

15.8.4 Application of Membrane Technologies to Macroalgae

Membrane technologies can be applied to seaweed as part of a cascading biorefinery process to maximise valorisation of all components. A combination of membrane technologies can be employed to isolate algal proteins with the same principles of molecular weight cut-offs (MWCOs) used in the dairy industry, as discussed earlier. UF could then be used to isolate proteins and other macromolecules between 1 and 200 kDa, similar to the way it is used in the dairy industry to generate enriched fractions less than 10, 5, 3, and 1 kDa. NF could then be used to remove monovalent salts to minimise osmotic pressure, followed by RO to reduce fluid volume (Kumar et al. 2013; Bleakley and Hayes 2017).

Algal proteins including lectins and phycobiliproteins are often exploited for several industrial applications. Lectins are most commonly extracted from macroalgal sources, while phycobiliproteins are typically isolated from microalgae (Hayes 2013). Phycobiliproteins, especially phycoerythrin, can constitute a significant proportion of the overall protein content in red algae, with levels of 1.2% (total dw) reported for *P. palmata* (Wang et al. 2010). Lectins are found at similar levels and a yield of 1% lectins was obtained from *Eucheuma serra* (Rhodophyta) (Bleakley and Hayes 2017).

15.8.5 Application of Membrane Technologies in the Dairy Industry

In the dairy industry, membrane filtration systems have been used to recover organic compounds, suspended solids, colour, and turbidity from anaerobically treated dairy wastewater (Zielinska and Galik 2017). Combinations of direct MF, UF, MF and UF, and a combination of UF with coagulation using two conventional coagulants were investigated. The installation with ceramic membranes was operated at a pressure of 0.15 MPa (MF) and 0.3 MPa (UF). Chemical oxygen demand (COD) removal was 89% ± 2% in MF and 95% ± 1% in UF and 99% in microfiltration-ultrafiltration (MF-UF). This study concluded that use of MF-UF was most appropriate for removal of pollutants/compounds and that MF-UF also resulted in 80% higher average permeate flux in this two-stage system. The most effective strategy to mitigate membrane fouling was to use MF as a pretreatment before# UF.

15.9 Novel Proteins

15.9.1 Extraction of Proteins from Insects

Insect protein is highly nutritious and represents a notable alternative protein source. Bußler et al. (2016) recently reviewed the protein extractability and technofunctionality of insect flour fractions recovered from *Tenebrio molitor* and *Hermetia illucens.* To isolate proteins from the flours, protein solubility must first be optimised by changing the pH, ionic strength, and temperature of the extraction solvent. The processing parameters required for protein isolation from these insect species were similar to those required for plant proteins, enabling the use of traditional methods of protein processing. The proteins appeared to be highly soluble at alkaline pH values. In terms of ensuring the safety of the final insect protein products, Bußler and colleagues recommended several physical separation methods and thermal treatments as a strategy to ensure safety (Bußler et al. 2016). Several patents exist that look at the topic of protein recovery from various insects.

Protein production from insect sources can also result in the generation of the valuable by-product chitin. Chitin provides the strong exoskeleton to insects and other animals, including prawns and crabs, and is a valuable compound that has several applications in the agricultural, food, pharmaceutical, medical, and functional food industries (Hayes et al. 2008).

15.9.2 Fish Wastewater Proteins

Fish industry wastewater contains potentially valuable molecules for the food sector, including proteins, flavours, and other compounds, and recovering these may lead to significant productivity gains (Masse et al. 2008). Afonso and Bórquez (2002) looked at the performance of an integrated process based on a combination of MF and UF membranes for the recovery of proteins from the effluent of a fishmeal plant. MF was found to reduce the oil, grease, and suspended matter content and the use of a UF ceramic membrane (Carbosep M2, tubular, 15 kDa MWCO) reduced the organic load of the MF permeate and allowed the recovery of valuable raw materials, including proteins.

A conceptual process for the treatment of $10\,m^3\,h^{-1}$ of fishmeal effluent was designed. The plant generated $1\,m^3\,h^{-1}$ of concentrate containing $170\,g\,L^{-1}$ of solids and $112\,g\,L^{-1}$ of proteins. An economic assessment of the integrated MF-UF process for production of 544 tonnes per year of fishmeal (66% protein content) yielded a net present worth of 160 000$, a 17% rate of return and an eight-year return on investment, indicating the feasibility of the process for protein recovery and pollution reduction (Afonso et al. 2004). Polymer membranes are usually made from cellulose acetate, polyvinylidene fluoride, or other substances.

The wastewaters generated in fishmeal production (average flow rates of $1100\,m^3\,h^{-1}$ for a plant capacity of 100 tonnes fish/h) contain a high organic load, and therefore should not be discharged directly into the sea without effective treatment. However, these effluents contain a large amount of potentially valuable proteins, which can be concentrated by UF and recycled into the fishmeal process, improving its quality and the economic benefits from the raw material, while the treated water can be discharged into the sea or reused in the plant. Due to the high concentration of suspended matter in these effluents, MF pretreatment is required. Two types of effluent from a fishmeal plant in Talcahuano, Chile, were characterised and microfiltrated with a Whatman filter No. 1. A mineral tubular membrane, Carbosep M2 (MWCO = 15 kDa, ID = 6 mm, L = 1.2 m), was used in the UF experiments. The operating conditions were optimised in total recirculation mode, and the subsequent concentration experiments were carried out at 4 bar, $4\,m\,s^{-1}$, ambient temperature and natural pH. The results show that UF reduced the organic load from the fishmeal wastewaters and allowed the recovery of valuable raw materials, including proteins (Afonso and Bórquez 2002). Although the membrane undergoes severe fouling, it can easily be cleaned using caustic washing. Separation and treatment techniques for process waters generated in the food industry are shown in Table 15.1.

15.9.3 Characterisation of Fish Processing Wastewater Proteins – Methods

Details of how wastewaters may be sampled and characterised were provided in the PIPE report (Baron et al. 2015). Herring wastewaters were sampled at three time points: March and September 2012 and September 2013. Samples were collected from four different types of process waters: refrigerated seawater (RSW), storage water (SW), processing water (PW), and salt brines (SBs). Samples were assessed for dry matter, pH and ionic strength, total proteins and polypeptide profiles, total and free amino acid content, total fatty acids, trace elements and total volatile base nitrogen and thiobarbituric reactive substances (TBARSs), and enzyme activities. The protein content of all samples was analysed using the bicinchoninic acid assay (BCA), using bovine serum

Table 15.1 Separation and treatment techniques for process water in the food industry.

Membrane method	Product	Recovery of fractions	Reference
Membrane and ultrafiltration	Fishmeal	Recovery and concentration of protein	Afonso et al. 2004
Ultrafiltration	Fish brine, meal, surimi, fish	Recovery and fractionation of protein and lipids	Afonso and Bórquez 2002
Reverse osmosis	Fishmeal	Preconcentration of solids	Matthiasson and Sivik 1978
Decanter and anaerobic treatment	Tuna	Treatment of wastewater	Afonso and Bórquez 2002
Deammonification and anaerobic carbon removal	Fish	Energy-efficient treatment of wastewater	Trautmann et al. 2011

albumin (BSA) as a standard. The polypeptide profile was evaluated using SDS-PAGE. The total free and total amino acid content measurement was performed on selected samples as described by Farvin et al. (2010). Briefly, 500 μL of each sample was subjected to microwave-assisted acid hydrolysis in sealed ampoules; the method did not allow for detection of methionine, tryptophan, and cysteine.

15.10 Conclusion

The development of infrastructure to reduce the amount of protein that ends up in the water following food processing could yield economically valuable biomolecules for potential application in human food, functional foods, pharmaceutical, and medical applications. Indeed, currently, the recovery of mammalian blood proteins is of great interest and the technology exists in this sector to recover blood in a hygienic manner so that the blood is considered food grade under European Union (EU) legislation. Recovery and storage of the blood from fish at a registered process facility under hygienic conditions would help to reduce the biological oxygen demand (BOD) and COD and environmental pollution from marine processing. Recovered proteins could be fractionated into serum albumin and a haemoglobin fraction using basic centrifugation at temperatures of 4 °C and then dried using spray-drying or freeze-drying technologies that would result in protein products suitable for use as food, pharma, and functional food ingredients. Development of blood recovery techniques from fish during processing is an area that is currently under investigation in Iceland and Norway, by companies including Skaginn 3X and Marel (personal communication with Skaginn 3X).

It is recommended that in order to reduce BOD/COD in bloodwaters prior to treatment with a dissolved air flotation (DAF) system, bloodwaters that are not collected using food-grade methods under EU legislation should be collected in bulk and stored at 4 °C and then fractionated and dried immediately using MWCO ceramic membranes and UF membranes. UF/MWCO membranes with cut-off values of 10 000 Da can be used to generate permeate fractions which can then be dried and used as protein ingredients in feed manufacture. The retentate fractions and proteins greater than 10 000 Da could be used for non-food applications such as bioenergy creation or fertiliser production.

Conversion of recovered 'waste' proteins from food processing into health beneficial bioactive peptides would also result in products suitable for use in the pet food industry as the humanisation of pets has promoted a growing animal health and welfare market trend. However, much research is required in order to make a health claim on human food products and claims must be substantiated and scientifically validated in line with European Food Safety Authority (EFSA) requirements and EU legislation.

References

Afonso, M.D. and Bórquez, R. (2002). Review of the treatment of a seafood processing wastewater and recovery of proteins therein by membrane separation processes – prospects of the ultrafiltration of wastewater for the fish meal industry. *Desalination* 142: 29–45.

Afonso, M.D., Ferrer, J., and Bórquez, R. (2004). An economic assessment of protein recovery from fish meal effluents by ultrafiltration. *Trends in Food Science and Technology* 15: 506–512.

Augustin, M.A. and Udabage, P. (2007). Influence of processing on functionality of milk and dairy proteins. *Advances in Food and Nutrition Research* 53: 1–38.

Barbarino, E. and Lourenco, S.O. (2005). An evaluation of methods for extraction and quantification of proteins from marine macro and microalgae. *Journal of Applied Phycology* 17 (5): 447–460.

Baron, C.P., Osteberg, C., Kemp, C. et al. (2015). *PIPE – Pelagic Industry Processing Effluents Innovative and Sustainable Solutions.* Oslo: Nordic Innovation Publications.

Bleakley, S. and Hayes, M. (2017). Algal proteins: extraction, application and challenges concerning production. *Foods* 26 (6): 5.

Buβler, S., Rumpold, B.A., Jander, E. et al. (2016). Recovery and techno-functionality of flours and proteins from two edible insect species: meal worm (*Tenebrio molitor*) and black soldier fly (*Hermetia illucens*) larvae. *Heliyon* 2 (12): e00218.

Dutournié, P., Limousy, L., Anquet, I.J., and Déon, S. (2017). Modification of the selectivity properties of tubular ceramic membranes after alkaline treatment. *Membranes* 7 (4): 65.

Farvin, K.H.S., Baron, C.P., Nielsen, N.S. et al. (2010). Antioxidant activity of yoghurt peptides: part 2 – characterisation of peptide fractions. *Food Chemistry* 123: 1090–1097.

Fleurence, J., Le Couer, C., Mabeau, S. et al. (1995). Comparison of different extractive procedures for proteins from the edible seaweeds *Ulva rigida* and *Ulva rotundata. Journal of Applied Phycology* 7: 577–582.

Galanakis, C.M., Castro-Munoz, R., Cassano, A., and Conidi, C. (2016). Recovery of high-added-value compounds from food waste by membrane technology. In: *Membrane Technology for Biorefining* (ed. A. Figoli, A. Cassano and A. Basile), 189–215. Cambridge: Woodhead Publishing.

Hayes, M. (2013). Biological activities of proteins and marine-derived peptides from byproducts and seaweeds. In: *Marine Proteins and Peptides: Biological Activities and Applications* (ed. S.-K. Kim). Chichester: Wiley.

Hayes, M., Carney, B., Slater, J., and Brúck, W. (2008). Mining marine shellfish wastes for bioactive molecules: chitin and chitosan: part A: extraction. *Biotechnology Journal* 3 (7): 871–877.

Kadam, S.U., Tiwari, B.K., and O'Donnell, C.P. (2013). Application of novel extraction technologies for bioactives from marine algae. *Journal of Agricultural and Food Chemistry* 61 (20): 4667–4675.

Kadam, S.U., Álvarez, C., Tiwari, B.K., and O'Donnell, C.P. (2017). Extraction and characterisation of proteins from Irish brown seaweed *Ascophyllum nodosum. Food Research International* 99: 1021–1027.

Kumar, P., Sharma, N., Ranjan, R. et al. (2013). Perspective of membrane technology in the dairy industry: a review. *Asian-Australasian Journal of Animal Sciences* 26 (9): 1347–1358.

Massé, L., Massé, D.I., and Pellerin, Y. (2008). The effect of pH on the separation of manure nutrients with reverse osmosis membranes. *Journal of Membrane Sciences* 325: 914–919.

Mathiasson, E. and Sivik, B. (1978). New methods of recovering fish processing waters.(in Swedish). Livsmedelsteknik. 20 (5): 241–244.

Muro, C., Riera, F., and Fernández, A. (2013). Advancements in the fractionation of milk biopeptides by means of membrane processes. In: *Bioactive Food Peptides in Health and Disease.* Intech. doi: 10.5772/53674.

Mussatto, S.I. (2015). Microwave assisted extraction of Fucoidan from microalgae. *Methods in Molecular Biology* 1308: 151–157.

Postma, P.R., Pataro, G., Capitoli, M. et al. (2016). Selective extraction of intracellular components from the microalga *Chlorella vulgaris* by combined pulsed electric field temperature treatment. *Bioresource Technology* 203: 80–88.

Töpfl, S. 2006. Pulsed electric fields (PEF) for permeabilisation of cell membranes in food and bioprocessing – applications, process and equipment design and cost analysis. Doctoral thesis submitted to Technische Universität Berlin. doi: 10.14279/depositonce-1441

Trautmann, N., Beier, M., Phan, L.-C., and Rosenwinkel, K.-H. (2011). A new approach to energy efficient treatment of waste water produced by the fish industry in Vietnam. *Water Science and Technology* 64: 279–285.

Wang, T., Jónsdóttir, R., Kristinsson, H.G. et al. (2010). Enzyme enhanced extraction of antioxidant ingredients from red algae *Palmaria palmata*. *LWT – Food Science and Technology* 43: 1387–1393.

Wang, Z., Yu, H., Xia, F. et al. (2012). Novel GO-blended PVDF ultrafiltration membranes. *Desalination* 299: 50–54.

Zielinska, M. and Galik, M. (2017). Use of ceramic membranes in membrane filtration supported by coagulation for the treatment of dairy wastewater. *Water, Air, and Soil Pollution* 228: 173.

16

The Role of Immunoglobulins from Bovine Colostrum and Milk in Human Health Promotion

Shane Feeney[1,2]*, Sinead T. Morrin*[1]*, Lokesh Joshi*[2]*, and Rita M. Hickey*[1]

[1] *Teagasc Food Research Centre, Moorepark, Fermoy, Co. Cork, Ireland*
[2] *Advanced Glycoscience Research Cluster, National Centre for Biomedical Engineering Science, National University of Ireland Galway, Galway, Ireland*

16.1 Introduction

Immunoglobulins are antibodies that belong to a family of globular proteins and comprise a group of glycoproteins present in the serum and tissue fluids of all mammals. There are five classes of immunoglobulins: IgG, IgA, IgM, IgD, and IgE. Each class is characterised by variations in size, electrical charges, amino acid content, and carbohydrate content (Atkinson et al. 2006). The monomeric basic structure of all Igs appears similar, being composed of four subunits of polypeptides, forming a Y-shaped molecule (Figure 16.1). This structure consists of two identical Fab 'arms' and an Fc stem. The Fab regions contain the antigen-binding sites, while the elimination of the antigen is mediated by the Fc through the activation of host immune mechanisms (Recio et al. 2009). IgG occurs as monomeric Ig (~160 kDa), while IgA (~370 kDa) and IgM (~1000 kDa) form polymeric Ig. IgG can also be present in two subclasses, called IgG1 and IgG2, IgG1 being the most abundant Ig isotype in bovine milk and colostrum, accounting for approximately 80% of the total Ig content (Atkinson et al. 2006; Gapper et al. 2007; Hurley and Theil 2013; Bogahawaththa et al. 2017).

All IgGs contain glycan chains at conserved positions in the constant regions of the heavy chain (Figure 16.1). In IgG, *N*-glycosylation is conserved at Asn297 (Recio et al. 2009) of the Fc region (Takimori et al. 2011). In fact, the constant regions of the Fc stem interact via their attached glycan chains (Deisenhofer 1981). Takimori et al. (2011) investigated IgG glycosylation over lactation and identified complex type *N*-glycans with core fucosylation throughout lactation. However, there was a notable decrease in the level of IgG sialylation postpartum. The glycan chains of milk Igs may protect the Ig protein from digestion by proteolytic enzymes, allowing the intact or only partially digested Ig to reach the intestine for absorption into the blood (O'Riordan et al. 2014). The glycosylation of the Fc portion of IgG is also thought to keep the heavy chains in an open conformation, contributing to the binding of the Fc to Fc-gamma receptors (Anthony and Ravetch 2010).

Novel Proteins for Food, Pharmaceuticals, and Agriculture: Sources, Applications, and Advances, First Edition. Edited by Maria Hayes.

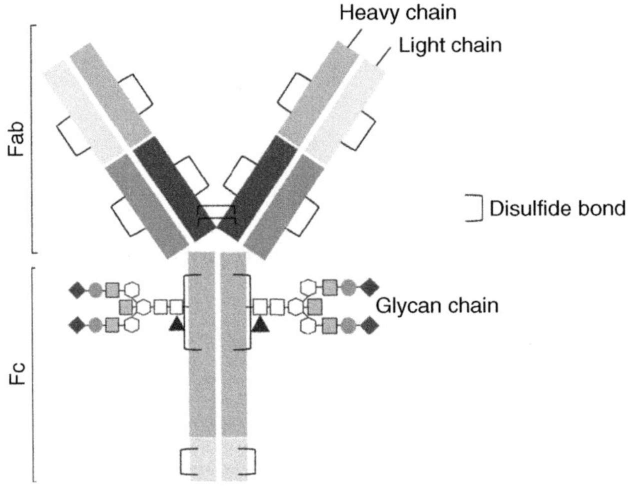

Figure 16.1 Generalised structure of an immunoglobulin.

Immunoglobulin concentrations in bovine colostrum are influenced by many factors. For instance, delayed milking after parturition leads to decreases in Ig concentrations due to dilution effects of increasing colostrum volumes (Pritchett et al. 1991; Moore et al. 2009). Older cows are reported to have higher colostrum Ig concentrations, due to the longer exposure time to farm-specific pathogens (Meganck et al. 2012). Some studies have shown that season has no influence on colostrum Ig concentrations, except in periods of serious heat stress (Nardone et al. 1997; Morin et al. 2010; Eihvalde et al. 2012). Shortening of the dry period to 28 days was shown not to decrease colostrum Ig concentrations compared to cows with a dry period of 56 days (Annen et al. 2004; Watters et al. 2008; Eihvalde et al. 2012). Recently, Verweij et al. (2014) found that continuous milking resulted in a decrease in colostrum Ig concentrations when compared with colostrum samples of cows that underwent a traditional dry period of at least 42 days. Table 16.1 provides the concentrations of different Ig classes in bovine colostrum and milk.

The biological function of antibodies is to bind antigens and thereby facilitate their removal from the body. Each antibody recognises an epitope, a small portion of the

Table 16.1 Immunoglobulin composition in bovine colostrum and milk.

Immunoglobulin	Molecular mass (kDa)	Colostrum (g/L)	Milk (g/L)
IgG1	146–163	15–180	0.35
IgG2	146–154	1–3	0.02–0.12
IgM	900	5	0.04–1.0
IgA	385–430	3.5	0.05–0.14

Source: Adapted from Pagnoncelli et al. (2017).

surface of a large molecule, such as a polysaccharide or protein (Borghesi et al. 2014). Therefore, the main role of milk Igs is to agglutinate bacteria, neutralise toxins, inactivate viruses, and provide an environment favourable for the growth of beneficial bacteria (McWilliams 1986; Kvistgaard et al. 2004; Bojsen et al. 2007). In other words, milk Igs give the offspring an immunological protection against infections in the gastrointestinal tract and respiratory system. Not taking into account colostrum, the total amounts of Igs in bovine milk are lower than in human milk, and levels are significantly lower or absent in infant formula. This could partly account for the increased incidence of infections in formula-fed infants compared with breast-fed infants (Dewey et al. 1995), which supports the concept of supplementing formulas with Igs. High concentrations of antibodies against a particular pathogen may be achieved by immunising cows with the pathogen or its antigens.

Through a strategy of vaccination and established purification procedures in the dairy industry, large-scale manufacture of various IgGs is possible (Kramski et al. 2012). Such hyperimmune milk products have been tested in the prophylaxis and treatment of various infections (Hurley and Theil 2013). This chapter reviews knowledge on the isolation and applications of bovine milk immunoglobulins.

16.2 Digestion of Immunoglobulins

Milk Igs are degraded in the human GI tract into $F(ab')_2$, Fab, and Fc fragments by the intestinal proteolytic enzymes pepsin, trypsin, chymotrypsin, carboxypeptidase, and elastase (Marnila and Korhonen 2002). Susceptibility to digestion varies with Ig species and subtype, with bovine IgG1, for example, being more readily digested by pepsin compared to IgG2 (Butler and Kennedy 1978). Even after digestion, Igs can retain bioactivity as long as their Fab dimer and monomer fragments are not denatured (Roos et al. 1995). The reader is referred to a very comprehensive review of the human studies which detail the survivability of orally administered Ig preparations (Jasion and Burnett 2015). Twelve of the fifteen studies the authors reviewed clearly demonstrated that orally administered Ig (particularly IgGs), many from bovine colostrum and milk, can survive gastric exposure and resist proteolytic digestion in the stomach and intestinal tract. The stability of the IgG is based on the structural properties of the molecule, including the intra- and interchain disulfide bonds, the post-translationally added glycans, and the three-dimensional folded domains (Jasion and Burnett 2015).

The passage of ingested Igs across intestinal cells of the neonate can be specific or non-specific, depending on the species, and it occurs only for a certain period after birth. The gut of the suckling calf is non-selective and open to the transport of a variety of macromolecules from 12 to 36 hours after birth (Staley and Bush 1985), corresponding to the time of the highest IgG concentrations in colostrum. Although the main Ig class of bovine colostrum is IgG1, the passive transport of other Ig classes can occur, although in smaller amounts, and is also considered important for the calf. The Fc receptors on the enterocytes of the small intestine mediate selective binding and transcytosis of IgG until the day of weaning. Also, in the human infant, colostral Igs are absorbed in small quantities for a short period of 18–24 hours after birth (Hurley and Theil 2011; Marnila and Korhonen 2011).

16.3 Applications and Functionality of Immunoglobulins

16.3.1 Protection against Infections

The use of bovine colostral and milk immunoglobulins to provide passive immunity against diseases in other species, especially in humans, has received much interest. Here, the cow's immune system is used to produce antigen-specific antibodies that are secreted in colostrum and milk where they are then used to protect consumers against a specific disease. The cows are hyperimmunised against one or more antigens representing pathogens of bacterial or viral origin. Such products may include immunoglobulin with no alteration to its concentration or partial immunoglobulin isolation or concentration in a whey protein concentrate (Hurley and Theil 2011). An overview of the efficacy of bovine colostral and immunoglobulin preparations is presented in Table 16.2.

16.3.1.1 *Escherichia coli*

Enterotoxigenic *Escherichia coli* (ETEC) is a common bacterial pathogen associated with traveller's diarrhoea and infant bacterial diarrhoea in developing countries. Current preventive treatments are limited and not ideal as long-term solutions (Steele et al. 2013).

Passive oral administration of hyperimmune bovine milk Ig has been experimentally examined as a preventive or therapeutic treatment against ETEC infection (Tacket et al. 1988; Freedman et al. 1998). Notably, hyperimmune IgG generated against a cocktail of inactivated whole-cell ETEC and against purified colonisation factor antigen I (CFA/I) fimbriae conferred 76–100% protection against challenge with a prototype CFA/I-ETEC strain in volunteers (Tacket et al. 1988; Freedman et al. 1998; Otto et al. 2011), corroborating other lines of evidence that anticolonisation factor immunity is associated with protection against ETEC diarrhoea (Evans et al. 1982; Rao et al. 2005). In addition, Tawfeek et al. (2003) tested the efficacy of anti-ETEC hyperimmune bovine colostral (HBC) antibodies in preventing diarrhoea in infants, and found a significant reduction in incidence and duration of diarrhoea in HBC-fed infants compared with formula-fed, but not breast-fed, infants. These HBC products were the precursors of Travelan, an anti-ETEC HBC product manufactured by Immuron Ltd. and commercially available for prophylaxis of traveller's diarrhoea in the United States, Australia, and Canada (Immuron Ltd, Blackburn North, Victoria, Australia). A similar newer product, designated IMM-124E, is produced by Immuron (www.immuron.com).

Sears et al. (2017) investigated several commercial HBC products formulated to reduce the risk of ETEC-induced diarrhoea, including Travelan and IMM-124E. The products contained high levels of IgG specific for multiple ETEC antigens, including O-polysaccharide 78 and CFA/I present in the administered vaccines. Antimicrobial activity was measured *in vitro* using novel functional assays. The IgG present in the products greatly reduced ETEC motility in soft agar and promoted complement-mediated lysis *in vitro*.

Other types of *E. coli* have also been investigated with Ig products (Mietens et al. 1979; Huppertz et al. 1999; Takemura et al. 2002, 2003, 2004). An Ig-enriched colostrum product, IMMULAC™, made from the colostrum of non-immunised cows, was found to reduce the mortality of mice orally challenged with enterohaemorrhagic *E. coli* (EHEC) O157: H7 (Funatogawa et al. 2002).

Table 16.2 Efficacy of bovine immune colostrum, milk, or IgG against microbial infections.

Micro-organism	Symptoms	Material	References
E. coli (ETEC and EHEC)	Diarrhoea	HBC, Ig-enriched colostrum	Tacket et al. (1988), Freedman et al. (1998), Funatogawa et al. (2002), Tawfeek et al. (2003), Takemura et al. (2002, 2003, 2004), Otto et al. (2011), Sears et al. (2017)
Helicobacter pylori	Malnutrition and diarrhoea	HBC, milk containing anti-*H. pylori* polyclonal bovine Igs	Czinn et al. (1993), Casswall et al. (1998), Opekun et al. (1999), Marnila et al. (2003), Attallah et al. (2009), Gorrell and Robins-Browne (2009), Hu et al. (2015)
Clostridium difficile		HBC	Lyerly et al. (1991), Naaber et al. (1996), van Dissel et al. (2005), Numan et al. (2007), Mattila et al. (2008), Sponseller et al. (2014)
Shigella	Shigellosis (diarrhoea)	HBC	Tacket et al. (1992), Ashraf et al. (2001), Sears et al. (2017)
Cryptosporidium	Diarrhoea	HBC	Tzipori et al. (1986, 1987), Shield et al. (1993), Greenberg and Cello (1996), Okhuysen et al. (1998)
Streptococci	Dental caries	Colostrum from immunised cows	Michalek and McGhee (1977), Filler et al. (1991), Loimaranta et al. (1997, 1998a,b, 1999), Oho et al. (1999), Shimazaki et al. (2001), Wei et al. (2002), Ramezanalizadeh et al. (2017)
Rotavirus	Diarrhoea	HBC	Brüssow et al. (1987), Hilpert et al. (1987), Davidson et al. (1989), Turner and Kelsey (1993), Mitra et al. (1995), Sarker et al. (1998), Pant et al. (2007), Günaydın et al. (2014)
Respiratory syncytial virus	Cold-like	Bovine milk-derived IgG	den Hartog et al. (2014)
Human herpes virus	Blistering (in the mouth or on the genitals)	Immune milk	Masuzawa et al. (2016)

EHEC, enterohaemorrhagic *E. coli*; ETEC, enterotoxigenic *E. coli*; HBC, hyperimmune bovine colostrum; Ig, immunoglobulin.

16.3.1.2 *Helicobacter pylori*

Helicobacter pylori causes gastric infection in children and adults, and is especially prevalent in children in developing countries, where it is associated with chronic malnutrition and diarrhoea. Many animal studies have shown that milk containing bovine antibody against *H. pylori* reduces bacterial load, thus preventing and even eradicating *H. pylori* infection (Czinn et al. 1993; Marnila et al. 2003; Attallah et al. 2009; Gorrell and Robins-Browne 2009).

However, until recently, the available clinical studies focused on bovine antibodies in milk have not shown *H. pylori* eradication. For instance, Opekun et al. (1999) evaluated healthy, adult *H. pylori*-infected patients who were given an oral suspension of a HBC cocktail with several IgG isotypes. Patients were assessed based on a urea breath test or a histological *H. pylori* count. The results indicated that the cocktail had no significant effect on either measurement. In a double-blind study, *H. pylori*-positive infants were administered either HBC or a colostrum placebo from a non-immunised cow for 30 days (Casswall et al. 1998). None of the infants tested negative by the urea breath test after the treatment, leading the authors to speculate that HBC may not contain sufficient concentrations of immunoglobulin to successfully treat the intracellular stage of *H. pylori* infection, since this study used the highest practical dose (Casswall et al. 1998).

Recently, Hu et al. (2015) performed a clinical study to evaluate the efficacy and safety of specific anti-*H. pylori* polyclonal bovine antibodies in milk for the clearance of intragastric *Helicobacter* infection in O blood group subpopulations. Of the 30 antibody-treated subjects, 13 tested negative by the urea breath test, whereas none of the 30 placebo-treated subjects became urea breath test negative after the medication. Of the 13 patients who tested negative, three became positive again at the end of the follow-up. Both intention to treat and per protocol analysis indicated a significant difference in the clearance rate of infected patients between the groups treated with bovine antibody-containing milk and placebo ($P = 0.001$, $P < 0.05$) and no significant difference in adverse effects ($P > 0.05$ all). The authors concluded that bovine antibody-based oral immunotherapy appears to be safe and has a significant clearance effect on intragastric *H. pylori* in this group of adults.

16.3.1.3 *Clostridium difficile*

Clostridium difficile is a spore-forming, gram-positive bacterium that has emerged as a leading cause of infectious diarrhoea. *C. difficile*-associated diarrhoea primarily affects aged patients as a consequence of antimicrobial therapy, often during hospitalisation (Johnson 2009). The use of HBC against *C. difficile* infection has produced promising preliminary results and has been reviewed by Steele et al. (2013). Anti-*C. difficile* HBC was shown to be effective in inhibiting adhesion of *C. difficile* bacteria to enterocyte-like Caco-2 cells *in vitro* (Naaber et al. 1996). In addition, Lyerly et al. (1991) found that concentrated HBC IgG was effective at neutralising toxin A and toxin B of *C. difficile* in a tissue culture assay and protected hamsters against *C. difficile* infection, particularly when combined with Enfamil milk replacer. IgG was also found to survive passage through healthy adult human gastrointestinal tracts and retain its neutralisation activity when protected by enteric capsules or given with sodium bicarbonate and omeprazole (Kelly et al. 1997; Warny et al. 1999).

Van Dissel et al. (2005) used a preparation from cows immunised against *C. difficile* toxins and whole-cell *C. difficile* as a supportive treatment for two weeks after a standard antibiotic treatment in an uncontrolled cohort study. Nine of 16 patients had a history of relapsing *C. difficile* diarrhoea. After the regimen, with the exception of one patient, *C. difficile* toxins had disappeared from faeces and during the follow-up period of 11 months, none of the patients had another episode of *C. difficile* diarrhoea. Anti-*C. difficile* HBC has also been found to be as effective at treating mild to moderate recurring *C. difficile* infection as metronidazole (Numan et al. 2007; Mattila et al. 2008).

In a recent study by Sponseller et al. (2014), a pregnant dairy cow was repeatedly immunised with recombinant mutants of toxins A and B, and the resultant HBC was evaluated in gnotobiotic piglets with diarrhoea caused by *C. difficile* infection. Control piglets received non-immune colostrum and developed moderate to severe diarrhoea and colitis. In contrast, HBC-treated piglets had mild or no diarrhoea and mild or no colitis. Lyophilisation had no detectable impact on HBC efficacy and the HBC had no discernible effect on the composition of normal human gut microbiota in the porcine intestinal tract. The authors concluded that HBC provides an oral, cost-effective, and safe alternative to antibiotic therapy for *C. difficile* infection. By preserving intestinal microbiota, HBC may be more efficacious than antibiotics.

Recently, Hillreiner et al. (2017), using a gene expression profiling approach, found that gene expression of proinflammatory cytokines was altered between the high and low responder cows when challenged with *C. difficile*. This research could aid in the identification of cows with a very effective immune response and therefore the ability to produce greater amounts of Igs in their milk.

16.3.1.4 *Shigella*

Shigellosis, caused by various *Shigella* species, remains a common cause of diarrhoea in young children in developing countries, in certain communal care settings, and in travellers and military personnel (Steele et al. 2013). While antibiotics are the standard treatment, immunological approaches, including HBC, have been investigated for both treatment and prevention.

Tacket et al. (1992) investigated the prophylactic efficacy of immunoglobulin concentrate from HBC in healthy adult volunteers and found that the product significantly reduced the incidence of diarrhoea resulting from challenge with *Shigella flexneri*. However, no difference in diarrhoea or other symptoms in children with stools positive for *Shigella dysenteriae* was found whether treated with bovine colostrum from cows immunised against *S. dysenteriae* or with colostrum from non-immunised cows (Ashraf et al. 2001). Interestingly, the Igs in the HBC product IMM-124E, produced by Immuron, were recently shown to recognise *Shigella* polysaccharides in addition to ETEC, as mentioned above (Sears et al. 2017). This is probably due to the presence of shared epitopes.

16.3.1.5 *Cryptosporidium*

Cryptosporidium species generally cause self-limiting diarrhoea in healthy individuals, but can cause chronic, wasting diarrhoea in malnourished and immunocompromised populations. Cryptosporidiosis was recognised as an important gastrointestinal infection in HIV-infected humans in the 1980s (Steele et al. 2013). HBCs have proven effective *in vivo* in the prevention of infection induced by this parasite in both immunocompromised (Tzipori et al. 1987; Greenberg and Cello 1996) and healthy humans (Okhuysen et al. 1998). For instance, Tzipori et al. described the effective treatment of cryptosporidiosis in immunocompromised children and adults using orally administered HBC (Tzipori et al. 1986, 1987). Greenberg and Cello (1996) later demonstrated that immunoglobulin concentrate derived from HBC was effective at treating diarrhoea in *C. parvum*-infected AIDS patients. Similarly, in a child with AIDS who had severe diarrhoea caused by cryptosporidiosis, administration of a commercial hyperimmune bovine colostrum preparation with anticryptosporidial activity improved the diarrhoea and eliminated the parasite (Shield et al. 1993). Okhuysen et al. (1998) found that HBC was also effective in the prevention of cryptosporidiosis in healthy volunteers when administered before challenge with *C. parvum* spores.

In contrast, non-immune colostrum and non-specific bovine Ig concentrate have been shown to afford only marginal protection against the parasite, emphasising the importance of using specific antibodies (Saxon and Weinstein 1987; Rump et al. 1992; Plettenberg et al. 1993).

16.3.1.6 Streptococci

Bovine milk antibodies specific to human dental caries-promoting streptococci have been shown to reduce the amounts of these bacteria in human dental plaque (Filler et al. 1991; Loimaranta et al. 1999) and to protect gnotobiotic rats against dental caries (Michalek and McGhee 1977). A bovine whey preparation of colostrum from cows immunised with caries-inducing bacterial strains (*Streptococcus mutans* and *S. sobrinus*), and containing over 60% immunoglobulin of which 80% was IgG1, has been used in several studies evaluating its effect on caries-producing bacteria. The colostral whey preparation reduced adherence of *S. mutans in vitro* and caused aggregation of suspended bacteria (Loimaranta et al. 1998a; Wei et al. 2002), as well as inhibiting glucose uptake by the test organism (Loimaranta et al. 1997, 1998b). Moreover, mouth rinsing with immune milk collected from cows immunised with a fusion protein representing two of the major factors implicated in oral colonisation by *S. mutans* inhibited recolonisation of saliva and plaque by the bacteria (Oho et al. 1999; Shimazaki et al. 2001).

A recent study by Ramezanalizadeh et al. (2017) demonstrated that the colostrum of hyperimmune cows reduced the attachment of *S. mutans* and *S. sobrinus* by 69% and 43% respectively, using an *in vitro* microtitre plate method, and also reduced bacterial growth.

16.3.1.7 Rotavirus

Rotavirus is the primary cause of non-bacterial, acute diarrhoea in infants and young children six months to two years of age in both developed and developing countries. No effective vaccine is available to date, so reduction of duration and severity of symptoms in infants is an important public health concern (Steele et al. 2013).

Brüssow et al. (1987) gave hyperimmune colostrum for 10 days to children of 3–15 months of age. Fifty-five were in the trial group while 65 were in the placebo group. Nine (14%) of the 65 on placebo developed rotavirus diarrhoea, whereas none in the trial group developed diarrhoea. Hilpert et al. (1987) treated infants hospitalised with acute diarrhoea with anti-rotavirus immunoglobulin concentrate without a significant decrease in duration of diarrhoea or excretion of virus. However, Mitra et al. (1995) showed a decrease in the duration of diarrhoea and a reduction in stool output when treating rotavirus-infected infants with HBC. In addition, two groups tested the efficacy of passive immunisation of healthy infants with anti-rotavirus HBC and demonstrated a reduction in incidence and duration of diarrhoea in treated infants (Davidson et al. 1989; Turner and Kelsey 1993).

In 1998, Sarker et al. performed a double-blind placebo-controlled trial involving 80 children with rotavirus diarrhoea. The children were randomly assigned to receive orally either 10 g of immunised bovine colostrum (containing 3.6 g of anti-rotavirus antibodies) daily for four days or the same amount of a placebo preparation. Children who received the immunised bovine colostrum had significantly less daily and total stool output and frequency, and required a smaller amount of oral rehydration solution than did children who received the placebo. Clearance of rotavirus from the stool was also earlier in the immunised bovine colostrum group compared with the placebo group.

Pant et al. (2007) evaluated the therapeutic potential of different probiotics and their combination with HBC against rotavirus in a mouse model of rotavirus diarrhoea. A combination of HBC with *Lactobacillus rhamnosus* GG reduced diarrhoea outcome measures significantly, prevented histopathological changes, and reduced virus load in the intestine. In order to further improve this form of treatment, *L. rhamnosus* GG was engineered to display surface-expressed IgG-binding domains of protein G (GB1, GB2, and GB3) which capture HBC-derived IgG antibodies (HBC-IgG) and thus target rotavirus. The combination therapy was significantly more effective in reducing the prevalence, severity, and duration of diarrhoea in comparison to HBC alone or a combination of wild-type *L. rhamnosus* GG and HBC in mice (Gunaydin et al. 2014).

16.3.1.8 Respiratory Syncytial Virus

Respiratory syncytial virus (RSV) infection is a major cause of death in the first year of life, with especially high mortality rates in African and Asian countries (Black et al. 2003; Nair et al. 2010; Lozano et al. 2012).

In 2014, den Hartog et al. investigated whether bovine milk-derived IgG could modulate immune responses against human RSV. The group investigated the specificity and functional relevance of bovine IgG against RSV and other human respiratory pathogens, the ability of bovine IgG to bind to human Fcc receptors, and the induction of effector functions in human myeloid cells. Bovine IgG was found to recognise human RSV, influenza haemagglutinin and *Haemophilus influenzae*. IgG also bound to FccRII on neutrophils, monocytes, and macrophages, but not to FccRI and FccRIII, and could bind simultaneously to human RSV and human FccRII on neutrophils. In addition, human neutrophils and dendritic cells internalised pathogens that were opsonised with bovine IgG. Bovine IgG was also shown to prevent infection of HEp2 cells by human RSV. These findings are based on various *in vitro* assays and require *in vivo* studies to formally demonstrate that bovine milk-derived IgG can contribute to protection against this respiratory pathogen.

16.3.1.9 Human Herpes Virus

Bovine herpes virus type-1 (BHV-1) naturally infects cattle worldwide, and its antibody is found in milk. Moreover, BHV-1 glycoprotein K, the major antigen, exhibits substantial homology with human herpes virus simplex 1 (HSV-1) glycoprotein K. On the basis of this, Masuzawa et al. (2016) hypothesised that BHV-1 antibody exists in immune milk and may suppress HSV-1 activity. The authors found that immune milk has protective activity against human HSV-1 by opsonic activity owing to its high concentration of HSV-1-reactive IgG.

16.3.2 Other Health Benefits

Some early studies on immune milk involved its use in the treatment of rheumatoid arthritis and hay fever (Campbell and Petersen 1963). In 1988, a patent was published on the treatment and prevention of rheumatoid arthritis using IgG antibody obtained from the milk of immunised cows (Stolle and Beck 1988). Immune milk preparations have also been studied for the prevention and treatment of high blood cholesterol and high blood pressure (Golay et al. 1990; Sharpe et al. 1994). Sharpe et al. (1994) compared the hypocholestrolaemic and blood pressure effects of ordinary skim milk powder and skim milk powder produced from hyperimmunised cows. The subjects consumed 90 g of control or immune milk daily in a randomised, double-blind crossover study. The immune milk caused a significant reduction in total and low-density lipoprotein (LDL) cholesterol levels with no change in high-density lipoprotein (HDL) cholesterol or triacylglycerols. Similar results were observed using a blood-derived bovine IgG fraction in a placebo-controlled trial performed by Earnest et al. (2005).

An animal model utilising leptin-deficient *ob/ob* mice documented the effects of supplementation with IgG-rich colostrum from cows immunised against enterotoxigenic *E. coli* or the purified anti-LPS IgG alone on blood lipids, glucose tolerance, and hepatic steatosis for eight weeks (Adar et al. 2012). Reductions in circulating triglycerides (but not total cholesterol) and hepatic fat were noted following supplementation with either the IgG-rich colostrum or the purified IgG, in addition to improved glucose tolerance following supplementation with the purified IgG, suggesting the potential for bovine immunoglobulin supplementation as a means to affect indices of metabolic disease.

Immune milk preparations have also been studied for the prevention and treatment of oral submucous fibrosis (OSF). OSF is a chronic, insidious disease affecting the oral cavity and sometimes the pharynx. It is characterised by the progressive build-up of constricting bands of collagen in the cheeks and adjacent structures of the mouth, which can cause problems with speech and swallowing and severely restrict mouth opening and tongue movement, leading to difficulty in eating (Singh 2015). Tai et al. (2001) performed a preliminary study in which 26 OSF patients were given an immune milk treatment (45 g of immune milk powder twice a day) for three months and oral habit intervention was included in the experimental group. Another 20 OSF patients received only oral habit intervention, which served as the control group. The authors reported significant improvement in mouth opening and symptoms among the 26 OSMF patients who received immune milk treatment in comparison to the control group.

During and after major surgical procedures, endotoxaemia and acute phase reactions may occur due to intestinal endotoxins reaching the bloodstream by translocation of bacterial products (Buttenschoen et al. 2001). A number of placebo-controlled studies (Bölke et al. 2001, 2002; Friedrich et al. 2002) verified that prophylactic oral application of Ig-enriched colostrum milk preparation diminishes perioperative endotoxaemia, prevents reduction of endotoxin-neutralising capacity, and reduces postoperative C-reactive protein levels, suggesting a stabilisation of the gut barrier (Orth et al. 2013). This effect may be caused by immunoglobulin transportation by the neonatal receptor (FcRn) of the mucosal epithelium. These studies were performed on patients undergoing abdominal and cardiac surgery.

16.4 Isolation of Immunoglobulins

Isolation of Igs from colostral whey has been accomplished traditionally by precipitation methods with either ammonium sulfate or ethanol followed by several other chromatography methods. These methods yield highly pure IgG fractions but are more appropriate for small-scale applications than large-scale production (El-Loly 2007).

Other challenges which arise in the isolation of Igs from colostrum are their complex composition and the sensitivity of Igs to heat treatments (Chen and Chang 1998). Rapid developments in separation technologies and application of membrane separation methods have made possible the industrial-scale isolation of immunoglobulins from various streams, including whey and colostrum (Korhonen 2004; Korhonen and Pihlanto 2007).

Membrane filtration and chromatography have been employed alone or in combination. For instance, microfiltration combined with ultrafiltration of bovine colostrum led to an IgG/total solids purity of more than 90% (Piot et al. 2004). Mehra and Kelly (2004) used a 0.1 μM membrane and selective membrane separation using pH manipulation to produce an Ig-rich preparation from cheese whey, with a protein composition similar to that produced from colostrum whey. Al-Mashikhi et al. (1988) successfully isolated IgG from acid and cheddar whey using an ultrafiltration membrane combined with immobilised metal chelate affinity chromatography, producing purities of 77% and 53%, respectively. In 1998, Akita and Li-Chan developed immune affinity columns using specific egg yolk IgY against bovine IgG1 and IgG2 and isolated pure IgG1 and IgG2 from cheddar cheese whey or colostrum. Fukumoto et al. (1994) developed a method to isolate Igs from cheese whey using ultrafiltration and immobilised metal affinity chromatography (IMAC). IMAC was found to give higher recovery and purity of IgG from concentrated whey when compared to ion exchange chromatography.

In 1998, Korhonen et al. used a multistep process with a series of membrane methods (ultrafiltration, microfiltration, reverse osmosis) followed by a cation exchange resin used as a molecular sieve for concentration of Igs from colostral whey, which resulted in 40–75% of enrichment of Ig. Indeed, several chromatographic techniques can be used for purification of Igs from bovine milk fluids, including size exclusion (Pessela et al. 2006), ion exchange (Hahn et al. 1998; Bai et al. 2000; Doultani et al. 2004; Wongchuphan et al. 2011; Dong et al. 2013; Matlschweiger et al. 2017), IMAC (Fukumoto et al. 1994), expanded bed adsorption (Nielsen et al. 2002; Du et al. 2014), metal chelate (Kaneko et al. 1985; Al-Mashikhi et al. 1988) and hydroxyapatite (Rossano et al. 2001). More recently, hydrophobic charge induction chromatography has been used to isolate and purify bovine IgG with high purity (Wu et al. 2015). Serial chromatographic techniques have also been employed to separate bovine IgG. For instance, Qi et al. (2001) developed a high-performance liquid chromatography (HPLC) method that was based on coupled diethylaminoethanol (DEAE) anion exchange and protein G affinity columns for isolation of IgG, while Wu and Xu (2009) purified IgG from bovine colostrum with serial cation-anion exchange chromatography.

Affinity chromatography is undoubtedly the most widely used method for antibody purification because of its high selectivity (Bottomley et al. 1995; Kim and Li-Chan 1998; Özkara et al. 2003; Yang et al. 2009; Barroso et al. 2010). With affinity chromatography, the IgG can be purified by virtue of its specific binding properties to an immobilised ligand (Liu et al. 2010). For instance, lectins have been used to isolate Igs, with the lectin jacalin (known for its ability to bind O-linked oligosaccharides) employed to obtain a colostral Ig population rich in oligosaccharide chains (Porto et al. 2007). This preparation contained 10% of the total colostral Ig and was typically three to six times more resistant to pepsin digestion than the Ig contained in the fraction that was not bound to jacalin, which presumably did not contain O-glycans. The most commonly used affinity binders, however, are the protein A or protein G ligands, which strongly bind to the Fc regions of IgG, IgM, IgA, and IgD (Zettlitz 2010; Darcy et al. 2011). However, these ligands have been linked with several inherent problems for process development, mainly because of cost and loss of activity during the sanitisation conditions applied during column regeneration (Billakanti et al. 2014). Another major problem encountered when using bacterial proteins such as protein A in affinity purification of Igs is contamination of the purified samples with traces of these proteins, also known as ligand leakage (Choe et al. 2016).

Novel small molecules have been developed to mimic the binding properties of bacterial proteins to Igs and are reviewed by Choe et al. (2016). Short peptides are known to selectively and strongly bind to a great variety of biomolecules. Such peptides have potential advantages in chromatography processes because they can be more stable and less immunogenic than protein ligands and their production cost is low (Fassina et al. 1996; Yang et al. 2005). Moreover, these synthetic affinity ligands extend their selectivity to other classes of immunoglobulins (IgG, IgE, IgM, and IgA), unlike conventional protein A/G ligands (Palombo et al. 1998a,b,c; Yang et al. 2009).

Recently, Billakanti et al. (2014) employed a hexamer peptide (HWRGWV) affinity matrix for the isolation of bovine immunoglobulins from various dairy streams (skim milk, acid whey, and colostrum). Bound immunoglobulins were recovered with a purity of >85% in a single step. Methods such as this hold promise as an alternative to conventional proteinA/G chromatography for direct capture of immunoglobulins from streams containing relatively high immunoglobulin concentrations such as colostrum, transgenic, or hyperimmune milk.

16.5 Detection of Immunoglobulins

Before and after isolation of Igs, a prerequisite for any potential application in the infant or adult human diet is their reliable quantitative determination in milk and colostrum, or indeed any foods or dietary supplements incorporating Ig derived from bovine sources. Analytical methods involving immunodiffusion (Akita and Li-Chan 1998; Li et al. 2008), immunoturbidimetry (Jin et al. 2012), HPLC (Merin et al. 2001; Ferreira 2007), protein G affinity chromatography (Holland et al. 2011), and enzyme-linked immunosorbent methods (Kummer et al. 1992) have been used successfully for the detection of IgG. Methods that detect multiple proteins, such as thermally addressed immunosorbent assays (Ma et al. 2010), have also been developed. Rapid methods that may be integrated into milking systems, such as surface plasmon resonance (SPR)-based immunosensors, have also been evaluated (Crosson et al. 2010).

Until relatively recently, however, no standard method for the quantification of bovine IgG in bovine colostrum or milk powders and dietary supplements containing colostrum of bovine origin has been recognised by the Association of Analytical Communities (AOAC) International or any regulatory authority (Gapper et al. 2007). A method based on an affinity liquid chromatography with a protein G column for the determination of IgG in bovine dairy products (Copestake et al. 2006) underwent an AOAC collaborative trial in 2010 (Abernethy and Otter 2010). Eleven laboratories provided data for the study and assayed blind duplicates of six materials. The method with casein removal had adequate reproducibility for measuring IgG in colostrum-derived powders. In 2013, nine laboratories participated in an AOAC collaborative study to determine bovine IgG levels in selected dairy powders and dietary supplements by SPR methodology (Gapper 2013). In this study, IgG was determined by diluting samples and then dispensing on a 96-well plate. Analysis was performed with an instrument utilising SPR detection over a surface functionalised with affinity-purified, polyclonal goat anti-bovine IgG antibody with regeneration between injections. The results of the SPR study were adopted as an Official First Action Method by AOAC International and complement existing methodology recognised by international regulatory authorities.

16.6 Effect of Processing on Immunoglobulins

The expression of biological activity of any protein is associated with its three-dimensional conformational structure. Maintenance of this biological integrity is of the utmost importance for inclusion of Igs in functional foods. Igs are considered relatively heat labile, in which antigen-binding sites are more heat sensitive than the other areas (Bogahawaththa et al. 2017). Exposure of Ig to heat treatments can directly alter the conformational integrity, which in turn influences Ig functionality.

Many authors have suggested that minimum exposure to heat treatments should be achieved when manufacturing Ig-based functional foods or immune supplements (Gapper et al. 2007; Hurley and Theil 2013; Bogahawaththa et al. 2017). Studies have shown that the structure of Ig and its functions are usually affected by heat treatments of above 65 °C (Calmettes et al. 1991; Li et al. 2005; Gapper et al. 2007). Such treatments have resulted in reduced antigen-binding activity of bovine IgG (Dominguez et al. 1997, 2001). However, bovine IgG was shown to have higher heat stability in milk than in phosphate-buffered saline (PBS) (Li-Chan et al. 1995) and in colostrum than in either whey or PBS (Chen and Chang 1998). IgG is the most thermo-stable and IgM is the most thermo-labile of the Igs found in bovine milk (Mainer et al. 1997).

Donahue et al. (2012) demonstrated that batch heat treatment of colostrum at 60 °C for 60 minutes can be successfully conducted on commercial dairy farms to decrease colostrum microbial counts while maintaining colostrum IgG concentrations. Also, Gelsinger et al. (2014) reported that heat treatment (60 °C for 30 minutes) of colostrum containing approximately 50–100 mg IgG/mL increases absorption of IgG from colostrum in neonatal calves. Bogahawaththa et al. (2017) designed a study to investigate the thermal denaturation of native bovine IgG and its association with three other native whey proteins (beta-lactoglobulin, alpha-lactalbumin, and bovine serum albumin). The authors concluded that the presence of other whey proteins did not contribute to thermal stability of IgG at 100 °C.

Ig stability is also affected by pH and studies on isolated immunoglobulin over a pH range indicate that bovine IgG isolated from milk is stable for several hours at 37 °C and neutral pH, but stability is significantly reduced at a pH of below 4.0 or above 10.0 (Shimizu et al. 1993; Chen and Chang 1998). The negative effect of pH on IgG stability, even in the range of 4.5–6.5, is augmented under elevated temperature conditions (Dominguez et al. 2001; Gao et al. 2010).

16.7 Conclusion

As a result of the progress made in the separation and isolation of Igs from bovine colostrum and cheese whey, IgG products can be manufactured at large scale. The use of immune milks and Ig concentrates as dietary supplements most probably leaves the normal gut microflora intact, and also provides an alternative strategy to preventing infections in the face of the emerging antibiotic resistance crisis.

Successful development of products such as novel immunotherapeutic agents could revolutionise current management of infectious disease, yielding improved outcomes for patients and sustaining public health. Additional studies are required before definitive conclusions can be drawn and claimed health benefits must be substantiated. It may be that synergistic effects of Igs with probiotics and other milk bioactives such as lactoferrin and glycomacropeptide will provide new opportunities for nutritional intervention in the prevention of disease.

References

Abernethy, G. and Otter, D. (2010). Determination of immunoglobulin G in bovine colostrum and milk powders, and in dietary supplements of bovine origin by protein G affinity liquid chromatography: collaborative study. *Journal of AOAC International* 93 (2): 622–627.

Adar, T., Ben Ya'acov, A., Lalazar, G. et al. (2012). Oral administration of immunoglobulin G-enhanced colostrum alleviates insulin resistance and liver injury and is associated with alterations in natural killer T cells. *Clinical and Experimental Immunology* 167 (2): 252–260.

Akita, E. and Li-Chan, E. (1998). Isolation of bovine immunoglobulin G subclasses from milk, colostrum, and whey using immobilized egg yolk antibodies. *Journal of Dairy Science* 81 (1): 54–63.

Al-Mashikhi, S.A., Li-Chan, E., and Nakai, S. (1988). Separation of immunoglobulins and lactoferrin from cheese whey by chelating chromatography. *Journal of Dairy Science* 71 (7): 1747–1755.

Annen, E., Collier, R., McGuire, M., and Vicini, J. (2004). Effects of dry period length on milk yield and mammary epithelial cells. *Journal of Dairy Science* 87: E66–E76.

Anthony, R.M. and Ravetch, J.V. (2010). A novel role for the IgG Fc glycan: the anti-inflammatory activity of sialylated IgG Fcs. *Journal of Clinical Immunology* 30: S9–S14.

Ashraf, H., Mahalanabis, D., Mitra, A. et al. (2001). Hyperimmune bovine colostrum in the treatment of shigellosis in children: a double-blind, randomized, controlled trial. *Acta Paediatrica* 90 (12): 1373–1378.

Atkinson, D.E., Boyd, R.D.H., and Sibley, C.P. (2006). Placental transfer. In: *Knobil and Neill's Physiology of Reproduction*, 3e (ed. J.D. Neill), 2787–2846. St Louis: Elsevier.

Attallah, A.M., Abbas, A.T., Ismail, H. et al. (2009). Efficacy of passive immunization with IgY antibodies to a 58-kDa *H. pylori* antigen on severe gastritis in BALB/c mouse model. *Journal of Immunoassay and Immunochemistry* 30 (4): 359–377.

Bai, L., Burman, S., and Gledhill, L. (2000). Development of ion exchange chromatography methods for monoclonal antibodies. *Journal of Pharmaceutical and Biomedical Analysis* 22 (3): 605–611.

Barroso, T., Temtem, M., Hussain, A. et al. (2010). Preparation and characterization of a cellulose affinity membrane for human immunoglobulin G (IgG) purification. *Journal of Membrane Science* 348 (1–2): 224–230.

Billakanti, J.M., Fee, C.J., Naik, A.D., and Carbonell, R.G. (2014). Application of peptide chromatography for the isolation of antibodies from bovine skim milk, acid whey and colostrum. *Food and Bioproducts Processing* 92 (2): 199–207.

Black, R.E., Morris, S.S., and Bryce, J. (2003). Where and why are 10 million children dying every year? *Lancet* 361 (9376): 2226–2234.

Bogahawaththa, D., Chandrapala, J., and Vasiljevic, T. (2017). Thermal denaturation of bovine immunoglobulin G and its association with other whey proteins. *Food Hydrocolloids* 72: 350–357.

Bojsen, A., Buesa, J., Montava, R. et al. (2007). Inhibitory activities of bovine macromolecular whey proteins on rotavirus infections *in vitro* and *in vivo*. *Journal of Dairy Science* 90 (1): 66–74.

Bölke, E., Jehle, P., Orth, K. et al. (2001). Changes of gut barrier function during anesthesia and cardiac surgery. *Angiology* 52 (7): 477–482.

Bölke, E., Orth, K., Jehle, P.M. et al. (2002). Enteral application of an immunoglobulin-enriched colostrum milk preparation for reducing endotoxin translocation and acute phase response in patients undergoing coronary bypass surgery – a randomized placebo-controlled pilot trial. *Wiener Klinische Wochenschrift* 114 (21–22): 923–928.

Borghesi, J., Mario, L.C., Rodrigues, M.N. et al. (2014). Immunoglobulin transport during gestation in domestic animals and humans – a review. *Open Journal of Animal Sciences* 4 (05): 323.

Bottomley, S.P., Sutton, B.J., and Gore, M.G. (1995). Elution of human IgG from affinity columns containing immobilised variants of protein A. *Journal of Immunological Methods* 182 (2): 185–192.

Brüssow, H., Hilpert, H., Walther, I. et al. (1987). Bovine milk immunoglobulins for passive immunity to infantile rotavirus gastroenteritis. *Journal of Clinical Microbiology* 25 (6): 982–986.

Butler, J.E. and Kennedy, N. (1978). The differential enzyme susceptibility of bovine immunoglobulin G1 and immunoglobulin G2 to pepsin and papain. *Biochimica et Biophysica Acta – Protein Structure* 535 (1): 125–137.

Buttenschoen, K., Buttenschoen, D.C., Berger, D. et al. (2001). Endotoxemia and acute-phase proteins in major abdominal surgery. *American Journal of Surgery* 181 (1): 36–43.

Calmettes, P., Cser, L., and Rajnavolgyi, E. (1991). Temperature and Ph-dependence of immunoglobulin-G conformation. *Archives of Biochemistry and Biophysics* 291 (2): 277–283.

Campbell, B. and Petersen, W.E. (1963). Immune milk – a historical survey. *Dairy Science Abstracts* 25 (9): 345–358.

Casswall, T., Sarker, S., Albert, M. et al. (1998). Treatment of *Helicobacter pylori* infection in infants in rural Bangladesh with oral immunoglobulins from hyperimmune bovine colostrum. *Alimentary Pharmacology and Therapeutics* 12 (6): 563–568.

Chen, C.-C. and Chang, H.-M. (1998). Effect of thermal protectants on the stability of bovine milk immunoglobulin G. *Journal of Agricultural and Food Chemistry* 46 (9): 3570–3576.

Choe, W., Durgannavar, T.A., and Chung, S.J. (2016). Fc-binding ligands of immunoglobulin G: an overview of high affinity proteins and peptides. *Materials* 9 (12): 994.

Copestake, D.E., Indyk, H.E., and Otter, D.E. (2006). Affinity liquid chromatography method for the quantification of immunoglobulin G in bovine colostrum powders. *Journal of AOAC International* 89 (5): 1249–1256.

Crosson, C., Thomas, D., and Rossi, C. (2010). Quantification of immunoglobulin G in bovine and caprine milk using a surface plasmon resonance-based immunosensor. *Journal of Agricultural and Food Chemistry* 58 (6): 3259–3264.

Czinn, S.J., Cai, A., and Nedrud, J.G. (1993). Protection of germ-free mice from infection by *Helicobacter felis* after active oral or passive IgA immunization. *Vaccine* 11 (6): 637–642.

Darcy, E., Leonard, P., Fitzgerald, J. et al. (2011). Purification of antibodies using affinity chromatography. *Methods in Molecular Biology* 681: 369–382.

Davidson, G., Daniels, E., Nunan, H. et al. (1989). Passive immunisation of children with bovine colostrum containing antibodies to human rotavirus. *Lancet* 334 (8665): 709–712.

Deisenhofer, J. (1981). Crystallographic refinement and atomic models of a human Fc fragment and its complex with fragment B of protein A from *Staphylococcus aureus* at 2.9- and 2.8-A resolution. *Biochemistry* 20: 2361–2370.

den Hartog, G., Jacobino, S., Bont, L. et al. (2014). Specificity and effector functions of human RSV-specific IgG from bovine milk. *PLoS One* 9 (11): e112047.

Dewey, K.G., Heinig, M.J., and Nommsen-Rivers, L.A. (1995). Differences in morbidity between breast-fed and formula-fed infants. *Journal of Pediatrics* 126 (5): 696–702.

Dominguez, E., Perez, M.D., and Calvo, M. (1997). Effect of heat treatment on the antigen-binding activity of anti-peroxidase immunoglobulins in bovine colostrum. *Journal of Dairy Science* 80 (12): 3182–3187.

Dominguez, E., Perez, M.D., Puyol, P. et al. (2001). Effect of pH on antigen-binding activity of IgG from bovine colostrum upon heating. *Journal of Dairy Research* 68 (3): 511–518.

Donahue, M., Godden, S.M., Bey, R. et al. (2012). Heat treatment of colostrum on commercial dairy farms decreases colostrum microbial counts while maintaining colostrum immunoglobulin G concentrations. *Journal of Dairy Science* 95 (5): 2697–2702.

Dong, S.S., Chen, L., Dai, B. et al. (2013). Isolation of immunoglobulin G from bovine milk whey by poly(hydroxyethyl methacrylate)-based anion-exchange cryogel. *Journal of Separation Science* 36 (15): 2387–2393.

Doultani, S., Turhan, K.N., and Etzel, M.R. (2004). Fractionation of proteins from whey using cation exchange chromatography. *Process Biochemistry* 39 (11): 1737–1743.

Du, Q.Y., Lin, D.Q., Zhang, Q.L., and Yao, S.J. (2014). An integrated expanded bed adsorption process for lactoferrin and immunoglobulin G purification from crude sweet whey. *Journal of Chromatography B* 947: 201–207.

Earnest, C.P., Jordan, A.N., Safir, M. et al. (2005). Cholesterol-lowering effects of bovine serum immunoglobulin in participants with mild hypercholesterolemia. *American Journal of Clinical Nutrition* 81 (4): 792–798.

Eihvalde, I., Kairiša, D., and Zagorska, J. (2012). Analysis of factors influencing immunoglobulin concentration in colostrum of dairy cows. *Lucrări Ştiinţifice-Seria Zootehnie* 57: 256–259.

El-Loly, M. (2007). Bovine milk immunoglobulins in relation to human health. *International Journal of Dairy Science* 2 (3): 183–195.

Evans, D., de la Cabada, F., and Evans, D. (1982). Correlation between intestinal immune response to colonization factor antigen/I and acquired resistance to enterotoxigenic

Escherichia coli diarrhea in an adult rabbit model. *European Journal of Clinical Microbiology* 1 (3): 178–185.

Fassina, G., Verdoliva, A., Odierna, M.R. et al. (1996). Protein A mimetic peptide ligand for affinity purification of antibodies. *Journal of Molecular Recognition* 9 (5–6): 564–569.

Ferreira, I.M. (2007). Chromatographic separation and quantification of major human milk proteins. *Journal of Liquid Chromatography and Related Technologies* 30 (4): 499–507.

Filler, S., Gregory, R., Michalek, S. et al. (1991). Effect of immune bovine milk on *Streptococcus mutans* in human dental plaque. *Archives of Oral Biology* 36 (1): 41–47.

Freedman, D.J., Tacket, C.O., Delehanty, A. et al. (1998). Milk immunoglobulin with specific activity against purified colonization factor antigens can protect against oral challenge with enterotoxigenic *Escherichia coli*. *Journal of Infectious Diseases* 177 (3): 662–667.

Friedrich, I., Silber, R., Baumann, B. et al. (2002). IgM-enriched immunoglobulin preparation for immunoprophylaxis in cardiac surgery. *European Journal of Medical Research* 7 (12): 544–549.

Fukumoto, L., Li-Chan, E., Kwan, L., and Nakai, S. (1994). Isolation of immunoglobulins from cheese whey using ultrafiltration and immobilized metal affinity chromatography. *Food Research International* 27 (4): 335–348.

Funatogawa, K., Ide, T., Kirikae, F. et al. (2002). Use of immunoglobulin enriched bovine colostrum against oral challenge with enterohaemorrhagic *Escherichia coli* O157: H7 in mice. *Microbiology and Immunology* 46 (11): 761–766.

Gao, W., Chen, L., Xu, L.B., and Huang, X.H. (2010). Specific IgG activity against diarrheagenic bacteria in bovine immune milk and effect of pH on its antigen-binding activity upon heating. *Journal of Dairy Research* 77 (2): 220–224.

Gapper, L. (2013). Determination of bovine immunoglobulin G in bovine colostrum powders, bovine milk powders, and dietary supplements containing bovine colostrum products by an automated direct immunoassay with optical biosensor: collaborative study. *Journal of AOAC International* 96 (5): 1026–1032.

Gapper, L.W., Copestake, D.E., Otter, D.E., and Indyk, H.E. (2007). Analysis of bovine immunoglobulin G in milk, colostrum and dietary supplements: a review. *Analytical and Bioanalytical Chemistry* 389 (1): 93–109.

Gelsinger, S.L., Gray, S.M., Jones, C.M., and Heinrichs, A.J. (2014). Heat treatment of colostrum increases immunoglobulin G absorption efficiency in high-, medium-, and low-quality colostrum. *Journal of Dairy Science* 97 (4): 2355–2360.

Golay, A., Ferrara, J.M., Felber, J.P., and Schneider, H. (1990). Cholesterol-lowering effect of skim milk from immunized cows in hypercholesterolemic patients. *American Journal of Clinical Nutrition* 52 (6): 1014–1019.

Gorrell, R.J. and Robins-Browne, R.M. (2009). Antibody-mediated protection against infection with *Helicobacter pylori* in a suckling mouse model of passive immunity. *Infection and Immunity* 77 (11): 5116–5129.

Greenberg, P.D. and Cello, J.P. (1996). Treatment of severe diarrhea caused by *Cryptosporidium parvum* with oral bovine immunoglobulin concentrate in patients with AIDS. *Journal of Acquired Immune Deficiency Syndromes* 13 (4): 348–354.

Gunaydin, G., Zhang, R., Hammarstrom, L., and Marcotte, H. (2014). Engineered *Lactobacillus rhamnosus* GG expressing IgG-binding domains of protein G: capture of hyperimmune bovine colostrum antibodies and protection against diarrhea in a mouse pup rotavirus infection model. *Vaccine* 32 (4): 470–477.

Hahn, R., Schulz, P., Schaupp, C., and Jungbauer, A. (1998). Bovine whey fractionation based on cation-exchange chromatography1. *Journal of Chromatography A* 795 (2): 277–287.

Hillreiner, M., Schmautz, C., Ballweg, I. et al. (2017). Gene expression profiling in pbMEC – in search of molecular biomarkers to predict immunoglobulin production in bovine milk. *BMC Veterinary Research* 13 (1): 369.

Hilpert, H., Briissow, H., Mietens, C. et al. (1987). Use of bovine milk concentrate containing antibody to rotavirus to treat rotavirus gastroenteritis in infants. *Journal of Infectious Diseases* 156 (1): 158–166.

Holland, P.T., Cargill, A., Selwood, A.I. et al. (2011). Determination of soluble immunoglobulin G in bovine colostrum products by protein G affinity chromatography – turbidity correction and method validation. *Journal of Agricultural and Food Chemistry* 59 (10): 5248–5256.

Hu, D., Zhang, F., Zhou, J. et al. (2015). The clearance effect of bovine anti-*Helicobacter pylori* antibody-containing milk in O blood group *Helicobacter pylori*-infected patients: a randomized double-blind clinical trial. *Journal of Translational Medicine* 13 (1): 205.

Huppertz, H.-I., Rutkowski, S., Busch, D.H. et al. (1999). Bovine colostrum ameliorates diarrhea in infection with diarrheagenic *Escherichia coli*, Shiga toxin-producing *E. coli*, and *E. coli* expressing intimin and hemolysin. *Journal of Pediatric Gastroenterology and Nutrition* 29 (4): 452–456.

Hurley, W. and Theil, P.K. (2013). Immunoglobulins in mammary secretions. In: *Advanced Dairy Chemistry: Volume 1A: Proteins: Basic Aspects*, 4e (ed. L.H. McSweeney and P.F. Fox), 275–294. Dordrecht: Springer.

Hurley, W.L. and Theil, P.K. (2011). Perspectives on immunoglobulins in colostrum and milk. *Nutrients* 3 (4): 442–474.

Jasion, V.S. and Burnett, B.P. (2015). Survival and digestibility of orally-administered immunoglobulin preparations containing IgG through the gastrointestinal tract in humans. *Nutrition Journal* 14 (1): 22.

Jin, M.J., Shao, H., Jiang, Z.M. et al. (2012). A reliable immunoturbidimetry method for immunoglobulin G in bovine colostrum. *Food and Agricultural Immunology* 23 (2): 133–144.

Johnson, S. (2009). Recurrent *Clostridium difficile* infection: a review of risk factors, treatments, and outcomes. *Journal of Infection* 58 (6): 403–410.

Kaneko, T., Wu, B., and Nakai, S. (1985). Selective concentration of bovine immunoglobulins and α-lactalbumin from acid whey using FeCl3. *Journal of Food Science* 50 (6): 1531–1536.

Kelly, C.P., Chetham, S., Keates, S. et al. (1997). Survival of anti-*Clostridium difficile* bovine immunoglobulin concentrate in the human gastrointestinal tract. *Antimicrobial Agents and Chemotherapy* 41 (2): 236–241.

Kim, H. and Li-Chan, E. (1998). Separation of immunoglobulin G from cheddar cheese whey by avidin-biotinylated IgY chromatography. *Journal of Food Science* 63 (3): 429–434.

Korhonen, H. (2004). Isolation of immunoglobulins from colostrum. *International Dairy Federation Bulletin* 389: 78–81.

Korhonen, H. and Pihlanto, A. (2007). Technological options for the production of health-promoting proteins and peptides derived from milk and colostrum. *Current Pharmaceutical Design* 13 (8): 829–843.

Korhonen, H., E. Syvaeoja, E. Vasara, T. Kosunen, and P. Marnila. 1998. Pharmaceutical composition, comprising complement proteins, for the treatment of *Helicobacter* infections and a method for the preparation of the composition. Patent application PCT (W098/00150).

Kramski, M., Center, R.J., Wheatley, A.K. et al. (2012). Hyperimmune bovine colostrum as a low-cost, large-scale source of antibodies with broad neutralizing activity for HIV-1 envelope with potential use in microbicides. *Antimicrobial Agents and Chemotherapy* 56 (8): 4310–4319.

Kummer, A., Kitts, D., Li-Chan, E. et al. (1992). Quantification of bovine IgG in milk using enzyme-linked immunosorbent assay. *Food and Agricultural Immunology* 4 (2): 93–102.

Kvistgaard, A., Pallesen, L., Arias, C. et al. (2004). Inhibitory effects of human and bovine milk constituents on rotavirus infections. *Journal of Dairy Science* 87 (12): 4088–4096.

Li-Chan, E., Kummer, A., Losso, J.N. et al. (1995). Stability of bovine immunoglobulins to thermal-treatment and processing. *Food Research International* 28 (1): 9–16.

Li, S.Q., Bomser, J.A., and Zhang, Q.H. (2005). Effects of pulsed electric fields and heat treatment on stability and secondary structure of bovine immunoglobulin G. *Journal of Agricultural and Food Chemistry* 53 (3): 663–670.

Li, Z., Liu, C., Li, M., and Wang, J. (2008). Modified single immunediffusion method and detection of the content of IgG in bovine colostrums. *China Dairy Cattle* 3: 41–44.

Liu, H.F., Ma, J., Winter, C., and Bayer, R. (2010). Recovery and purification process development for monoclonal antibody production. *mAbs* 2 (5): 480–499.

Loimaranta, V., Carlén, A., Olsson, J. et al. (1998a). Concentrated bovine colostral whey proteins from *Streptococcus mutans*/*Strep. sobrinus* immunized cows inhibit the adherence of *Strep. mutans* and promote the aggregation of mutans streptococci. *Journal of Dairy Research* 65 (4): 599–607.

Loimaranta, V., Laine, M., Soèderling, E. et al. (1999). Effects of bovine immune and non-immune whey preparations on the composition and pH response of human dental plaque. *European Journal of Oral Sciences* 107 (4): 244–250.

Loimaranta, V., Tenovuo, J., and Korhonen, H. (1998b). Combined inhibitory effect of bovine immune whey and peroxidase-generated hypothiocyanite against glucose uptake by *Streptococcus mutans*. *Molecular Oral Microbiology* 13 (6): 378–381.

Loimaranta, V., Tenovuo, J., Virtanen, S. et al. (1997). Generation of bovine immune colostrum against *Streptococcus mutans* and *Streptococcus sobrinus* and its effect on glucose uptake and extracellular polysaccharide formation by mutans streptococci. *Vaccine* 15 (11): 1261–1268.

Lozano, R., Naghavi, M., Foreman, K. et al. (2012). Global and regional mortality from 235 causes of death for 20 age groups in 1990 and 2010: a systematic analysis for the global burden of disease study 2010. *Lancet* 380 (9859): 2095–2128.

Lyerly, D.M., Bostwick, E., Binion, S., and Wilkins, T. (1991). Passive immunization of hamsters against disease caused by *Clostridium difficile* by use of bovine immunoglobulin G concentrate. *Infection and Immunity* 59 (6): 2215–2218.

Ma, L., Wang, C., Hong, Y. et al. (2010). Thermally addressed immunosorbent assay for multiplexed protein detections using phase change nanoparticles. *Analytical Chemistry* 82 (4): 1186–1190.

Mainer, G., Sanchez, L., Ena, J.M., and Calvo, M. (1997). Kinetic and thermodynamic parameters for heat denaturation of bovine milk IgG, IgA and IgM. *Journal of Food Science* 62 (5): 1034–1038.

Marnila, P. and Korhonen, H. (2002). Immunoglobulins. In: *Encyclopedia of Dairy Science*, vol. 3, 1950–1956. Amsterdam: Academic Press.
Marnila, P. and Korhonen, H. (2011). Milk proteins: immunoglobulins. In: *Encyclopedia of Dairy Sciences*, 2, nd edne, vol. 3 (ed. J.W. Fuquay, P.F. Fox and M.S. P.L.H.), 807–815. San Diego: Academic Press.
Marnila, P., Rokka, S., Rehnberg-Laiho, L. et al. (2003). Prevention and suppression of *Helicobacter felis* infection in mice using colostral preparation with specific antibodies. *Helicobacter* 8 (3): 192–201.
Masuzawa, M., Masuzawa, M., Arakawa, N. et al. (2016). Immune milk suppresses herpes simplex virus type 1. *Journal of Nutritional Science and Vitaminology* 62 (1): 67–70.
Matlschweiger, A., Engelmaier, H., Himmler, G., and Hahn, R. (2017). Secretory immunoglobulin purification from whey by chromatographic techniques. *Journal of Chromatography B* 1060: 53–62.
Mattila, E., Anttila, V.-J., Broas, M. et al. (2008). A randomized, double-blind study comparing *Clostridium difficile* immune whey and metronidazole for recurrent *Clostridium difficile*-associated diarrhoea: efficacy and safety data of a prematurely interrupted trial. *Scandinavian Journal of Infectious Diseases* 40 (9): 702–708.
McWilliams, M. (1986). *Nutrition for the Growing Years*. New York: Macmillan.
Meganck, V., Laureyns, J. and Opsomer, G. (2012). The importance of a proper colostrum management in modern cattle farms. *Vlaams Diergeneeskundig Tijdschrift*. 81: 373–381.
Mehra, R. and Kelly, P. (2004). Whey protein fractionation using cascade membrane filtration. *Bulletin of the International Dairy Federation* 389: 40–44.
Merin, U., Bernstein, S., Bloch-Damti, A. et al. (2001). A comparative study of milk serum proteins in camel (*Camelus dromedarius*) and bovine colostrum. *Livestock Production Science* 67 (3): 297–301.
Michalek, S.M. and McGhee, J.R. (1977). Effective immunity to dental caries: passive transfer to rats to antibodies to *Streptococcus mutans* elicits protection. *Infection and Immunity* 17 (3): 644–650.
Mietens, C., Keinhorst, H., Hilpert, H. et al. (1979). Treatment of infantile *E. coli* gastroenteritis with specific bovine anti-*E. coli* milk immunoglobulins. *European Journal of Pediatrics* 132 (4): 239–252.
Mitra, A., Mahalanabis, D., Ashraf, H. et al. (1995). Hyperimmune cow colostrum reduces diarrhoea due to rotavirus: a double-blind, controlled clinical trial. *Acta Paediatrica* 84 (9): 996–1001.
Moore, D., Taylor, J., Hartman, M., and Sischo, W. (2009). Quality assessments of waste milk at a calf ranch. *Journal of Dairy Science* 92 (7): 3503–3509.
Morin, D.E., Nelson, S.V., Reid, E.D. et al. (2010). Effect of colostral volume, interval between calving and first milking, and photoperiod on colostral IgG concentrations in dairy cows. *Journal of the American Veterinary Medical Association* 237 (4): 420–428.
Naaber, P., Lehto, E., Salminen, S., and Mikelsaar, M. (1996). Inhibition of adhesion of *Clostridium difficile* to Caco-2 cells. *Pathogens and Disease* 14 (4): 205–209.
Nair, H., Nokes, D.J., Gessner, B.D. et al. (2010). Global burden of acute lower respiratory infections due to respiratory syncytial virus in young children: a systematic review and meta-analysis. *Lancet* 375 (9725): 1545–1555.

Nardone, A., Lacetera, N., Bernabucci, U., and Ronchi, B. (1997). Composition of colostrum from dairy heifers exposed to high air temperatures during late pregnancy and the early postpartum period. *Journal of Dairy Science* 80 (5): 838–844.
Nielsen, W., Olander, M., and Lihme, A. (2002). Expanding the frontiers in separation technology. *Scandinavian Dairy Information* 2: 50–53.
Numan, S.C., Veldkamp, P., Kuijper, E.J. et al. (2007). *Clostridium difficile*-associated diarrhoea: bovine anti-*Clostridium difficile* whey protein to help aid the prevention of relapses. *Gut* 56 (6): 888–889.
O'Riordan, N., Kane, M., Joshi, L., and Hickey, R.M. (2014). Structural and functional characteristics of bovine milk protein glycosylation. *Glycobiology* 24 (3): 220–236.
Oho, T., Shimazaki, Y., Mitoma, M. et al. (1999). Bovine milk antibodies against cell surface protein antigen PAc-glucosyltransferase fusion protein suppress cell adhesion and alter glucan synthesis of *Streptococcus mutans*. *Journal of Nutrition* 129 (10): 1836–1841.
Okhuysen, P.C., Chappell, C.L., Crabb, J. et al. (1998). Prophylactic effect of bovine anti-*Cryptosporidium* hyperimmune colostrum immunoglobulin in healthy volunteers challenged with *Cryptosporidium parvum*. *Clinical Infectious Diseases* 26 (6): 1324–1329.
Opekun, A., El-Zaimaity, H., Osato, M. et al. (1999). Novel therapies for *Helicobacter pylori* infection. *Alimentary Pharmacology and Therapeutics* 13 (1): 35–42.
Orth, K., Knoefel, W.T., van Griensven, M. et al. (2013). Preventively enteral application of immunoglobulin enriched colostrums milk can modulate postoperative inflammatory response. *European Journal of Medical Research* 18 (1): 50.
Otto, W., Najnigier, B., Stelmasiak, T., and Robins-Browne, R.M. (2011). Randomized control trials using a tablet formulation of hyperimmune bovine colostrum to prevent diarrhea caused by enterotoxigenic *Escherichia coli* in volunteers. *Scandinavian Journal of Gastroenterology* 46 (7–8): 862–868.
Özkara, S., Yavuz, H., and Denizli, A. (2003). Purification of immunoglobulin G from human plasma by metal-chelate affinity chromatography. *Journal of Applied Polymer Science* 89 (6): 1567–1572.
Pagnoncelli, M.G.B., de Melo Pereira, G.V., Fernandes, M.J. et al. (2017). Milk immunoglobulins and their implications for health promotion. In: *Nutrients in Dairy and their Implications on Health and Disease* (ed. R.R. Watson, R.J. Collier and V.R. Preedy), 87–96. San Diego: Academic Press.
Palombo, G., Falco, S.D., Tortora, M. et al. (1998a). A synthetic ligand for IgA affinity purification. *Journal of Molecular Recognition* 11 (1–6): 243–246.
Palombo, G., Rossi, M., Cassani, G., and Fassina, G. (1998b). Affinity purification of mouse monoclonal IgE using a protein a mimetic ligand (TG19318) immobilized on solid supports. *Journal of Molecular Recognition* 11 (16): 247–249.
Palombo, G., Verdoliva, A., and Fassina, G. (1998c). Affinity purification of immunoglobulin M using a novel synthetic ligand. *Journal of Chromatography B: Biomedical Sciences and Applications* 715 (1): 137–145.
Pant, N., Marcotte, H., Brussow, H. et al. (2007). Effective prophylaxis against rotavirus diarrhea using a combination of *Lactobacillus rhamnosus* GG and antibodies. *BMC Microbiology* 7: 86.
Pessela, B.C., Torres, R., Batalla, P. et al. (2006). Simple purification of immunoglobulins from whey proteins concentrate. *Biotechnology Progress* 22 (2): 590–594.
Piot, M., Fauquant, J., Madec, M.-N., and Maubois, J.-L. (2004). Preparation of serocolostrum by membrane microfiltration. *Le Lait* 84 (4): 333–341.

Plettenberg, A., Stoehr, A., Stellbrink, H.-J. et al. (1993). A preparation from bovine colostrum in the treatment of HIV-positive patients with chronic diarrhea. *Clinical Investigator* 71 (1): 42–45.

Porto, A., Oliveira, L., Ferraz, L. et al. (2007). Isolation of bovine immunoglobulins resistant to peptic digestion: new perspectives in the prevention of failure in passive immunization of neonatal calves. *Journal of Dairy Science* 90 (2): 955–962.

Pritchett, L.C., Gay, C.C., Besser, T.E., and Hancock, D.D. (1991). Management and production factors influencing immunoglobulin G1 concentration in colostrum from Holstein Cows. *Journal of Dairy Science* 74 (7): 2336–2341.

Qi, Y., Yan, Z., and Huang, J. (2001). Chromatography on DEAE ion-exchange and protein G affinity columns in tandem for the separation and purification of proteins. *Journal of Biochemical and Biophysical Methods* 49 (1–3): 263–273.

Ramezanalizadeh, F., Aliasghari, A., Khorasgani, M.R. et al. (2017). Evaluation of hyperimmune colostrum production in bovine against cariogenic streptococci and its impact on growth and bacterial biofilm formation. *Journal of Dental Medicine* 29 (4): 237–246.

Rao, M.R., Wierzba, T.F., Savarino, S.J. et al. (2005). Serologic correlates of protection against enterotoxigenic *Escherichia coli* diarrhea. *Journal of Infectious Diseases* 191 (4): 562–570.

Recio, I., Moreno, F.J., and López-Fandiño, R. (2009). Glycosylated dairy components: their roles in nature and ways to make use of their biofunctionality in dairy products. In: *Dairy Derived Ingredients – Food and Nutraceutical Uses* (ed. M. Corredig), 170–211. Cambridge: Woodhead Publishing.

Roos, N., Mahe, S., Benamouzig, R. et al. (1995). 15N-labeled immunoglobulins from bovine colostrum are partially resistant to digestion in human intestine. *Journal of Nutrition* 125 (5): 1238–1244.

Rossano, R., d'Elia, A., and Riccio, P. (2001). One-step separation from lactose: recovery and purification of major cheese-whey proteins by hydroxyapatite – a flexible procedure suitable for small-and medium-scale preparations. *Protein Expression and Purification* 21 (1): 165–169.

Rump, J., Arndt, R., Arnold, A. et al. (1992). Treatment of diarrhoea in human immunodeficiency virus-infected patients with immunoglobulins from bovine colostrum. *Clinical Investigator* 70 (7): 588–594.

Sarker, S.A., Casswall, T.H., Mahalanabis, D. et al. (1998). Successful treatment of rotavirus diarrhea in children with immunoglobulin from immunized bovine colostrum. *Pediatric Infectious Disease Journal* 17 (12): 1149–1154.

Saxon, A. and Weinstein, W. (1987). Oral-administration of bovine colostrum anti-Cryptosporidia antibody fails to alter the course of human cryptosporidiosis. *Journal of Parasitology* 73 (2): 413–415.

Sears, K.T., Tennant, S.M., Reymann, M.K. et al. (2017). Bioactive immune components of anti-diarrheagenic enterotoxigenic *Escherichia coli* hyperimmune bovine colostrum products. *Clinical and Vaccine Immunology* 24 (8): e00186–e00216.

Sharpe, S.J., Gamble, G.D., and Sharpe, D.N. (1994). Cholesterol-lowering and blood pressure effects of immune milk. *American Journal of Clinical Nutrition* 59 (4): 929–934.

Shield, J., Melville, C., Novelli, V. et al. (1993). Bovine colostrum immunoglobulin concentrate for cryptosporidiosis in AIDS. *Archives of Disease in Childhood* 69 (4): 451–453.

Shimazaki, Y., Mitoma, M., Oho, T. et al. (2001). Passive immunization with milk produced from an immunized cow prevents oral recolonization by *Streptococcus mutans*. *Clinical and Diagnostic Laboratory Immunology* 8 (6): 1136–1139.

Shimizu, M., Nagashima, H., and Hashimoto, K. (1993). Comparative studies on molecular stability of immunoglobulin-G from different species. *Comparative Biochemistry and Physiology Part B* 106 (2): 255–261.

Singh, P. (2015). Recent concepts of etiopathogenesis and management of oral submucous fibrosis: a review of literature. *Journal of Evolution of Medical and Dental Sciences* 4 (78): 13728–13741.

Sponseller, J.K., Steele, J.A., Schmidt, D.J. et al. (2014). Hyperimmune bovine colostrum as a novel therapy to combat *Clostridium difficile* infection. *Journal of Infectious Diseases* 211 (8): 1334–1341.

Staley, T. and Bush, L. (1985). Receptor mechanisms of the neonatal intestine and their relationship to immunoglobulin absorption and disease. *Journal of Dairy Science* 68 (1): 184–205.

Steele, J., Sponseller, J., Schmidt, D. et al. (2013). Hyperimmune bovine colostrum for treatment of GI infections: a review and update on *Clostridium difficile*. *Human Vaccines and Immunotherapeutics* 9 (7): 1565–1568.

Stolle, R. J. and L. R. Beck. 1988. Prevention and treatment of rheumatoid arthritis. US Patent 4,732,757.

Tacket, C.O., Binion, S.B., Bostwick, E. et al. (1992). Efficacy of bovine milk immunoglobulin concentrate in preventing illness after *Shigella flexneri* challenge. *American Journal of Tropical Medicine and Hygiene* 47 (3): 276–283.

Tacket, C.O., Losonsky, G., Link, H. et al. (1988). Protection by milk immunoglobulin concentrate against oral challenge with enterotoxigenic *Escherichia coli*. *New England Journal of Medicine* 318 (19): 1240–1243.

Tai, Y., Liu, B., Wang, J. et al. (2001). Oral administration of milk from cows immunized with human intestinal bacteria leads to significant improvements of symptoms and signs in patients with oral submucous fibrosis. *Journal of Oral Pathology and Medicine* 30 (10): 618–625.

Takemura, K., Hogan, J., Lin, J., and Smith, K. (2002). Efficacy of immunization with ferric citrate receptor FecA from *Escherichia coli* on induced coliform mastitis. *Journal of Dairy Science* 85 (4): 774–781.

Takemura, K., Hogan, J., and Smith, K. (2003). Effect of immunoglobulin G from cows immunized with ferric citrate receptor (FecA) on iron uptake by *Escherichia coli*. *Journal of Dairy Science* 86 (1): 133–137.

Takemura, K., Hogan, J., and Smith, K. (2004). Growth responses of *Escherichia coli* to immunoglobulin G from cows immunized with ferric citrate receptor, FecA. *Journal of Dairy Science* 87 (2): 316–320.

Takimori, S., Shimaoka, H., Furukawa, J.I. et al. (2011). Alteration of the N-glycome of bovine milk glycoproteins during early lactation. *FEBS Journal* 278 (19): 3769–3781.

Tawfeek, H.I., Najim, N.H., and Al-Mashikhi, S. (2003). Efficacy of an infant formula containing anti-*Escherichia coli* colostral antibodies from hyperimmunized cows in preventing diarrhea in infants and children: a field trial. *International Journal of Infectious Diseases* 7 (2): 120–128.

Turner, R.B. and Kelsey, D.K. (1993). Passive immunization for prevention of rotavirus illness in healthy infants. *Pediatric Infectious Disease Journal* 12 (9): 718–722.

Tzipori, S., Roberton, D., and Chapman, C. (1986). Remission of diarrhoea due to cryptosporidiosis in an immunodeficient child treated with hyperimmune bovine colostrum. *British Medical Journal* 293 (6557): 1276–1277.

Tzipori, S., Roberton, D., Cooper, D., and White, L. (1987). Chronic cryptosporidial diarrhoea and hyperimmune cow colostrum. *Lancet* 330 (8554): 344–345.

Van Dissel, J.T., de Groot, N., Hensgens, C.M. et al. (2005). Bovine antibody-enriched whey to aid in the prevention of a relapse of *Clostridium difficile*-associated diarrhoea: preclinical and preliminary clinical data. *Journal of Medical Microbiology* 54 (2): 197–205.

Verweij, J., Koets, A., and Eisenberg, S. (2014). Effect of continuous milking on immunoglobulin concentrations in bovine colostrum. *Veterinary Immunology and Immunopathology* 160 (3–4): 225–229.

Warny, M., Fatimi, A., Bostwick, E. et al. (1999). Bovine immunoglobulin concentrate-*Clostridium difficile* retains *C difficile* toxin neutralising activity after passage through the human stomach and small intestine. *Gut* 44 (2): 212–217.

Watters, R., Guenther, J., Brickner, A. et al. (2008). Effects of dry period length on milk production and health of dairy cattle. *Journal of Dairy Science* 91 (7): 2595–2603.

Wei, H., Loimaranta, V., Tenovuo, J. et al. (2002). Stability and activity of specific antibodies against *Streptococcus mutans* and *Streptococcus sobrinus* in bovine milk fermented with *Lactobacillus rhamnosus* strain GG or treated at ultra-high temperature. *Molecular Oral Microbiology* 17 (1): 9–15.

Wongchuphan, R., Tey, B.T., Tan, W.S. et al. (2011). Purification of rabbit polyclonal immunoglobulin G using anion exchangers. *Process Biochemistry* 46 (1): 101–107.

Wu, M.-B. and Xu, Y.-J. (2009). Isolation and purification of lactoferrin and immunoglobulin G from bovine colostrum with serial cation-anion exchange chromatography. *Biotechnology and Bioprocess Engineering* 14 (2): 155–160.

Wu, M., Zhang, F., Liang, Y. et al. (2015). Isolation and purification of immunoglobulin G from bovine colostrums by hydrophobic charge-induction chromatography. *Journal of Dairy Science* 98 (5): 2973–2981.

Yang, H., Gurgel, P.V., and Carbonell, R.G. (2005). Hexamer peptide affinity resins that bind the Fc region of human immunoglobulin G. *Chemical Biology and Drug Design* 66 (s1): 120–137.

Yang, H.O., Gurgel, P.V., and Carbonell, R.G. (2009). Purification of human immunoglobulin G via Fc-specific small peptide ligand affinity chromatography. *Journal of Chromatography A* 1216 (6): 910–918.

Zettlitz, K.A. (2010). Protein A/G chromatography. In: *Antibody Engineering* (ed. R. Kontermann and S. Dübel), 531–535. Berlin: Springer.

Index

Novel Proteins for Food, Pharmaceuticals, and Agriculture: Sources, Applications, and Advances, First Edition. Edited by Maria Hayes.

q

14970450-1-Text.pdf TPS: 244x170 Plates: 0 Spine: 16.20

Printed and bound by CPI Group (UK) Ltd, Croydon, CR0 4YY

27/09/2026

14970450-0001